QUALITY STRATEGY FOR RESEARCH AND DEVELOPMENT

WILEY SERIES IN SYSTEMS ENGINEERING AND MANAGEMENT

Andrew P. Sage, Editor

A complete list of the titles in this series appears at the end of this volume.

QUALITY STRATEGY FOR RESEARCH AND DEVELOPMENT

Ming-Li Shiu
Jui-Chin Jiang
Mao-Hsiung Tu

WILEY

Published by John Wiley & Sons, Inc., Hoboken, New Jersey
Published simultaneously in Canada

Library of Congress Cataloging-in-Publication Data:

Shiu, Ming-Li, 1978–
Quality strategy for research and development / Ming-Li Shiu, Jui-Chin Jiang, Mao-Hsiung Tu.
pages cm
Includes index.
Includes bibliographical references and index.
ISBN 978-1-118-48763-1 (hardback)
1. Quality control. 2. Engineering–Management. 3. Systems engineering. I. Jiang, Jui-Chin, 1959– II. Tu, Mao-Hsiung, 1955– III. Title.
TS156.S4727 2013
658.4'013–dc23

2013004130

10 9 8 7 6 5 4 3 2 1

To our dear families for their immense love

CONTENTS

FOREWORD

Dr. Mingli Shiu and I have known each other for many years since 2004 when he was a visiting researcher in Japan. At an occasion for discussing quality function deployment (QFD) with him, I shared my viewpoints and case studies about the latest development of QFD then and expected his success in making further progress. I am pleased and consider that the publishing of this book is a wonderful achievement for him and for the field of QFD.

Overall, this book offer insights into the development of QFD over the past several decades, and further renews and expands it based on a better perception of QFD and a better integration of QFD's technology deployment and robust engineering.

In Japan, QFD is widely recognized as a comprehensive system for simultaneously assuring product quality and work quality. This is because QFD, broadly defined, stands for the combination of quality deployment and (narrowly defined) quality function deployment (which means the deployment of job functions that create quality). However, some misunderstandings about the essence of QFD were gradually engendered in the international popularization of QFD, such as: (1) quality deployment is equivalent to "quality chart" (house of quality) and (2) QFD is equivalent to quality deployment. This book can be viewed as one of a few English-version books that clarify what QFD essentially is and how it is powered by reinforcing the technology deployment through updating the approach of robust engineering. Connecting the technology deployment of QFD and robust technology development is a good evolutionary strategy. To realize such strategy, the commonly seen execution approach of separating QFD projects from the corporate operating system should be avoided. This book also provides the process for simultaneously implementing the evolutionary QFD and corporate new product development activities.

I would like to strongly recommend this book for the new viewpoint on QFD, better comprehension of QFD, and integration of technology deployment and robust engineering.

Yoji Akao
Chairman, International Council for QFD

PREFACE

This book embodies the experiences of the authors in researching and coaching the methods of strategic quality management.

The strategic quality management methods presented by the authors is to transform the mental and operating models of an organization (in this book, the intended target is an R&D organization) to be more customer-focused, rationalized, and efficient, and to reinforce quality management capital. In practice, we are more concerned with the methods for achieving these strategic objectives; meanwhile, these methods are expected to be applied to various fields of technology or product broadly in the long term.

Today, although the philosophy of total quality management (TQM) may be well known throughout the whole organization, in practice, the field that applies the most quality management concepts and methods and can also keep up with the newest development of theories is still manufacturing. Regarding R&D, which could obtain the biggest cost-and-effect impact from applying quality management, the application of quality methodologies is fewer. Quality can be classified into two types: customer-driven quality and engineered quality. For customer-driven quality, marketing and R&D personnel have to be able to truly grasp customer-expressed and latent requirements to realize various products used to satisfy customers and deliver attractive quality. On the other hand, R&D personnel have to, at the earliest stage of product development, propose the proactive prevention countermeasures for "unknown items" (various unknown usage conditions and sources of variability in market) to make improvements before problems arise (i.e., ensuring engineered quality).

In organizations which are mutually qualified to be competitors, the team intelligence of their R&D personnel is similar, so it is difficult for a company to distinguish itself and establish a competitive advantage. Although it seems as if organizations compete in products and technologies, actually, they compete in the effectiveness of methodologies adopted by R&D as well as the R&D paradigm formed by extensively applying these effective methodologies in the organization. The most commonly seen approach to quality assurance

(QA) in practice is as follows: "Quality is attained by tests and inspections"—that is, by discovering quality problems with the help of tests and inspections (i.e., debug approach) and then analyzing cause-and-effect relationships to solve the problems. However, the biggest value provided by QA practitioners should be to propose prevention countermeasures for unknown items, in order to make improvements before quality problems arise and to assure the R&D paradigm is customer-focused, rational, and efficient. In this book, we introduce the methodologies and integration strategy that can effectively realize customer-driven quality and engineered quality at the upstream stages of the corporate value chain (i.e., marketing process, technology development, and product design). By applying and promoting these concepts and methods, an R&D organization can shift its conventional paradigm, and outperform the competition or even escape from the competition.

In this book, the quality strategy for effectively realizing customer-focused, rationalized, and efficient R&D is to shift the current paradigm of R&D through the implementation of QFD (quality function deployment) and RE (robust engineering). The four key tools of QFD are quality deployment, technology deployment, cost deployment, and reliability deployment; and the focuses of RE are ideal function formulation, dynamic signal-to-noise ratio, and orthogonal array. This book provides actionable knowledge, strategy, and implementation steps on the strength of systems thinking and integrated perspectives.

We have to thank the wisdom and efforts of many quality-field scholars, experts, and practitioners who inspired us during the process of writing this book. They have developed many effective and efficient methodologies, practical approaches, and useful philosophies and concepts that enable us to introduce, expand, or develop the strategies and methods mentioned in this book. We hope, following the examples, that our efforts will help to establish a world in which people can enjoy higher quality. We especially thank the Quality gurus, Dr. Noriaki Kano and Dr. Yoji Akao, for their guidance and kindheartedness in various aspects when Ming-Li Shiu was in Japan, and even now, when he is in Taiwan.

Moreover, we would like to express our sincerest thanks to all the people who guided us, inspired us, encouraged us, gave us opportunities to understand the theories and practices, supported us, energized us, and directly contributed to the birth of this book. They are Mr. Mark Ko, Ms. Yi-Ching Chang, Mr. Simon Chang, Mr. Alan Wu, and all of our dearest families, including Ming-Li Shiu's infant daughter who was born in December 2012.

Also, we offer our heartfelt appreciation to the following individuals for their professional advice and excellent support, thereby enabling this book to be presented to the public in a form superior to the original manuscript: Professor Andrew P. Sage of George Mason University; George J. Telecki and Kari Capone at John Wiley & Sons; Wiley's editorial, marketing, publicity, and sales teams; and the composition and project management teams at Toppan Best-set Premedia.

The pursuit of quality is an endless journey for individuals, enterprises, the society, and the world. We must always keep endeavoring to ensure quality in our world.

Ming-Li Shiu
Jui-Chin Jiang
Mao-Hsiung Tu

Taipei, Taiwan
March 2013

1 Introduction to Quality by Design

1.1 WHAT IS QUALITY?

"Quality" is one of the most important foundations of trading, whose modern concepts have had a long history since their development in the 1930s. To develop a stable trade relationship, a company must be dedicated to ensure product quality so that their product can be used to (a) "exchange" the money and resources required by the company's operation and growth and (b) create more added value through the accumulation of such resources. A company expects to continually exchange more resources with the market and do more businesses to pursue sustainable operation and growth. In the past decades, many quality scholars and experts have tried to develop better philosophies regarding quality, along with more effective quality assurance (QA) methods to make their society and the world more prosperous and thereby enable human beings to enjoy a better quality of life. Quality, cost, and productivity are the most important factors contributing to the success of a company. Kondo (1995) considered that among the three key management indicators, quality is distinguished from the other two by the following two features:

1. The history of quality (in other words, the relationship between quality and human beings) is far longer than that of either cost or productivity.
2. Quality is the only one of the three indicators to be of common concern to both manufacturer and customer.

Kondo (2000) indicated that quality has a very long history of about 1.7 million years, and he expressed the fact that we feel a deeper connection with it than with either cost or productivity. Hence, improving quality is more easily sympathized with and accepted, and it is harder to refuse than a call to cut cost or to raise productivity. While we may emphasize "quality first" and stress the importance of establishing a "quality culture," for example, we do not commonly use the terms "cost culture" or "productivity culture." Therefore, quality is the center of integrated management (Kondo, 2002). Although

Quality Strategy for Research and Development, First Edition. Ming-Li Shiu, Jui-Chin Jiang, and Mao-Hsiung Tu.

quality concepts have been developed for a long time, it was not until the 1930s that Dr. Walter A. Shewhart explored quality's significance in modern industry and its control methods with a completely new perspective. After Dr. W. Edwards Deming (a student of Dr. Shewhart) in the 1950s applied the philosophies and methods of quality control to help post-war Japan to win quality leadership in the global market, quality philosophies and various methods spread rapidly worldwide and developed swiftly.

What is quality? "Quality" is a term that can be defined and explained from many different angles. If people have closer definitions about quality, just like the way we define "cars" and "televisions," then quality is concrete and easy to be understood. However, the definition of quality is diverse with various levels and contents, making quality seem abstract, and thus our recognition about it would vary with circumstances. Kano et al. (1984) described that quality bi-dimensional in terms of "objectivity" and "subjectivity," and they pointed out that Aristotle (384–322 BC) first expressed this viewpoint clearly in his book *Metaphysics*. Making the comparison between product A and product B for instance, if product A is evaluated as having a better quality characteristic value than that of product B, then such approach used to evaluate the difference of "quality level" according to defined quality characteristics (or performance indicators) is "objective." If product A and product B are evaluated as "good" or "bad" according to people's different subjectively valued properties (e.g., likes and dislikes, preferences, values), then people recognize quality by using the "subjective" approach. With such concept, we know that the development of the definitions or concepts of quality such as "zero defects," "freedom from deficiencies," and "six sigma" is based on the quality recognition of objectivity, while the definitions or concepts of quality such as "product features," "customer satisfaction," and "customer perception" are based on the quality recognition of subjectivity. The "bi-dimension of quality" provides an effective and easy-to-understand framework to help us know, classify, explain, and analyze the qualities defined and described from various different angles.

Based on the above, when we speak of quality, we can classify it into two types as follows (Taguchi et al., 2000):

1. *Customer-Driven Quality.* What the customer wants—for example, function, appearance, color, and so on. The way we improve customer-driven quality is to fulfill the customer's expressed and latent requirements.
2. *Engineered Quality.* Freedom from what the customer does not want—for example, noise, vibrations, failures, pollution, and so on. The way we improve engineered quality is to lower the variability around an ideal function caused by various sources of variability.

Customer-driven quality is related to market segmentation and product planning; a company determines the business scope through a product's positioning planning and develops the products that can meet the requirements of a customer group within this scope. After marketing and sales personnel identify the market segment and determine product features and portfolios, R&D personnel have to, according to that product planning, develop the products that can be robust against various customer usage conditions so as not to have functional variability in applications. A product has functional variability during customer use, which means that the customer has to bear a certain degree of losses (including invisible loss such as worry and bother as well as visible loss such as time and money). Therefore, the objective of enhancing engineered quality is to reduce societal loss after products are shipped. In the historical development of quality management, customer-driven quality (i.e., planning and development of "functions") is emphasized; but in the twenty-first century, engineered quality (i.e., reduction of functional variability, optimization of "functionality"), which is closely related to lowering functional variability, noise, pollution, and societal loss, plays a very important role in the future development of quality management.

Table 1.1 summarizes the differences between customer-driven quality and engineered quality. Here, we use two well-known definitions of quality to enhance the understanding about the two types of quality: Customer-driven quality means "fitness for use"; engineered quality means "the societal loss a product causes after being shipped." The so-called "fitness for use" by Juran (1992) includes two meanings:

1. *Product Features.* In the eyes of customers, the better the product features, the higher the quality.

TABLE 1.1. Two Types of Quality

Item	Customer-Driven Quality	Engineered Quality
Concept	Features what customer wants (e.g., function per se, features, color and appearance)	Problems customer does not want (e.g., functional variability, defects, failures, noise and vibrations)
Useful definition	Fitness for use (by Dr. Joseph M. Juran)	The loss a product causes to society after being shipped (by Dr. Genichi Taguchi)
Definition component	• Product features • Freedom from deficiencies	• Loss caused by variability of function • Loss caused by harmful side effects
Quality implication	Known or foreseeable items	Unknown or unforeseeable items
Design focus	Function design	Functionality design
Evaluation criteria	Customer satisfaction	Signal-to-noise (SN) ratio
Core method	Quality function deployment	Robust engineering

2. *Freedom from Deficiencies.* In the eyes of customers, the fewer the deficiencies, the better the quality.

As for Taguchi, he considers that a quality problem is concerned with reducing the loss a product causes to society after being shipped. In the context of this definition of quality, loss should be restricted to two categories, other than any losses caused by functions per se:

1. Loss caused by variability of function.
2. Loss caused by harmful side effects.

Therefore, ensuring that a design will perform its intended functions without variability, as well as cause little loss through harmful side effects, is a quality issue. These losses can be viewed as "unknown or unforeseeable items"; that is, the customer's usage conditions are diverse and have various possibilities, so it's very hard to predict product's functional performance and the loss caused by functional variability under various unpredictable or unknown usage conditions. Therefore, in the everyday practice of R&D we usually need verification tests with many samples and ample time to ensure the performance of products. Even so, the central issue of quality by design is to deal with such unpredictable circumstances and make the product's functional performance under various usage conditions achieve the state where the variability is very small. When a product can perform its intended functions without variability, we say that the product has good functionality, and the design methods that make the product achieve functionality are called "functionality design." An indicator called signal-to-noise (SN) ratio is the criteria used to evaluate how well the functionality developed. We will interpret its meaning and applicability in Part II of this book.

Relatively speaking, to achieve customer-driven quality is to deal with "known or foreseeable items" because the clarification of customer requirements or specifications is the task that must be sufficiently done in product planning. If the requirements or specifications cannot be clarified, we can't proceed to the product design stage. Customers use a product to accomplish certain tasks through product functions; hence, at the product planning stage, what the we need to clarify is "what the customer wants" or "what product functions the customer would need to accomplish certain tasks." Customer satisfaction is the criterion used to evaluate how well the function is designed. A design method that is helpful to plan product function and convert that function into the items of technology and product development as well as set quality assurance (QA) points is called quality function deployment (QFD). QFD is an important method to implement function design and quality deployment; and robust engineering (RE) is an effective method to implement functionality design. Both of them are the core focuses of this book. We will introduce QFD and RE, respectively, in Part I and Part II of this book, and we will interpret their integration model and implementation process after that.

1.2 WHY QUALITY BY DESIGN?

The pursuit of quality has emerged as a constant and dominant theme in management thinking since the 1930s. Although the initial theory emerged from American sources, the early industrial applications were predominantly conducted in Japanese companies (Beckford, 1998). At first, Japanese companies used quality control in manufacturing and inspection areas; but in the mid-1950s, quality control spread company-wide, which made quality control a management tool. Total quality management (TQM) is the tool formed by integrating all the concepts, facilitation programs, and techniques of quality management for implementing company-wide quality management.

The ultimate objective of TQM is customer satisfaction. Companies must, at the appropriate time and price, provide products with the quality that satisfies customers, thereby exchanging the needed resources for subsistence and operation. Through accumulated resources, companies create more added values and obtain more resources, so as to pursue sustainable development and growth. If a company is viewed as a collection of various planning and operational activities, all these activities can be represented using a value chain (Porter, 1985). As quality management is extended to the whole company, assurance activities of quality push toward the upper reaches of the value chain. In other words, during technology development and new product development (NPD), quality must be "built in" and "designed in." This is because the quality built on these stages has the maximum return and benefit in cost reduction and customer satisfaction. It far surpasses the improvement effects brought about, after design release, by relying on promotion efforts in selling or detection and modification in manufacturing.

1.3 HOW TO DESIGN FOR QUALITY

As mentioned above, we classify quality into two types: customer-driven quality and engineered quality. Therefore, the design methodologies needed to realize the quality of the two types are also different.

In late 1960s, when Japanese products started the period from "designed by imitation" to "developed by originality," the so-called "customer-driven quality" becomes important; that quality is the quality built into the product according to the voice of the customer (VOC). Also, customer-driven quality can be further described based on the definition of quality, "fitness for use," defined by Juran (1992). The "fitness for use" defined by Juran includes both meanings of "product features" and "freedom from deficiencies"; that is, it expresses that customer satisfaction is determined by how the product achieves the balance between more features and fewer deficiencies, and that balance state is quality. To achieve customer-driven quality and improve customer satisfaction, a quality assurance (QA) method called quality function deployment (QFD) applied to NPD was developed; it can effectively convert

VOC into quality characteristics as well as their design targets, as well as deploy various subsystems, components, parts, and process elements and facilitate the interrelationships needed to achieve those targets. The so-called quality function deployment (broadly defined) is actually the short form of both "quality deployment" and "quality function deployment (narrowly defined)"; the former is the method used to ensure product quality, and the latter is used to ensure work quality and process quality ("quality function" means the job functions that create quality). QFD is considered as a comprehensive approach to function design and quality deployment. It simultaneously focuses on function design as well as how to systematically deploy the designed function and the quality we want to assure into the whole NPD process and realize them.

On the other hand, to achieve engineered quality and reduce various losses brought by functional variability, a QA method called robust engineering (RE) (also known as quality engineering) applied to technology development and NPD was developed; it is a methodology of functionality evaluation and design which can make the functionality of technology and product approach its ideal state as much as possible and further avoid various quality problems arisen at the downstream. In other words, RE is to optimize functionality.

In Section 1.4, we interpret the relationship between QFD and NPD in detail, and we describe QFD's importance and evolutionary strategy based on the reflections on "the development of QFD" and "the evolution of NPD philosophy." In Section 1.5, we interpret the importance of technology development and its relationship with RE strategy.

1.4 NEW PRODUCT DEVELOPMENT AND QFD

Nowadays, NPD's quality management relies upon the guidance of modern NPD philosophy and integrated application of quality tools and methodologies to deal with the following three major issues:

1. *Improving R&D Productivity.* Today's competition on the market is time-based competition. Products require variety and shorter time from development to market, yet their lifetime after being launched to the market is relatively short. Consequently, R&D productivity needs to be enhanced so as to deal with competition. R&D productivity is defined as the R&D capability that can (1) more easily meet market demands for product variety and higher quality levels, (2) more readily combine company's existing "technology seeds" with "customer needs" or further create new customer requirements, (3) more easily prevent potential quality problems through early prediction, (4) effectively more accumulate product-development knowledge in a systemic manner and effectively transfer to future generations, and (5) attain more leverage from critical resources and thereby reduce product development costs.

2. *Managing NPD Cycle in Supply Chain Environment.* The design of supply chain has given rise to a situation in which product development, manufacturing, and support activities are spread out over different areas and companies. Therefore, modern NPD philosophy has shifted from concurrent engineering (CE), emphasizing paralleling of activities, to integrated product and process development (IPPD), stressing the integration of product design activities and process design activities. The latter more unequivocally reflect major challenges of NPD management in today's supply chain environment.
3. *Creating Attractive Quality.* Attractive quality is defined as an unexpected new quality achieved by meeting customers' latent requirements and is synonymously used with customer delight (Kano, 2002). Kano (2002) observed that compared with competitors in the same target market, most product makers, in particular of mature products, usually have considerably similar capabilities in surveying the "voice of the customer" and converting customer expressed requirements into concrete product specifications to conduct development, and that therefore creating product differentiation and the customer's surprised response is not easy. Since 1984 Kano has been advocating the concept and process of attractive quality creation (AQC) in order to serve as a quality strategy for NPD to realize customer delight (Kano et al., 1984). Garvin (1987), who held a similar viewpoint, also considered that in order to achieve quality gains, managers need a new way of thinking, a conceptual bridge to the customer's vantage point.

The term *quality function deployment* (QFD) refers to a concept and methodology of NPD under the umbrella of TQM. QFD was conceived in Japan in the late 1960s, and Akao first presented the concept and method of QFD (Akao and Mazur, 2003). QFD is one of the few techniques that could potentially have a quality improvement impact throughout a company's product development process (Booker, 2003). Its objectives are to identify the customer, determine what the customer wants, and provide a way to meet the customer's desires (Maddux et al., 1991). QFD combines various design engineering and managerial tools to create a customer-driven approach to developing new products (Özgener, 2003). It is the most complete and convincing methodology that can identify and prioritize customer requirements and translate these requirements into appropriate company requirements at each stage of the product life cycle (Conti, 1989; Burke et al., 2002).

We, in view of the importance of having to consider the foregoing three major issues of NPD quality management, make reflections upon the development of QFD and the evolution of NPD philosophy (as described in Sections 1.4.1 and 1.4.2) to help us identify some key issues and further develop QFD's expanded system (as described in Chapter 3) for more effective NPD.

1.4.1 Reflections on the Development of QFD

QFD has been developed for 35 years, and the published literature on QFD reveals that it has achieved a state where it is well known in the academic literature of product development management and has been widely applied in many countries (Cheng, 2003). Although QFD has attracted great attention and its importance has been recognized, its popularity has not always translated into successful practice (Akao, 1988). The reflections on the development of QFD are described as follows:

1. *Recognition of QFD definitions, QFD concepts, and QFD's relationships with NPD needs to be reinforced.* After QFD was internationally spread and transferred, some misunderstandings about the essence of QFD were gradually engendered and can be found in various relevant articles and books. The common misunderstandings include: (1) quality deployment being equivalent to quality chart and (2) QFD being equivalent to quality deployment. And the result is an incomplete and ineffective QFD in practice.

These misperceptions have been noticed by some researchers (Fortuna, 1988; Cox, 1992; Cohen, 1995; Partovi, 1999; Bouchereau and Rowlands, 2000; Cristiano et al., 2000, 2001; Akao and Mazur, 2003; Booker, 2003; Akao, 2004). Akao (2004) has undertaken to reinforce QFD concepts and contents regarding the second-most common misconceptions about QFD. Aside from that, views held by other researchers are mostly no more than descriptions of phenomena observed with no concrete response to those issues. In addition, the main shortcomings most often pointed out (Zairi and Youssef, 1995; Prasad, 1998; Bouchereau and Rowlands, 2000) are also caused by misconceptions about basic concepts of QFD and its relationships with NPD.

2. *The universal QFD implementation roadmap which includes the matrix-making of QFD and the main activities of NPD, as well as the essential tool set used in that roadmap, need to be developed.* The most frequently seen form of QFD is presented by four major matrixes and the relationships between them (Sullivan, 1986; Hauser and Clausing, 1988; Prasad, 1998; Kathawala and Motwani, 1994; Cohen, 1995). The main activity in most current implementations of QFD is the generation of these four charts (Prasad, 1998). However, if we regard QFD as the QA system managing NPD, then the generic roadmap of QFD needs to include the matrix-making of QFD and the main activities of NPD so that it can perform a higher value of QFD toward NPD's QA management.

Furthermore, if we regard QFD as a framework that can integrates various design tools, it is worthwhile to explore which tools can be systematically organized as the essential tool set used in NPD and QFD.

3. *QFD's research and development in Japan and the West need to be fused.* It can be analyzed that there is a main difference between the development trends of QFD in Japan and the West over the past 35 years: the former's QFD, in concept and methodology, toward the direction of "how to add value to every work activity in NPD cycle to achieve quality assurance and

competitively improve product quality" (Mizuno and Akao, 1978; Kogure and Akao, 1983; Akao, 1990a, b; Akao, 1997; Shindo, 1998; Akao, 2001a, b; Akao and Mazur, 2003; Cheng, 2003; Akao, 2004), in contrast to the latter's QFD toward the direction of "how to integrate various design tools and methodologies as well as more powerful numerical analysis methods to achieve QA and competitively improve product quality" (Clausing, 1988; Ross, 1988; Clausing and Simpson, 1990; Clausing and Pugh, 1991; Bendell et al., 1993; Clausing, 1993; Masud and Dean, 1993; Wasserman et al., 1993; Armacost et al., 1994; Lu et al., 1994; Eureka and Ryan, 1995; Khoo and Ho, 1996; Zhang et al., 1996; Terninko, 1997; ReVelle et al., 1998; Zhou, 1998; Dawson and Askin, 1999; Temponi et al., 1999; Bouchereau and Rowlands, 2000; Kim et al., 2000; Creveling et al., 2002; Büyüközkan et al., 2004a, b; Ramasamy and Selladurai, 2004; Chen et al., 2005; Lai et al., 2005; Yan et al., 2005).

Though QFD originated in Japan, currently most literatures published in English lay particular stress on the latter mentioned above. This has made the international mainstream of QFD's research and development overlook integration with many researches and practices published in Japanese.

4. *Reinforcement of QFD's effectiveness needs a renewal in both its structure and contents.* Since 1994, Akao has advocated "development management" to facilitate the further advancement of QFD, and he also indicated that another important research topic should be a management technology for making QFD more effective inside an organization (Akao, 1997). The direction of QFD development advocated by Akao can be concretely viewed as how, in harmony with the dominant NPD philosophy, to develop a more process-oriented approach to implement QFD, in order to effectively support the corporate NPD cycle.

Nevertheless, most QFD research has focused on three areas that are less related to the achievement of this evolution objective. The three focused areas are: (1) scoring mechanics of the matrix of matrices model (Masud and Dean, 1993; Wasserman, 1993; Wasserman et al., 1993; Armacost et al., 1994; Lu et al., 1994; Khoo and Ho, 1996; Zhang et al., 1996; Franceschini and Rossetto, 1997; Moskowitz and Kim, 1997; Park and Kim, 1998; Zhou, 1998; Dawson and Askin, 1999; Temponi et al., 1999; Bouchereau and Rowlands, 2000; Kim et al., 2000; Franceschini, 2002; Shin et al., 2002; Büyüközkan et al., 2004a, b; Ramasamy and Selladurai, 2004; Chen et al., 2005; Lai et al., 2005); (2) case studies of different industrial product applications (Cohen, 1988; Hauser, 1993; Armacost et al., 1994; Ghobadian and Terry, 1995; Mrad, 1997); and (3) management applications—such as strategic planning, policy deployment, project selection, and so on (Sullivan, 1988; Lu et al., 1994; Crowe and Cheng, 1996; Partovi, 1999; Creveling et al., 2002; Hunt and Xavier, 2003; Killen et al., 2005; LePrevost and Mazur, 2005). There has been a lack of interest in a comprehensive renewal of QFD structure and contents.

5. *QFD needs to be adapted for use in today's supply chain environment.* The application of QFD to product development according to the business model of a company is achieved as follows: (1) in companies that emphasize own

product design and (2) in companies that emphasize fabrication engineering. The difference in business models derives from the division of labor in global supply chain.

Indeed, QFD was conceived when Japanese industries were advancing toward product development based on originality, and it was therefore not designed for application in a business model that emphasizes fabrication engineering. The difference between the two corporate NPD models causes QFD to vary in its application focus (Cheng, 2003), so that QFD can be practiced more effectively in a supply chain environment. However, as of yet there is no QFD reinforced for the NPD model of fabrication suppliers.

6. *The mathematics underlying the other deployments beyond the "quality chart" needs to be specified.* Much of the QFD research focuses too much on the mechanics of scoring, while their range is restricted to a quality chart only (Masud and Dean, 1993; Wasserman, 1993; Khoo and Ho, 1996; Zhang et al., 1996; Franceschini and Rossetto, 1997; Moskowitz and Kim, 1997; Park and Kim, 1998; Zhou, 1998; Dawson and Askin, 1999; Temponi et al., 1999; Franceschini, 2002; Shin et al., 2002; Büyüközkan et al., 2004a, b; Ramasamy and Selladurai, 2004; Chen et al., 2005; Lai et al., 2005). QFD is not a method for finding an optimal design algorithmically, but specifying the mathematics underlying the overall QFD cycle (including the deployments of subsystems, process, technology, cost, and reliability beyond the quality chart) can facilitate understanding of QFD process and thereby yield greater benefits from its application.

1.4.2 Reflections on the Evolution of NPD Philosophy

Evolution in NPD philosophy signifies the needs for a change in NPD operating patterns and for a new development in core methodology. The reflections on NPD philosophy are described as follows:

1. *Changes in emphasis in the evolution of NPD philosophy oblige the enhancement of the application depth of existing methodologies and their integration with other tools and methodologies.* The NPD philosophy of CE has changed the operating pattern of NPD from over-the-wall or sequential engineering to parallel processing or simultaneous engineering. It mainly emphasizes interdepartmental communications in the NPD cycle. Later, the new philosophy of IPPD, considered an extension of CE (Allen, 1990; Prasad, 1996; Magrab, 1997; Usher et al., 1998), further concretely described and emphasized the essence of CE on product and process design integration.

Although in terms of basic concepts the two philosophies do not differ significantly (Allen, 1990; Turino, 1992; Clausing, 1993; Parsaei and Sullivan, 1993; Syan and Menon, 1994; Prasad, 1996; Zhang et al., 1996; Magrab, 1997; Wang, 1997; Franceschini, 2002), their change in emphasis (i.e., from emphasizing interdepartmental communications to product and process design integration) signifies the need to enhance the application depth and integration of existing core methodologies (e.g., QFD, computer-aided design, computer-aided manufacturing, design for manufacturing and assembly, rapid prototyping, etc.) (Allen, 1990; Turino, 1992; Kusiak, 1993; Parsaei and Sullivan, 1993; Syan and Menon, 1994; Prasad, 1996; Wang, 1997; Franceschini, 2002). Take the application of QFD in CE as an example. Even if only the first form of QFD—quality chart (also known as the house of quality)—is used, interdepartmental communications can be effectively improved (Sullivan, 1986; Hauser and Clausing, 1988; Akao and Mazur, 2003) and thus the weaknesses of the over-the-wall approach can be significantly improved. Therefore, using the quality chart alone has become the most common method of QFD application in CE. However, in applying QFD in IPPD, the deployments of subsystems, components, parts, process, and their relationships beyond the quality chart must be further implemented, and their integration with other design tools and methodologies must be stressed to enable product and process design integration.

2. *Core methodologies must tie in with philosophical evolution to have new development.* Since 1994, Akao has advocated development management to urge proper recognition and further advancement of QFD (Akao, 1997). Nevertheless, though the new philosophy of IPPD, considered an extension of CE, places more emphasis on the use of a systematic way to conduct integrated product and process development, the main method used, QFD (Ross, 1988; Shina, 1991; Carter and Baker, 1992; Wheelwright and Clark, 1992; Clausing, 1993; Kusiak, 1993; Parsaei and Sullivan, 1993; ReVelle et al., 1998; Shina, 1994; Syan and Menon, 1994; Prasad, 1996; Magrab, 1997; Schmidt, 1997; Monplaisir and Singh, 2002; Wang, 1997; Usher et al., 1998; Franceschini, 2002), has not touched on how itself can be renewed to adapt to the differences in applications resulting from changes in the dominant NPD philosophy. This situation indicates that in dealing with the key issues of NPD, the structural integrity of QFD is considered capable of achieving different NPD philosophies. However, the evolutionary needs of QFD have been neglected. This is why QFD research has focused far more on case studies than on comprehensive renewal of QFD structure and contents.

3. *New design tools and methodologies need a common framework that can achieve integration with existing methodologies.* One NPD philosophy can predominantly affect the operating pattern of an NPD cycle. However, an NPD cycle cannot achieve good development management by relying on just one tool or methodology. Numerous and complex items are processed by NPD, including product planning, concept development, product design, process design, production planning, and management of quality, technology, cost, and reliability. As a result, design tools and methodologies generally apply, depending on the timing and range of NPD. Hence, to continue improving development management, in addition to adopting better new tools and methodologies, a common framework is required to integrate new and existing tools of NPD to create a synergistic whole. QFD can play a dual role, as both a design methodology and a framework integrating various design methodologies.

1.5 TECHNOLOGY DEVELOPMENT AND FUNCTIONALITY DESIGN

Since Prahalad and Hamel proposed the concept of *core competence* in 1990, top management has to recognize the concept of *corporation* as a portfolio of competencies and, based on the strategy of developing core competence, build a competitive advantage by going beyond the short-term standpoint of focusing on products. Prahalad and Hamel (1990) stated that "the corporation, like a tree, grows from its roots. Core products are nourished by competencies and engender business units, whose fruit are end products." Therefore, it's not hard for us to know that technology development capability is one of the most important factors to develop and keep reinforcing core competence, that is, the key to establish competitive advantage.

Simply speaking, technology development capability signifies the capability of developing the technology with high readiness before product planning and ensuring that the technology can be effectively applied to develop the products with high quality, low cost, and short time to market. For different design targets of different products, the technology must also have a speedy adaptability which can effectively fulfill them. Enhancing technology development capability is helpful for a company to gain the following benefits:

1. Reinforcing core competence to establish and keep developing the company's competitiveness.
2. Establishing technology intelligence capital to define an industry's technology framework and standards or constitute the technology development barrier adverse to the competitors.
3. Gaining technology leadership to escape from the price war.
4. Applying technology to develop core products, as well as seeking new business concepts based on them.

When we speak of quality in two different fields such as technology development and NPD, the word "quality" would signify different meanings. At the technology development stage, since the planning for technology application and product concept may not be conducted yet, technology development is defined as researching a technical system formed by energy transformation. The quality of that technology means that energy transformation has achieved its ideal state (we call the ideal state an "ideal function"). Therefore, while we develop a technology that can make the energy fully transform in accordance with its ideal function (without energy loss or variability), we may say that technology is high quality, that is, its functionality or functional robustness is good. On the contrary, if energy transformation varies around the ideal function, that technology would create problems in applications and cause some loss: The larger the variability, the worse the quality the technology development and the worse the applicability. The so-called technology development is defined as follows: By means of small scale experiments, test pieces and simulations to predict and develop good functionality. Most work of technology development is to resist the functional variability caused by various sources of variability, so R&D (research & development) work can be regarded as robustness development (RD) of technologies or products.

Ideal function is energy-related and an energy transformation; through that energy transformation, a certain function can be provided. However, in the real world there are various sources of variability which cause the loss of that energy transformation and thus the function cannot fully take effect and be performed. According to conservation of energy, the lost energy will be used to create an unintended function and then produce various defect phenomena. During technology development, the way to improve various defect phenomena of quality characteristics is "firefighting"; the way to study how to make energy transformation not affected by various sources of variability to fully perform its ideal function and further to avoid causing various defect phenomena of quality characteristics is "proactive prevention." The latter way is the optimization of technology. The so-called technology optimization in this book is *optimized design of functionality*, and the design method to optimize functionality can be regarded as *technology development strategy* itself which can be broadly applied to various technological fields. RE plays the strategic role of realizing that optimization. It focuses on the rationalization and efficiency of technology development (including a totally new technological area), so as to make R&D personnel, at the possible lowest cost, predict and early improve the functional performance of a technology under various sources of variability. Once the functional performance is in a robust state, the technology realizes high quality. We will introduce this method and its implementation process in Part II of this book.

1.6 OUTLINE OF THIS BOOK

This book's body of framework is divided into two parts: Optimizing design for function and optimizing design for functionality. The content of each chapter is introduced as follows:

1. Chapter 1 describes the concept of quality and how to design for quality.
2. Chapter 2 introduces the methodology, QFD, used to optimize design for function and the dominant approaches to it.
3. Based on Chapter 2, Chapter 3 develops an expanded system of QFD and its implementation process using the NPD approach.
4. Chapter 4 describes the common seen R&D paradigm in practice and its improvement opportunities.
5. Chapter 5 presents how to evaluate a technology and conduct comparative assessment of technologies.
6. Chapter 6 describes the methodology, robust engineering (RE), used to optimize design for functionality, and it presents case studies of robust technology development.

7. Chapter 7 describes the technical and managerial key success factors (KSFs) for managing R&D paradigm shift.
8. Chapter 8 develops an integration strategy for QFD, RE, and other breakthrough strategies, such as DFX (design for excellence), DFSS (design for Six Sigma), and BOS (blue ocean strategy), in order to realize the customer-focused, rationalized, and efficient R&D.

The linkage structure of each chapter of this book is shown in Figure 1.1.

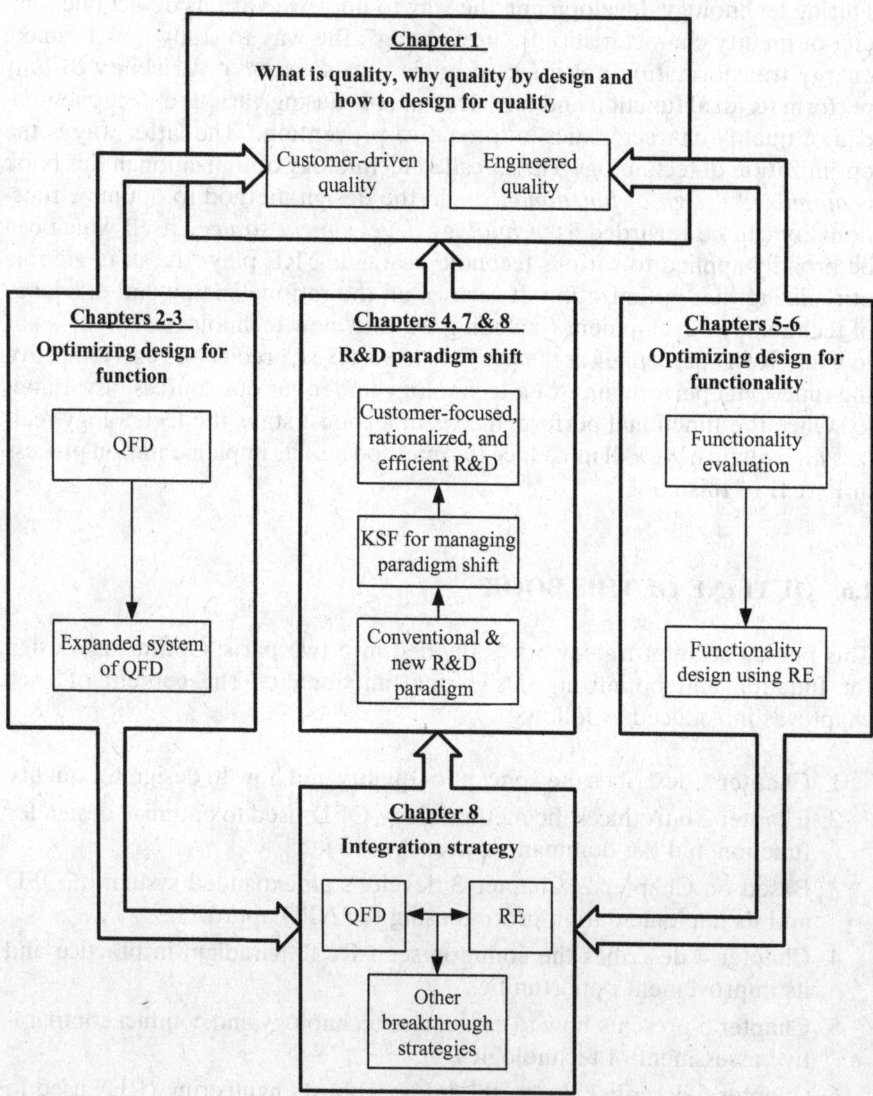

Figure 1.1. Framework of this book.

PART I
Optimizing Design for Function

2 Quality Function Deployment

2.1 HISTORICAL DEVELOPMENT AND DEFINITION OF QFD

Quality function deployment (QFD) was conceived in Japan in the late 1960s while Japanese industries were departing from their post-war mode of product development, which had essentially involved imitation and copying, and were moving to original product development. During this time, Akao first presented the concept and method of QFD. The development of QFD was motivated by two issues: (1) There was a recognition of the importance of designing quality into new products, but there was a lack of guidance on how this could be achieved; and (2) companies already reviewed the control points by using so-called quality assurance (QA) process sheets to inspect for quality, but these were produced at the manufacturing site after the new products were being produced (Akao and Mazur, 2003).

In 1966, a process assurance items table was presented by Oshiumi of the Bridgestone Tire Corporation. This table showed the links from the substitute quality characteristics, which were converted from true qualities to the process factors. Akao added a field called *design viewpoints* to the process assurance items table, but this approach was still inadequate in terms of setting design quality. This inadequacy was addressed by the creation of the *quality chart*, which was made public by the Kobe shipyards of Mitsubishi Heavy Industry. This systemized the true quality (customer's needs) in terms of functions, and it showed the relationship between these functions and the substitute quality characteristics (quality characteristics) (Akao and Mazur, 2003). All these ideas and developments were eventually integrated to produce quality deployment (QD). Akao et al. (1989) defined this as a methodology that

> . . . converts user demands into substitute quality characteristics (quality characteristics), determines the design quality of the finished good, and systematically deploys this quality into component quality, individual part quality and process elements and their relationships.

Quality Strategy for Research and Development, First Edition. Ming-Li Shiu, Jui-Chin Jiang, and Mao-Hsiung Tu.

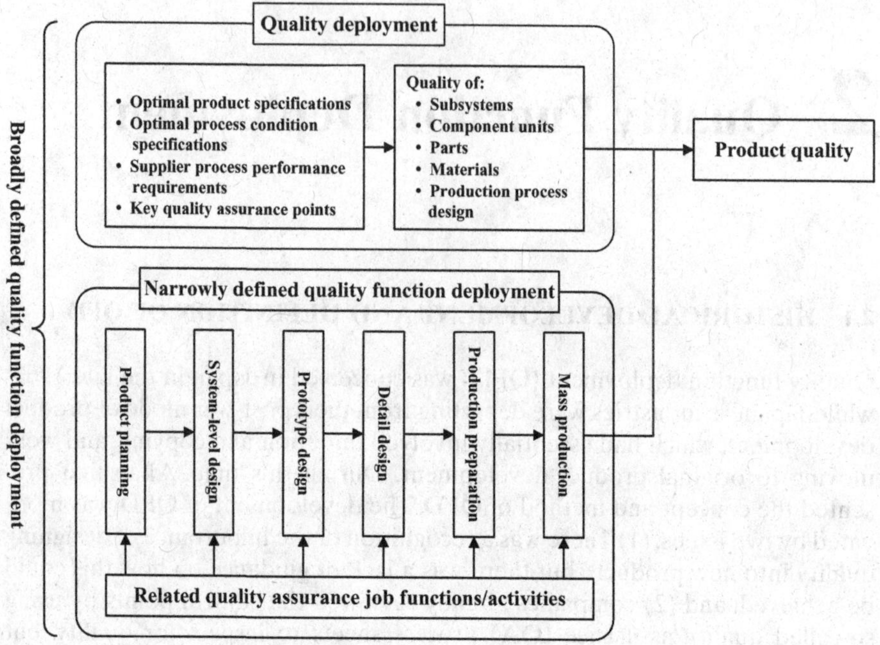

Figure 2.1. Broadly defined QFD.

Akao and Mazur (2003) also pointed out that there was another methodology merged into QFD. It is called narrowly defined QFD, which was described by Mizuno and Akao (1978) as

> step-by-step deployment of a job function or operation, that embodies quality, into their details through systemization of targets and means.

QFD, broadly defined, thus refers to the combination of quality deployment and narrowly defined QFD, as shown in Figure 2.1. Akao (2004) considered that the first focuses on the quality of a *product*, whereas the second focuses on the quality of the *work activities* required to achieve product quality.

Quality deployment is a means to convert customer's demands into design quality of the finished product, and it deploys the quality over the design targets and major QA points (which are the key points and control items in achieving selling propositions, preventing the recurrence of problem points of past product development, and preventing the occurrence of potential problem points of new designs) used at all levels of product architecture and during process design. Therefore, the purpose of quality deployment is to establish a quality network that can ensure the quality of a product itself. The network includes both (a) quality of physical elements—subsystems, component units, parts, and materials—that constitute a product architecture and (b) quality of production processes that manufacture and assemble semi-finished products into finished products, as illustrated in the quality deployment section in Figure 2.1.

Narrowly defined QFD is derived by extending the concept of value engineering (which was originally applied to defining the functions of a product) to the deployment of business process functions (Akao and Mazur, 2003). Therefore, the word "function" in QFD refers to the QA job functions of a new product development (NPD) cycle, rather than the functions of a product, and these job functions that can create product quality are called *quality functions*. The purpose of narrowly defined QFD is to establish a procedure network that is formed by various planning and operational QA activities and procedure flows needed to achieve product quality. This network includes QA activities in all phases of the product realization cycle, ranging from (a) product and technology development planning, system-level design, prototype design, and detail design to (b) production preparation and mass production, as illustrated in the narrowly defined QFD section in Figure 2.1.

To sum up, QFD is a system for translating the "voice of the customer" into appropriate company requirements at each stage, ranging from research through product design and development, to manufacture, to marketing and sales, and finally to distribution and disposal after usage (Burke et al., 2002). ReVelle et al. (1998) also pointed out that the name QFD expresses its true purpose, which is satisfying customers (*quality*) by translating their needs into a design and ensuring that all organizational units (*function*) work together to systematically break down their activities into finer and finer detail that can be quantified and controlled (*deployment*).

2.2 THE NATURE OF QFD

According to Kogure and Akao (1983), QFD provides an opportunity to employ a *design approach* that comprehensively views the entire picture of a system and finds means to achieve specific objectives to satisfy customer requirements, and it will be one of the important pillars supporting TQM together with the *analytical approach* that finds problem causes and manages to prevent recurrence of problems. Özgener (2003) also pointed out that QFD is about planning and problem prevention, not problem solving.

Ohfuji (1993) summarized that QFD is built upon three pairs of basic ideas or principles:

1. *Subdivision and Integration.* Analyzing customer requirements into technical characteristics, along with hierarchically synthesizing all technical elements to a quality product.
2. *Multidimensionality and Visualization.* Facilitating cross-functional communications and cooperation, along with using quality deployment charts to visualize the exchange of information and experiences.

3. *Holism and Partialism.* Not only building up systems with the intention of seeing a whole, but also logical thinking of relating and prioritizing system elements.

2.3 BENEFITS OF QFD

In concept and methodology, QFD supports an organization's TQM efforts by (ReVelle et al., 1998):

1. Strengthening the current development process through early identification of goals based on user needs, simultaneous focus on design and production, highlighting and prioritizing key issues, and enhanced communication and teamwork.
2. Providing an objective definition of product or service quality through (a) products and services that satisfy customer needs and expectations and (b) products and services that provide a competitive edge.
3. Facilitating team building and open communications because teams are enabled and empowered, and hidden agendas are eliminated.

Zairi and Youssef (1995) also believed that QFD is an essential pillar for a successful TQM implementation.

Many companies depend on their warranty programs, customer complaints, and inputs from their sales personnel to keep them in touch with their customer (Akao, 1988). The result is a focus on what is wrong with the existing product or service, with little or no attention being paid to what is right or what the customer really wants (Bouchereau and Rowlands, 2000). QFD provides an ideal opportunity to move away from "we know best what the customer wants" to a new culture of "let's hear the voice of the customer." In a sense it enables the organization to become very much proactive to quality problems rather than being reactive to them by waiting for customer complaints (Zairi and Youssef, 1995; Chan and Wu, 2002).

The ultimate benefits of QFD are higher R&D productivity, lower NPD costs, increased customer satisfaction, and increased market share. It has been well documented that the use of QFD can reduce the development time by 30–50 percent, the number of engineering changes by 30–50 percent, and the start-up and engineering costs by 20–60 percent, as well as increase customer satisfaction and reduce the warranty claims up to 50 percent (Fortuna, 1988; Clausing and Pugh, 1991; Hauser, 1993; Lockamy and Khurana, 1995; Zairi and Youssef, 1995; Swink, 2002). Also, QFD helps to formulate new product development plan and technology strategy, to identify specific competitive advantages and create the opportunity to supply niche products (Golomski, 1995), to improve designs and performance (Fortuna, 1988), and to retain product-development knowledge in a systematic manner so that it can be easily applied to future similar designs (Fortuna, 1988; Zairi and Youssef, 1995; Cheng, 2003).

2.4 TWO DOMINANT APPROACHES TO QFD

There are two dominant models that are similar but different approaches to QFD (ReVelle et al., 1998; Chan and Wu, 2002):

1. *Akao's Matrix of Matrices Model* (Mizuno and Akao, 1978; Akao et al., 1983; Akao, 1988; 1990b). It is also known as the basis of the GOAL/QPC (Growth Opportunity Alliance of Greater Lawrence/Quality Productivity Center) Matrix of Matrices developed by Bob King, which includes some 30 matrices (King, 1987), and as the more popular approach adopted in Japan (Cristiano et al., 2001).
2. *The Four-Matrix Model* (Sullivan, 1986; Hauser and Clausing, 1988). It is also known as the ASI (American Supplier Institute) Four-Matrix approach (Schmidt, 1997; ReVelle et al., 1998) and as the more common approach adopted in the Western Hemisphere (the United States in particular) (Cristiano et al., 2001; Chan and Wu, 2002).

Cohen (1995) briefly explained that the Four-Matrix Model is a blueprint for product development and covers basic product development steps, while the Matrix of Matrices Model is also designed for TQM and covers a host of activities—such as reliability planning, cost analysis, and manufacturing quality control—that are implicit or optional in the Four-Matrix Model.

2.4.1 Akao's Matrix of Matrices Model

Akao's approach to quality deployment (Mizuno and Akao, 1978) is shown in Figure 2.2; the procedure was established in four stages, eight phases, and 27 steps, as shown in Table 2.1.

In Figure 2.2, matrices or tables 1–17 are the major parts of quality deployment. And the arrow indicates links between some forms or fields: (1) Matrices 1–9 and data file 11 are useful for the system-level design of a product; (2) matrix 10 and tables 12, 13, 15, 16, and 17 are useful for the detail design and production deployment; and (3) table 14 is useful for the production ramp-up.

Ten years later, in Akao's book published in 1988 (English version published in 1990), Akao (1988) reorganized the quality deployment procedure as three stages and 27 steps, as shown in Table 2.2.

In 1983, Akao et al. constructed a comprehensive system of quality deployment—including technology, cost, and reliability—which can reflect quality, technology, cost, and reliability considerations simultaneously to

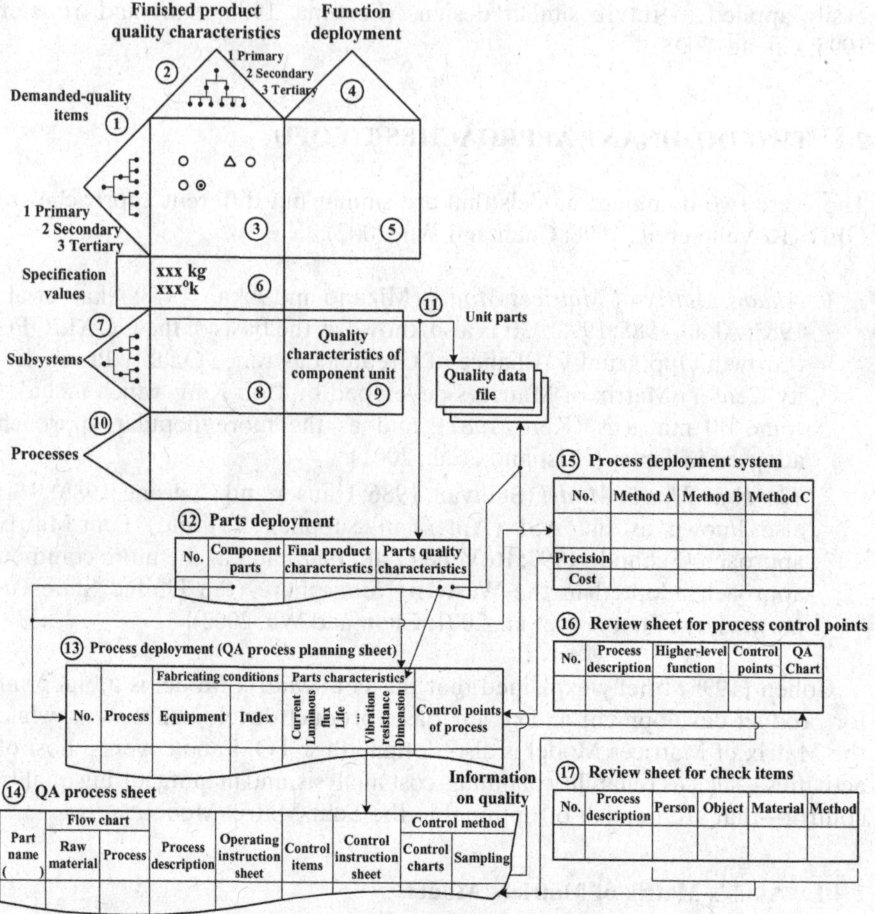

Figure 2.2. Quality deployment system (Mizuno and Akao, 1978).

ensure smooth product development (Akao et al., 1983). The comprehensive system is shown in Figure 2.3, and the objective of each deployment is as follows:

1. Quality deployment is to systematically deploy customers' demanded qualities.
2. Technology deployment is to extract any bottleneck-engineering (BNE) that hinders the realization of the quality and solve it at the earliest possible time.
3. Cost deployment is to achieve target cost while keeping a balance with quality.
4. Reliability deployment is to prevent failures and troubles and their effects through early prediction.

TABLE 2.1. Steps to a quality deployment system (Mizuno and Akao, 1978)

Set Planned Quality

I. Demanded-quality deployment
1. Determine what "thing" to produce
2. Grasp market information and make demanded-quality deployment chart
3. Conduct competitive analysis and determine selling points

II. Quality characteristics deployment
4. Make quality characteristics deployment chart
5. Conduct competitive analysis of quality characteristics and reliability
6. Make quality chart
7. Claims analysis
8. Set planned quality and design quality
9. Determine the evaluation for development

Deploy Intrinsic Technology and Set Design Quality

III. Deploy intrinsic technology
10. Make function deployment chart

IV. Deployment of subsystems
11. Make subsystems deployment chart
12. Analysis of claims, quality characteristics, reliability, product liability (PL), and cost
13. Determine design quality and select critical-to-safety parts and critical-to-function parts
14. Make improvement by the implementations of VE and FMEA
15. Determine quality evaluation items
16. Design reviews

Detailed Design and Production Deployment

V. Parts deployment
17. Make parts deployment chart

VI. Operation methods deployment
18. Operation methods research and operation methods deployment

VII. Process deployment
19. Deployment of process control points (make QC engineering planning tables)
20. Establish quality standards, operating standards and inspection criteria
21. Design reviews and prototype evaluation

Deployment to Initial-Stage Process Control

VIII. Deployment to manufacturing shop floor
22. Make QC engineering tables
23. Add process control points by reverse function deployment
24. Management by priority
25. Deployment to outside suppliers
26. Actively conduct cause analysis
27. Change operating model or feedback to next development

TABLE 2.2. Steps for Implementing Quality Deployment (Akao, 1988)

Setting the Planned Quality and Design Quality

1. First, survey the requirements of target customers. Grasp the expressed and latent requirements to determine what kind of "products" to make.
2. Understand the market and make the demanded-quality deployment chart.
3. Conduct a competitive analysis with the products of other companies, and determine the planned quality and selling points.
4. Determine the weight of each demanded quality.
5. List the quality elements and make the quality elements deployment chart.
6. Make the quality chart.
7. Conduct a competitive analysis of how the products of other companies satisfy each quality element.
8. Analyze customer complaints.
9. Determine the important quality elements.
10. Check the characterization of quality elements and concretely set the design quality.
11. Determine the quality confirmation methods and test methods.

Detailed Design and Production Preparation (Subsystems Deployment)

1. Deploy "units deployment chart" and the corresponding components deployment chart. Convert the finished product quality into quality characteristics of the components.
2. Clarify units, component functions, component quality characteristics and specifications.
   Regarding the existing products, note down the process capability data. If process capability indices are known, note down the values; if not, indicate the levels with A, B, C, etc.
3. Identify the quality assurance items, critical-to-function characteristics and critical-to-safety characteristics of each unit and its components.
   The so-called critical-to-function characteristics are those characteristics that would prevent product function from being performed if they do not satisfy the specifications. The critical-to-safety characteristics are those characteristics that would harm human life if they do not satisfy the specifications. The focal point of quality assurance (QA) is to clarify the two characteristics, mark them in the drawings and communicate the design purpose to the manufacturing site.
4. Select the inspection items for the units and their components.

Process Deployment

1. Conduct research on processing methods (processing methods deployment) and propose various methods that can improve process capability. Test these methods.

TABLE 2.2. (*Continued*)

2. Determine the optimal processing methods that can satisfy the necessary degree of precision at the lowest cost.
   To do this, compute the precision and cost of each processing method, and draw a precision–cost curve to determine which processing method can realize high precision and low cost.
3. Evaluate the outcomes of all of the previous steps, as well as the performance of trial products, to assess the feasibility of transferring to production.
4. Establish the component quality standards, inspection criteria, and purchasing standards. Decide "make" or "buy" and establish their standards.
5. Make the process planning chart (facility deployment).
   When determining the required facilities and process conditions, clarify the QA check points for each facility based on the two-dimensional matrix of finished product quality characteristics against facility conditions.
6. Make the QA process planning sheet for parts processing, that is, convert part quality characteristics into process control items.
7. Make the QA process sheet.
   Finish the QA process sheet which specifies who will measure the control items and check items in QA process planning sheet and who will be responsible for taking actions in case of occurring abnormal problems.
8. Make the QA process sheet for assembly processes.
   As above, establish the control mechanism for control items in assembly processes.
9. Standardize operation procedures.
   At this time, all the control items must be soundly included in operations management system.
10. Seek deployment to subcontractors and suppliers.
    Request the subcontractors and suppliers to submit QA process sheet linked to the quality characteristics of each part. If they are incapable of making it, educate them or make the QA process sheet by ourselves and instruct them according to it.
11. Conduct the proactive analysis for problem prevention.
    Analyze the design data deployed and gained from upstream research, prototyping, trial production and initial manufacturing.
12. Apply the results of analysis to design changes or feed them forward to the development of next product.

2.4.2 The Four-Matrix Model

The Four-Matrix Model logically divides an NPD cycle into four phases using four matrices. The main activity in most current implementations of QFD is the generation of charts corresponding to these four phases (Prasad, 1998).

As the term suggests, the Four-Matrix Model is constituted by four matrices. One of the most common matrices is called the house of quality (HOQ) (also

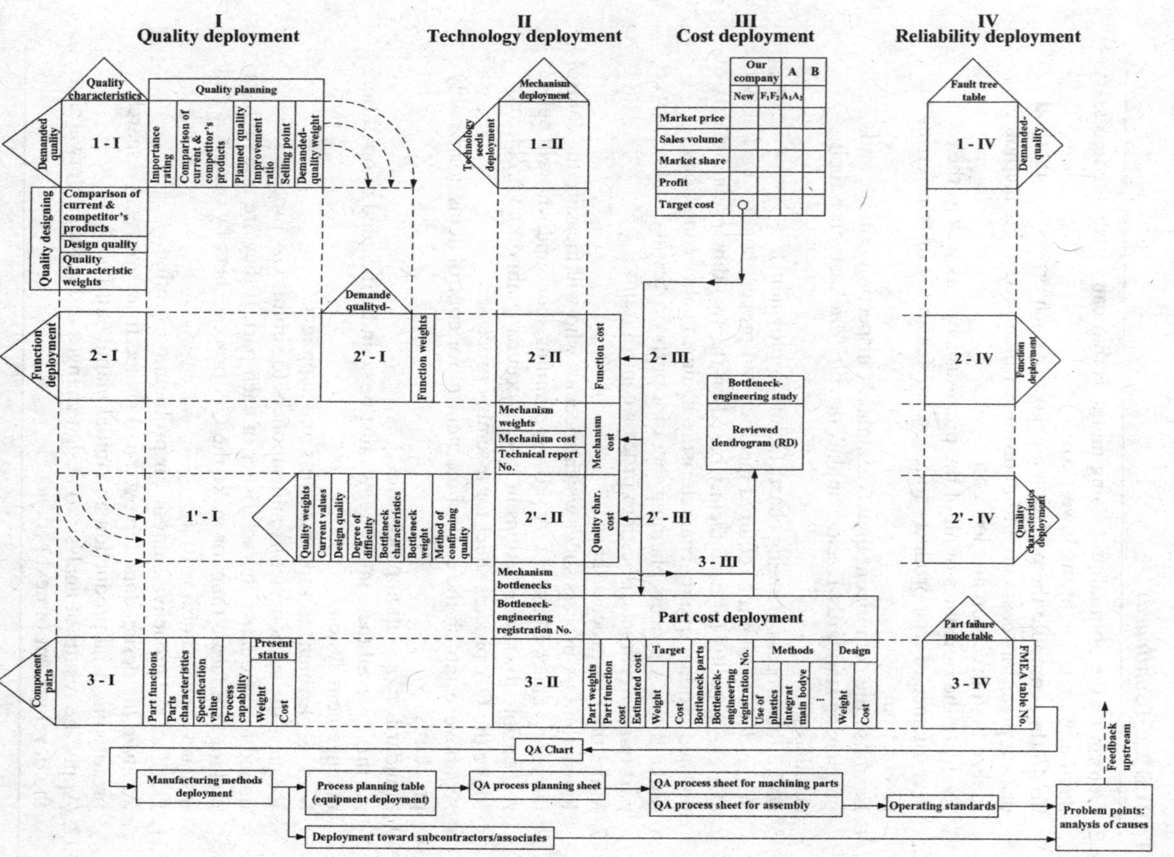

Figure 2.3. Quality deployment including technology, cost, and reliability (Akao et al., 1983).

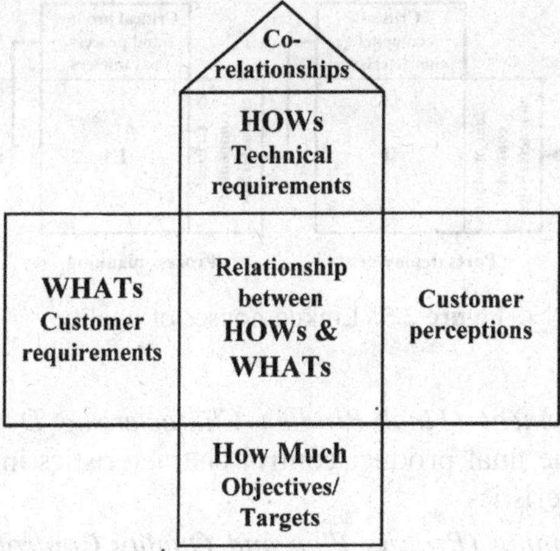

Figure 2.4. House of quality (Jiang et al., 2007).

known as the quality chart), as shown in Figure 2.4. This is a matrix that provides a map for the design process, as a construct for understanding customer requirements and establishing priorities of design requirements to satisfy them (Miguel, 2005).

HOQ can be used to provide four pieces of information: (1) WHAT is important to the customer? (2) HOW is it provided? (3) The relationships between the WHATs and HOWs. (4) How Much must be provided by the HOWs to satisfy the customer? (Ross, 1988). QFD utilizes a set of matrices to establish quality and deploy it through the NPD cycle. Therefore, other matrices are also used in a cause–effect relationship. These matrices relate the variables associated with one design phase to other variables associated with the subsequent design phase (Miguel, 2005). That is, the HOWs information in the first chart become the WHATs information for the second chart, and the HOWs in the second chart become the WHATs for the third chart, and the HOWs for the third chart become the WHATs for the fourth chart. QFD provides such a logical linkage between objectives (WHATs) and means (HOWs), along with a systematic method for developing or deploying the HOWs from the WHATs and for setting priorities (Hauser and Clausing, 1988; Cohen, 1995).

The Four-Matrix Model is usually depicted in a simple conceptual form as shown in Figure 2.5. These four matrices (also known as the "linked houses of quality") help to implicitly convey the voice of the customer through to manufacturing (Hauser and Clausing, 1988).

The four phases' respective objectives are more clearly described as follows (Sullivan, 1986; Kathawala and Motwani, 1994):

1. *Product Planning (Overall Customer Requirement Planning Matrix).* Translates the general customer requirements into specified final product control characteristics.

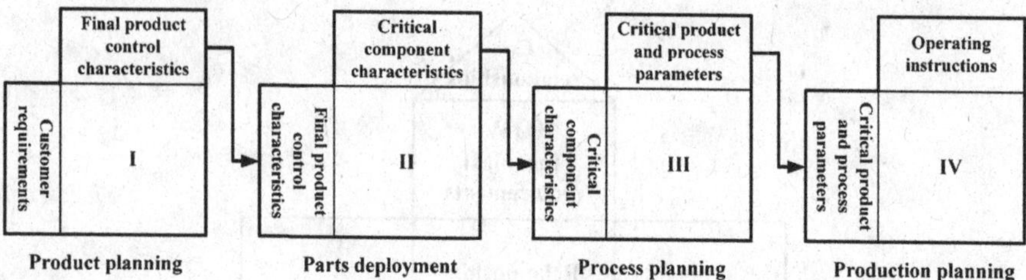

Figure 2.5. Linked houses of quality.

2. *Parts Deployment (Final Product Characteristic Deployment Matrix).* Translates the final product control characteristics into critical component characteristics.
3. *Process Planning (Process Plan and Quality Control Charts).* Identify critical product and process parameters, as well as develop check and control points for these parameters.
4. *Production Planning (Operating Instructions).* Identify operations to be performed by plant personnel to ensure that important parameters are achieved.

The overall QFD system based on these matrices traces a continuous flow of information from customer requirements to plant operating instructions, thus providing a common purpose of priorities and focus of attention (Sullivan, 1988).

2.5 SHORTCOMINGS OF QFD

While the structure provided by QFD can be significantly beneficial, it is not a simple tool to use. Many researchers considered that QFD is a complex and very time-consuming process to develop the QFD charts (Zairi and Youssef, 1995; Schmidt, 1997; Prasad, 1998; Bouchereau and Rowlands, 2000). Bouchereau and Rowlands (2000) considered that existing explanations of QFD assume that the customer requirements are constant—that is, either as unchanging over time or as the same for all customers at a given point. It does not address those situations where they are dynamic.

Prasad (1998) further indicated the major pitfalls of Akao's QFD approach as follows:

1. Conventional function deployment is mainly quality-focused and does not account for the complex PD^3 (product development, design, and delivery) process.
2. Conventional function deployment is a phased process that cannot be used for concurrent engineering.

3. Conventional function deployment is one-dimensional and difficult to address all aspects of total values management (TVM), such as X-ability, cost, tools and technology, responsiveness, and organization issues.

2.6 REVIEW COMMENTS ON QFD

The above content of this chapter offered an overview yet a basic knowledge of QFD. This section attempts to venture deeper into it from three aspects and extract some key issues to serve as focal points addressed in Part I of this book: (1) analyzing QFD's development trends and generation evolutions in Japan and the Western Hemisphere over the past 35 years, as well as discussing the misperceptions gradually produced when it spread and transferred internationally; (2) reviewing the shortcomings of QFD that have been pointed out, along with proposing the solutions, clarifications, or refutations for them; and (3) proposing an adaption model as reinforcement for practicing QFD in global supply chain.

2.6.1 Comments on QFD's Development Trends and Evolutions

Since Kogure and Akao published the first article that introduced QFD to the Western Hemisphere in 1983 (Kogure and Akao, 1983), QFD development has made tremendous progress in Japan and the Western Hemisphere (the United States in particular) alike and thereby strengthened QFD's body of knowledge.

Sullivan (1986) and Hauser and Clausing (1988) followed Kogure and Akao (1983) in putting a quality chart as their focal point when discussing QFD. They illustrated how to use a quality chart to convert customer's demands (WHATs) into substitute quality characteristics (HOWs) and determine their design targets (How Much) in great detail, but only used a simple conceptual model of "linked houses of quality" to show how a quality chart is combined with the following other deployments: subsystems deployment, parts deployment, and process deployment. Even so, they successfully made "house of quality" or "linked houses of quality" the most recognized QFD form today in the Western Hemisphere. In 1987, King proceeded to interpret QFD using the "30-matrix model" (King, 1987), which became the most complete QFD model made by Western researchers. The 30 matrices include:

1. Twenty matrices formed on purpose by the following deployments: (1) customer demands, (2) quality characteristics, (3) functions, (4) mechanisms first-level detail, (5) parts second-level detail, (6) new technology, (7) new concepts, (8) product failure modes, (9) parts failure modes, and (10) cost.

2. Based on the above 20 matrices, an additional 10 analysis tables and charts are needed: (1) value engineering, (2) product FTA (fault tree analysis) and FMEA (failure mode and effects analysis), (3) PDPC (process decision program chart) and RD factor analysis, (4) design improvement plan, (5) QA table, (6) equipment deployment, (7) process planning chart, (8) process FTA, (9) process FMEA, and (10) QC process chart.

Sullivan, King, and Clausing were arguably the most important figures in the initial stages of QFD development in the Western Hemisphere. When QFD was gradually known to researchers or practitioners in a number of fields, QFD combined with various design methodologies and numerical analysis methods, which also became a research trend afterward. These research areas can be divided into three main aspects as follows (Jiang et al., 2007):

1. QFD combined with TRIZ (Russian theory of inventive problem solving) (Terninko, 1997; ReVelle et al., 1998), Pugh concept selection (Clausing and Pugh, 1991; Clausing, 1993), and Taguchi methods (Clausing, 1988; Ross, 1988; Clausing and Simpson, 1990; Bendell et al., 1993; Clausing, 1993; Eureka and Ryan, 1995; Terninko, 1997; ReVelle et al., 1998; Bouchereau and Rowlands, 2000), in order to strengthen the effectiveness of QFD in product design and process design.
2. QFD combined with different numerical analysis methods, such as fuzzy sets (Masud and Dean, 1993; Wasserman et al., 1993; Khoo and Ho, 1996; Zhou, 1998; Temponi et al., 1999; Bouchereau and Rowlands, 2000; Kim et al., 2000; Büyüközkan et al., 2004a, b; Ramasamy and Selladurai, 2004; Chen et al., 2005), AHP (analytic hierarchy process) (Armacost et al., 1994; Lu et al., 1994), and neural network (Zhang et al., 1996; Bouchereau and Rowlands, 2000; Yan et al., 2005), in order to strengthen the accuracy of QFD in weight determination and numerical analysis.
3. The research of QFD application, such as QFD combined with strategic management, policy management, and project selection (Sullivan, 1988; Lu et al., 1994; Crowe and Cheng, 1996; Partovi, 1999; Hunt and Xavier, 2003; Killen et al., 2005; LePrevost and Mazur, 2005), and QFD's key successful factors in practice (Cox, 1992; Cristiano et al., 2001; Miguel, 2005).

With the development of "Six Sigma" (6σ) in recent years, the Western QFD has been consolidated into the core of the DFSS (design for Six Sigma) tool set. Some 6σ experts, such as Creveling et al. (2002), renamed QFD as "critical parameter management" (CPM). CPM process is summarized as follows:

1. Gather and identify critical voice of the customer needs, translate them into critical technical requirements, and then systematically (1) deploy the critical system requirements (critical function responses), critical

subsystem requirements (critical function parameters) and critical component requirements (critical-to-function specifications) by product architecture and (2) deploy the critical process and material requirements (critical process parameters and responses) by the manufacturing process.

2. Measure C_p/C_{pk} (process capability indices) values of the critical items of the deployed manufacturing processes, components, subsystems, and system, and take corrective actions during steady-state production and service.

The 6σ experts have made CPM as the axis when DMADV (design, measure, analyze, design, verify) activity is carried out. The core concept of DMADV implementation is to gradually "flow-down" new product's critical-to-quality characteristics (CTQs) to functional design, detailed design, and process control variables, and then "flow-up" capability to meet these requirements (Hahn et al., 2000). DFSS is considered a system by 6σ experts and facilitators, and an integrated approach is adopted to provide the applicable design tools and methodologies, as well as the required knowledge and skills for managing projects in NPD (Hahn et al., 2000; Antony and Banuelas, 2002; Creveling et al., 2002; Mader, 2002). As a result, when CPM is considered an extended model or extended application of QFD, a DFSS system may be deemed as a more integrated and more convincing body of knowledge of QFD.

On the other hand, in the same year when Kogure and Akao (1983) introduced QFD to the Western Hemisphere, Akao et al. (1983) published in Japan a comprehensive system of quality deployment—including technology, cost, and reliability—which can be considered the most complete quality deployment system and was recognized as a main form of QFD, different from the Western Hemisphere's "linked houses of quality" (also known as the "four-matrix model"). Though Akao (1988, 1990b) and Ohfuji et al. (1990, 1994) made publications successively about comprehensive quality deployment, these publications were in Japanese language. Therefore, the comprehensive model of quality deployment was not generally known in the Western Hemisphere, so the English literature related to the model was considerably limited.

In 1990 and 1996, the QFD Research Committee of JUSE (Japanese Union of Scientists and Engineers) discussed a series of subjects based upon comprehensive quality deployment and published them in *Quality Management* journal of JUSE (Akao et al., 1990; Akao, 1996). These subjects included (1) demanded-quality identification methods and its relationship with marketing, (2) *kansei* (sensibility), (3) quality deployment, (4) technology deployment, (5) cost deployment, (6) reliability deployment, (7) comprehensive QFD, (8) QFD in software development, (9) narrowly defined QFD, and (10) QFD as development management engineering. Moreover, Kanda (1995) and Mochimoto (1996, 1997) also proposed seven tools for new product planning (P7) and quality deployment for market pricing (QD_m) to strengthen QFD's effectiveness in quality planning and price design.

At about the same time, the committee conducted a four-year (1991–1994) research and discussion over narrowly defined QFD (Shindo, 1998). Since 1994, Akao began to promote the term *development management* (DM) (Akao, 1997) to materialize the concept of broadly defined QFD. It is defined as the use of quality deployment to ensure the design and manufacturing quality of new products and through narrowly defined QFD to establish various planning and operational QA activities and procedure flows needed to achieve product quality. DM aims to ensure product quality and the quality of work activities simultaneously and thereby strengthen the management of the overall system of NPD. These research and development trends affected the subsequent research of Japanese QFD and made the Japanese QFD particularly focused on how to use a systems approach rather than a project approach to manage NPD. The model, real-time database QFD (Rdb-QFD) can be considered representative of the Japanese QFD toward this direction, and its original concept was proposed by Ohfuji in 1996 (Shindo, 1998). Rdb-QFD refers to QFD practice based upon narrowly defined QFD (also known as the deployment of NPD's QA activities), and all quality deployment charts can be retrieved in real time from a computer database according to the needs in each phase of NPD. More specifically speaking, the concept of Rdb-QFD is as follow: Each quality deployment chart and its related product and process data are made and managed by the sales department, the R&D and engineering division, the production engineering division, the material procurement division, the manufacturing division, the QA division, and the after-sale service division, respectively according to business departmental functions; also, they are delivered, linked, and used within the organization in real time via the information system according to the need at each phase of the NPD cycle. Rdb-QFD emphasizes the implementation of QFD by cross-functional cooperation and operation, rather than task-oriented project teamwork, which enables the implementation of QFD to be fused into business practice so that it can facilitate the effectiveness of QFD implementation in organization.

On the same development track, in recent years there have been a great many case studies in Japan discussing how to establish a corporate quality management system (Akao, 2001a, b; Akao, 2004; Yoshizawa et al., 2004) by implementing QFD and to practice organizational knowledge management (Akao, 2002a, b; Akao, 2010; Akao and Inayoshi, 2003; Akao and Tsugawa, 2004; Akao and Kamimura, 2005; Akao and Kozawa, 2005; Akao and Takaomoto, 2005). Also, QFD has been classified as the following seven evolutions, e7-QFD for short, in terms of application areas (Nagai and Ohfuji, 2008; Ohfuji, 2010):

1. *Quality Assurance-QFD (QA-QFD).* Conventional QFD that focuses on product quality assurance.
2. *Job Function-QFD (Job-QFD).* QFD that emphasizes job function deployment for work quality.

3. *Taguchi Methods and TRIZ-QFD (TT-QFD).* QFD that integrates Taguchi Methods and TRIZ for design optimization and inventive problem solving.
4. *Statistical-QFD (Stat-QFD).* QFD that incorporates the use of statistical methods for effective market data analysis.
5. *Blue Ocean Strategy-QFD (BOS-QFD).* QFD that applies to new business creation and value innovation.
6. *Real-Time Database-QFD (Rdb-QFD).* QFD that incorporates computer system and real-time design database for computer-aided engineering.
7. *Sustainable Growth-QFD (Sus-QFD).* Wide utilization of QFD from the viewpoint of corporate management aiming for sustainable growth.

Additionally, Akao (2012) used a case study to demonstrate how QFD applied to improve the quality of motivation programs. These evolutions mentioned above represent the diversity of QFD and the expansibility for advancement in the future.

Figure 2.6 illuminates QFD developments in Japan and the Western Hemisphere over the past 35 years.

In general, after QFD was spread from Japan to the Western Hemisphere, some misunderstandings about the original QFD were gradually engendered and can be found in various relevant articles and books. The common misunderstandings include the following:

1. *Quality Deployment Being Equivalent to a Quality Chart.* Although quality deployment has been simplified as the form of "linked houses of quality" in the Western Hemisphere, the majority of QFD applications stop with the completion of the first-matrix quality chart, which is a chart used to convert customer's demands into substitute quality characteristics and determine their design targets (Fortuna, 1988; Cox, 1992; Cohen, 1995; Partovi, 1999; Bouchereau and Rowlands, 2000; Cristiano et al., 2000, 2001; Booker, 2003). Cox (1992) indicated that no more than 5 percent of Western companies go beyond the quality chart. However, most Japanese companies not only go beyond the quality chart to use the subsystems deployment chart, parts deployment chart, and process deployment chart (all of these are known as quality deployment), but also implement technology deployment, cost deployment, and reliability deployment (Cristiano et al., 2001).
2. *QFD Being Equivalent to Quality Deployment.* Broadly defined QFD refers to the combination of quality deployment and narrowly defined QFD. However, the acronym QFD has been used to mean quality deployment in itself, and narrowly defined QFD has been almost ignored by most Western QFD practitioners (Akao and Mazur, 2003; Akao, 2004). Hence, we can discover that the Western QFD developed in the research

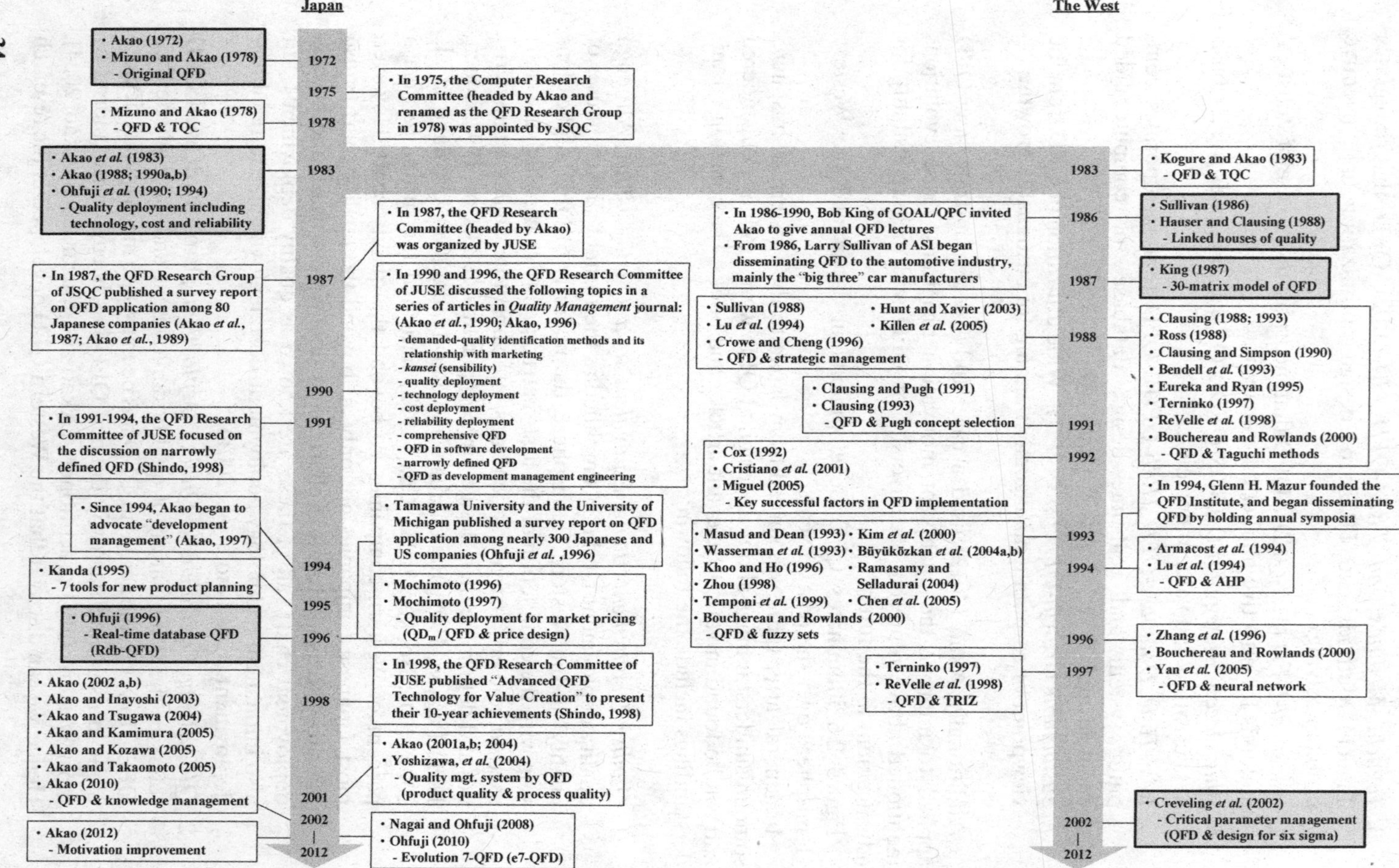

Figure 2.6. Overview of the development of QFD over the past 35 years.

direction of "how to integrate various design tools and methodologies as well as more powerful numerical analysis methods"—which was different from the Japanese QFD, which considers narrowly defined QFD to be the essential part for QFD to gain long-term buy-in, implementation, and compliance (Akao and Mazur, 2003).

With respect to the first misperception, apart from the fact that quality deployment includes (at least) a quality chart, individual part deployment, and process deployment, more should be said about comprehensive quality deployment. More than 20 years have passed since Akao et al. (1983) published a comprehensive system of quality deployment including technology, cost, and reliability, and 15 years have passed since its English publication (Akao, 1990a). Despite this passage of time, there have been very few English-language publications on the topic (Chan and Wu, 2002), but King's (1987) 30-matrix model deserves to be mentioned. "Linked houses of quality," as shown in Figure 2.5, is an often seen conceptual model of quality deployment. However, quality deployment is regarded as simply putting these houses into their logical sequence; in the absence of a thorough study of the connections among them, it is difficult to explain and deal with the deployments of technology, reliability, and cost that are necessarily involved in NPD. It is thus impossible to ensure that bottleneck-engineering (BNE) is discovered in the early phases of NPD, that potential failures or troubles are prevented, and that the target cost is achieved.

In addition, although narrowly defined QFD was combined with quality deployment to make up broadly defined QFD, it has been largely overlooked in the popularization of QFD. For this reason, Akao (2004) has stated that recognition of narrowly defined QFD should be reinforced.

Mistaking QFD for quality deployment has also caused difficulties in managing product-development knowledge. This has occurred because QFD is easily implemented by a project approach, rather than by a systems approach. The difference between the two is best represented by Nonaka and Takeuchi's (1995) hypertext organization model. This organizational structure enables an organization to create knowledge efficiently and continuously. As a computer operator moving easily through a hypertext document, organizational members can easily shift contexts—that is, get in and out of the various layers that exist within an organization—such as a project-team layer, a business-system layer, or a knowledge-based layer. A typical application of QFD is to build a cross-functional team and then launch the so-called QFD project on a new target product by completing various deployment charts and by using limited time and irregular operational limits. However, this approach (often characterized as QFD, although it is actually quality deployment) does not deal with the network of quality functions that are involved. In this situation, the effectiveness of the QFD that the company professes to implement is likely to be limited. Such an approach is inadequate because activities performed at the project-team layer cannot be effectively transferred into the business-system layer—with the result that accumulation of organizational

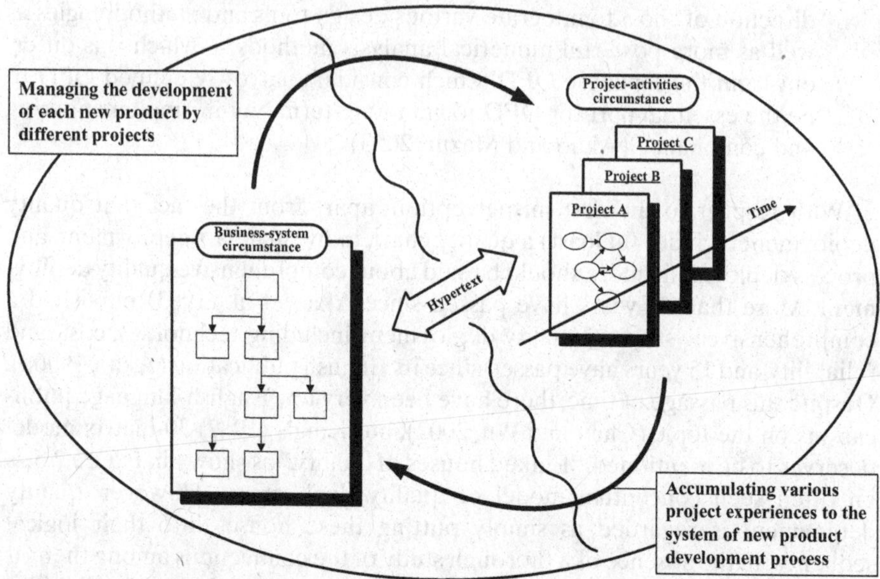

Figure 2.7. Conceptual model of hypertext NPD environment.

knowledge is difficult. Figure 2.7 shows the conceptual model of hypertext NPD environment.

We divide QFD of present and past in Japan and the Western Hemisphere (the United States in particular) into three generations in terms of evolution and give review comments, as shown in Table 2.3.

In the past, although some misperceptions about the essence of QFD did exist, Western QFD basically developed following the model of Japanese QFD. However, the difference of the current development of the so-called "third generation of QFD" between Japan and the Western Hemisphere can be considered a kind of change. It means that Japan and the Western Hemisphere have begun to develop their own QFD in different directions:

1. The current Japanese QFD, a real-time database QFD (Rdb-QFD) whose original concept was proposed by Ohfuji in 1996, is based on narrowly defined QFD (also known as the deployment of NPD's QA activities) for implementation, and all quality deployment charts can be retrieved in real time from a computer database according to the needs in each phase of NPD.
2. The Western QFD has now combined with TRIZ, the Kano model (an analysis for stratifying the importance of demanded qualities based on customer's perception), and Taguchi methods and have been developed by 6σ experts into a form of "critical parameter management" (CPM) that is used, in implementing DFSS activities, as the axis for the DMADV mode (define, measure, analyze, design, verify).

The main difference between the latest development of QFD in Japan and the Western Hemisphere is that the former's QFD, in concept and methodology, is toward the direction of "how to add value to every work activity in NPD cycle to achieve QA and competitively improve product quality" (i.e., to combine quality deployment and narrowly defined QFD to materialize the concept of broadly defined QFD), whereas the latter's QFD is toward the direction of "how to integrate various design tools and methodologies as well as more powerful numerical analysis methods to achieve QA and competitively improve product quality" (i.e., to focus on quality deployment in essence while neglecting narrowly defined QFD).

2.6.2 Comments on QFD's Shortcomings

This section will not only (1) give the clarifications or refutations, but also (2) extract some of the certain key issues and solutions against the QFD shortcomings pointed out by some researchers in the field:

1. *QFD Is Complex and Very Time-Consuming* (Zairi and Youssef, 1995; Schmidt, 1997; Prasad, 1998; Bouchereau and Rowlands, 2000). This is the most pointed-out shortcoming for QFD. The basic type structure of quality deployment already includes demanded-quality deployment, quality characteristics deployment, subsystems deployment, process deployment, and quality information exchanges among them. Implementation of a comprehensive system of quality deployment goes a step further to combine with technology deployment, cost deployment, and reliability deployment, and the resulting quality information network becomes more sophisticated. As a result, in the Western Hemisphere it is the first chart of quality deployment, the quality chart (house of quality), that is generally used because it is considered to have the biggest benefit. However, as known from the historical development of QFD, at first QFD was developed and expanded its size to prevent or resolve various problems in the practice of NPD, and then it was theorized into a reference model. QFD could not help being designed to be complicated due to the complexity of the practice of NPD per se, in order to prevent or address possible problems in the NPD cycle.

Another reason why QFD is considered to be complex and time-consuming is because the need for narrowly defined QFD to combine with quality deployment is neglected in the implementation of QFD, as described in Section 2.6.1. Consequently, QFD is usually practiced with a project approach rather than with a systems approach that combines with NPD cycle. Therefore, while QFD is considered to require additional project efforts (rather than practiced simultaneously in NPD cycle) and aims at the particular target product of the project (rather than be adapted and applied to future new products), the need to complete so many forms and create relationships among them would be misperceived as consuming too much time and sources of NPD, even affecting the time to market. In fact, as suggested by

TABLE 2.3. QFD Generation Evolutions in Japan and the Western Hemisphere

Item		First Generation	Second Generation	Third Generation
Japan	Developer and publication year	Akao (1972); Mizuno and Akao (1978)	Akao et al. (1983)	Ohfuji (1996)
	Representative model	Original QFD	Quality deployment including technology, cost, and reliability	Real-time database QFD (Rdb-QFD)
	Evolution characteristics	• Provides a systematic guidance on how to implement quality designing-in • Deploys the design targets and major QA points used in NPD prior to production start-up • Develops a model of "seeing a whole on one page" for NPD's QA management	• Comprehensively deploys quality, technology, cost and reliability in which NPD is bound to get involved	• Applies information technology (IT) to significantly enhance QFD's implementation efficiency • Deployment of NPD's QA activities (narrowly defined QFD) has become the axis in QFD implementation. All quality deployment charts can be retrieved in real time from a computer database according to the needs in each phase of NPD to materialize the concept of broadly defined QFD
	Potential vulnerabilities	• Not easy to specify how quality deployment integrates narrowly defined QFD in its implementation	• While developing the completeness of quality deployment, does not further strengthen narrowly defined QFD or specify how quality deployment integrates it in implementation	• Application of a real-time database not easy to facilitate QFD's content development and implementation effectiveness (though it can significantly enhance implementation efficiency)

The Western Hemisphere	Developer and publication year	Sullivan (1986); Hauser and Clausing (1988)	Bob King (1987)	6σ experts, such as Creveling et al. (2002)
	Representative model	Four linked houses of quality	Thirty-matrix model of QFD	Critical parameter management (CPM)
	Evolution characteristics	• The earliest QFD model made by Western researchers and the most recognized QFD form today in the Western Hemisphere	• The most complete (including technology, cost and reliability) quality deployment model made by Western researchers	• QFD combines with TRIZ, Kano model, and Taguchi methods • Implementation axis for the DMADV mode (define, measure, analyze, design, verify) of DFSS
	Potential vulnerabilities	• Oversimplifies QFD as a quality house (quality chart) or linked quality houses result in the losses of some other important quality information exchanges in the original QFD • Easily leads to the misunderstanding of quality deployment as a quality chart • Easily leads to the misunderstanding of QFD as quality deployment and causes its implementation to tend toward a project approach rather than a systems approach	• The quality deployment divided into 30 matrices loses the wholeness of the original comprehensive system as well as the connection among all deployments in macro view (though it can increase the understanding of content detail)	• Gives no consideration to the relationships among technology, cost, and reliability in which quality deployment and NPD are bound to get involved • Implementation of DMADV and CPM model tend toward a project approach rather than a systems approach

its definition and concept, QFD should be integrated into NPD cycle for implementation rather than be independent from it. As a result, the time needed for QFD actually is smaller than or equal to the time spent by NPD cycle. Besides, though the quality information network created by QFD is complicated, for development of similar products, most of them can continue to be used, and one only needs to focus on design change for priority management. That is, the "complexity" of QFD actually helps to diminish not only the complexity regarding R&D resulting from design uncertainty in new product, but also the schedule's delay. Cohen (1995) also indicated that experience has shown that the QFD process generally covers more ground faster than less structured methods. Not only can QFD reduce the overall development cycle by providing more complete planning, it also takes less time than less structured planning methods.

To sum up, the shortcomings of QFD's complexity and time-consuming nature are an inevitable result from practical needs rather than being derived from QFD methodology itself. Therefore, they do not come from QFD itself. Moreover, if the use of QFD can reduce the development time by 30–50 percent (Fortuna, 1988; Hauser, 1993; Lockamy and Khurana, 1995; Swink, 2002), criticizing QFD as very time-consuming would not make sense.

In order to address this kind of misperception about QFD, an active approach is to propose an NPD-process-oriented QFD implementation procedure—that is, to develop a QFD model that can achieve integration and simultaneous execution with the NPD cycle.

2. *QFD Assumes That the Customer Requirements Are Constant, and It Does Not Address Those Situations Where They Are Dynamic* (Bouchereau and Rowlands, 2000). The demanded qualities are the true qualities wanted by the customer. The customer's needs of true qualities almost never change along the time, that is, these fundamental requirements are constant; but the customer might dynamically change the focus of these demanded qualities along product evolution or time. The demanded-quality deployment chart in QFD is using the evaluation model of quality planning to deal with that situation; that is, when the focus of demanded qualities (i.e., importance evaluated according to the customer's stance) changes, the company's quality positioning planning also adapts along product-market strategy determined according to the company's stance, competitive analysis, and selling points determination which is not unchanged.

Thus, when customers' demanded qualities can be completely identified and properly described, the company grasps all the customer's fundamental requirements toward product quality. Hence, the demanded-quality deployment can offer a complete reference base and analysis framework of total customer needs. Companies can use it to proactively (1) select the most effective quality niches for the dynamic change of market and competitive environment and (2) create an unexpected new quality to delight the customers [also known as the strategy of "attractive quality creation" advocated by Kano (1994), (2002)].

In order to address this kind of misperception about QFD, an active approach is to promote the understanding of how to properly identify and describe the demanded qualities, in order to make the demanded-quality deployment more effective for the dynamic management of product's suitability to customer needs.

3. *QFD Is Mainly Quality-Focused and One-Dimensional without Addressing All Aspects of Total Values Management* (TVM) (Prasad, 1998). According to the study on quality theory (Kano et al., 1984), Aristotle (384–322 BC) may have been the first person to talk on the subject of quality in any systematic way. There are two aspects of quality along the flow of Aristotle's thoughts on quality: "different quality" and "good quality." Kano et al. (1984) also indicated that the discussions of quality have revolved around these two aspects since the time of Aristotle, despite differences in expression.

Since 1978, Akao quoted Juran as defining "quality" as "fitness for use" to describe the implications of "quality" in QFD (Mizuno and Akao, 1978). Juran's "fitness for use" includes the meanings of (1) "product features" trended toward "different quality" and (2) "freedom from deficiencies" trended toward "good quality." Hence, "quality" deployed by QFD not only refers to "good or bad generally meant by quality," but also refers to the differentiation of multiple dimensions like X-abilities. That is, when "quality" is recognized as "good or bad" while neglecting its implication of "differentiation," QFD is prone to be also recognized as merely deployed "quality" (i.e., good or bad) without giving regard to other dimensions of deployment (i.e., X-abilities).

Moreover, another reason why the necessity of addressing all aspects of TVM is proposed is because the word "function" in QFD was narrowly explained as "the functions of a product," rather than as "the QA job functions of a NPD cycle."

In order to address this kind of misperception about QFD, an active approach is to (1) develop an objective deployment structure of "quality" to deploy complete dimensions of "quality," which helps at the outset of product quality deployment to plan and design for excellence and completeness, and (2) establish a QA function deployment system for NPD to reinforce the recognition of narrowly defined QFD.

4. *QFD Is a Phased Process That Cannot Be Used for Concurrent Engineering* (Prasad, 1998).

When QFD is implemented by combining with NPD cycle, the linkage between it and NPD activities is generated according to the needs in each phase of the NPD cycle. Hence, when certain NPD activities with sequential relationships need to combine with QFD contents to generate some quality information, QFD also must be implemented in accordance with the sequence of its activities. In contrast, under conditions that some NPD activities can operate simultaneously, QFD's contents are simultaneously generated by different NPD activities. That is, QFD's contents are generated in coordination with the sequential or simultaneous operations of NPD activities. It is not restricted by the characteristic of the so-called phased process.

In order to address this kind of misperception about QFD, an active approach is to clarify the interrelationship between QFD and the NPD cycle—that is, to reinforce the recognition of how QFD simultaneously implemented with the NPD cycle.

2.6.3 Comments on QFD's Applications

The application of QFD to product development according to the business model of a company is achieved as follows:

1. In companies that emphasize their own product design, the internal activities involve the setting of product specifications. The internal process begins with demanded-quality deployment, and it makes use of planned quality information to determine critical substitute quality characteristics and design quality.
2. In companies that emphasize fabrication engineering, the product development activities begin with acceptance of a contract that clearly sets product specifications (substitute quality characteristics and design quality having already been determined). Demanded-quality deployment and quality planning before the setting of product specifications are part of the internal activities of the customer's company.

The above observations derive from the division of labor in the global supply chain. Generally speaking, Western and Japanese companies give priority to product design and marketing, whereas non-Japanese Asian countries are often perceived to be the "world's factory" (for reasons of manufacturing capability and cost advantage). This might explain why countries in the Western Hemisphere put particular emphasis on the quality chart when it comes to QFD. Indeed, QFD was conceived when Japanese industries were advancing toward product development based on originality, and it was therefore not designed for application in a business model that emphasizes fabrication engineering. The quality chart is the first chart of quality deployment. Therefore, the company may determine if it needs QFD according to it. In terms of the use of the quality chart from the above corporate NPD model, a majority of fabrication suppliers may consider that they do not need a quality chart and do not even need to practice QFD. Unlike own-design companies who consider core competence the ability of having "customer knowledge," fabrication suppliers consider it the ability of having "engineering knowledge." Therefore, their sources of innovation also differ: Own-design companies' innovation is driven by the "market pull" approach, whereas fabrication suppliers' innovation is driven by "technology push" approach. This difference causes QFD to vary in its application focus under the two different NPD models, so that QFD can be practiced more effectively in a supply chain environment. The comparison between own-design companies and fabrication suppliers is shown in Table 2.4.

TABLE 2.4. Comparison Between Own-Design Companies and Fabrication Suppliers

Item	Own Design Companies	Fabrication Suppliers
NPD model	NPD activities begin with actively studying customer's needs or receiving his commission to develop product planning and design concept, and based on the design concept, product development and production is then conducted	NPD activities begin with receiving finished product specifications commissioned by customer for manufacturing. Product development and production is then conducted based on the specifications
Quality focus	Quality differentiation	Quality goodness
Quality level that can be achieved	Having the most opportunity to create "attractive quality"	Even conformance to specifications is merely ensuring "must-be quality"
Core competence	Customer knowledge (marketing and product design capabilities)	Engineering knowledge (process design and manufacturing capabilities)
Product innovation driver	Market "pull"	Technology "push"

The important characteristics of QFD reinforced for application in fabrication companies are described as follows. With these characteristics, QFD will be more widely accepted and applied by fabrication suppliers, so as to be more effectively practiced in the global supply chain (Jiang et al., 2007c):

1. Though the NPD model of fabrication suppliers does not include product specifications determined using demanded-quality deployment and planned quality information, a new QFD must be equipped with an analytical mechanism that can conduct technical feasibility evaluation and design change management for customer-proposed specifications.
2. Because fabrication suppliers are professional manufacturing process designers and manufacturers, they must carry out a more effective process deployment approach than the original QFD to conduct technical feasibility evaluation and design change management of process design in process design and production.
3. The proactive prevention of potential design problems can be done by adequately grasping negative quality information and implementing failure mode and effects analysis (FMEA). However, the original QFD did not emphasize the implementation of the process design FMEA, so it is necessary to add the analytical mechanism to the new QFD.

4. Establish a deployment structure for narrowly defined QFD to deploy NPD's relevant QA activities such that quality deployment charts can be integrated more effectively with NPD cycle.

2.7 CONCLUDING REMARKS

In view of the above reviews, the extracted key issues can then be summarized as follows and dealt with by an expanded system of QFD as described in Chapter 3:

1. To comprehensively renew QFD's four aspects of quality, technology, cost, and reliability, in order to reinforce its quality designing-in methods.
2. To develop a more NPD-process-oriented approach to implement QFD, in order to effectively support the corporate NPD cycle.

Besides, we consider that the most important insight for QFD is in understanding the duality of theory and practice:

1. In discussing the quality assurance of NPD, QFD created an expression form of "seeing a whole on one page" for quality deployment. In addition, QFD uses it to (a) introduce how to convert customer's demands into design quality systematically and (b) deploy the quality over the design targets and major QA points used throughout the NPD cycle. Though narrowly defined QFD can be deemed as the NPD's QA practice itself in a company, quality deployment introduced by the expression form mentioned above is far more effective than narrowly defined QFD in facilitating researchers and learners to understand the most complicated NPD cycle in a company. That is, quality deployment can offer up a more general structure in discussing NPD, and cases of different companies can be discussed by means of attaching to it. This may partially explain why QFD is generally recognized as quality deployment.
2. QFD's practice must revolve around narrowly defined QFD, while quality deployment is applied to reinforce it. That is, in practice, QFD is implemented by establishing NPD's QA system and its activity lists as well as integrating the needed quality deployment charts and quality tools into the NPD system.

3 Expanded System of QFD

3.1 OVERVIEW OF EQFD SYSTEM AND ITS IMPLEMENTATION PROCESS

This chapter describes EQFD (expanded QFD) developed based on the structure of broadly defined QFD. QFD, broadly defined, refers to the combination of quality deployment and narrowly defined QFD. Quality deployment is viewed as the "breakdown structure of product quality" or "network of quality" used to ensure the quality of a product itself, and narrowly defined QFD is viewed as the "breakdown structure of job functions that create quality" or "network of quality functions" used to ensure the quality of the work activities required to achieve product quality.

Based upon this, we intend to renew the deployment structures of QFD's quality deployment and QA function deployment by developing an EQFD system and its implementation process. The concept is shown in Figure 3.1.

The EQFD system developed by this book, as shown in Figure 3.2, is a comprehensive system developed based on the four deployments (quality, technology, cost, and reliability) structure of the original QFD (Mizuno and Akao, 1978; Akao et al., 1983) and by renewing its contents and analytic mechanisms. Quality deployment and reliability deployment are used to realize customers' demanded qualities and failure-free qualities, while cost deployment and technology deployment are used to determine the allocated cost that can balance with quality and identify the product and process design changes and bottleneck-engineering (BNE).

In Figure 3.2, the codes I, II, III, and IV, respectively, stand for quality deployment, technology deployment, cost deployment, and reliability deployment; and codes 1, 2, 3, and 4 are, respectively, four applied scopes that deploy at different time points: finished product, semi-finished product, process, and shop floor management. As for making of quality deployment charts, it is conducted in general from codes I to IV and from codes 1 to 4.

Besides proposing an EQFD system by renewing the original QFD, we also intend to develop a process that can simultaneously execute EQFD and NPD

Quality Strategy for Research and Development, First Edition. Ming-Li Shiu, Jui-Chin Jiang, and Mao-Hsiung Tu.

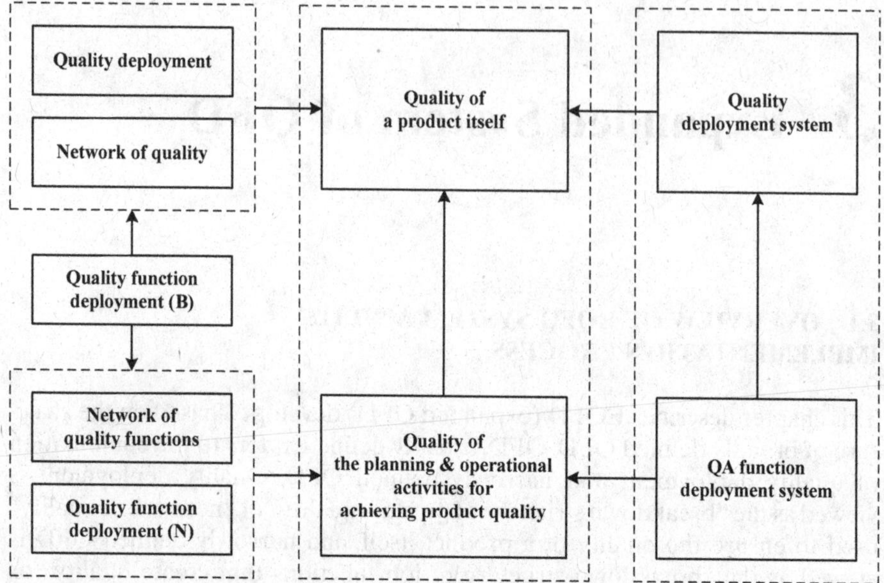

Figure 3.1. Conceptual framework of QFD.

cycle—that is, to develop an implementation process for how to use an NPD approach to establish the EQFD system shown in Figure 3.2 step by step. Though EQFD is suitable for application in both the NPD model of own-design companies and fabrication suppliers, we take the NPD cycle of electronics companies that emphasize fabrication engineering, for example, to illustrate how NPD integrates EQFD, and additionally we clarify how to design the important characteristics mentioned in Section 2.6.3 into the EQFD implementation process.

As for the NPD cycle of electronics fabrication companies, it generally refers to five phases of request for quotation (RFQ), prototype design, engineering verification test (EVT), design verification test (DVT) and production verification test (PVT) covered by product design and production preparations. The purposes of each phase are shown in Table 3.1, and the quality progress of new product experienced these phases is shown in Figure 3.3.

The EQFD implementation process is formed by four stages, eight phases, and 36 steps. The four stages refer to business/product and technology development planning, product design, production preparations, and mass production; the eight phases refer to the five phases from RFQ to PVT and the related business and product planning, technology development planning, and shop floor real-time management and abnormality management. As for the 36 steps, they are the combination of the activities conducted in practice of eight phases and the contents developed from the requirements of EQFD in implementation, as shown in Table 3.2.

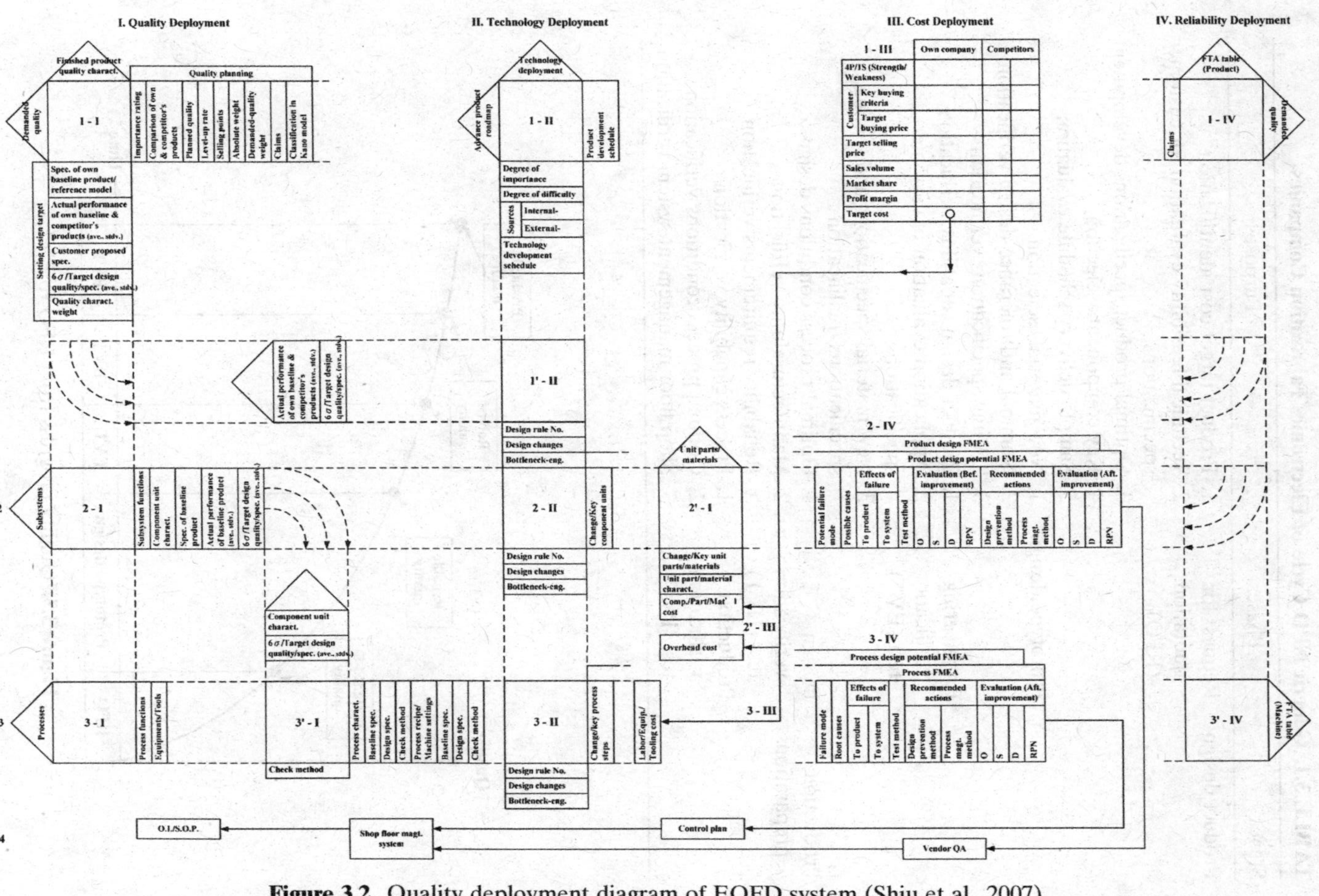

Figure 3.2. Quality deployment diagram of EQFD system (Shiu et al., 2007).

TABLE 3.1. Generic NPD Cycle of Electronics Fabrication Companies

Stages	Phases	Purposes
Product design	Request for quotation (RFQ)	1. Product target cost identification 2. Technical feasibility evaluation for customer specifications 3. Optimal product specifications development 4. Key components selection 5. Sample delivery schedule evaluation
	Prototype design	1. New materials selection 2. Nominal and tolerance design verification 3. Tooling specifications verification
	Engineering verification test (EVT)	1. Product design capability verification 2. Application evaluation of customer specifications 3. New material specifications and tooling specifications qualification
Production preparations	Design verification test (DVT)	1. Optimal process conditions design 2. Manufacturability verification 3. Reliability requirements verification
	Production verification test (PVT)	1. Process capability verification 2. Optimal process conditions verification 3. Shop floor management system verification

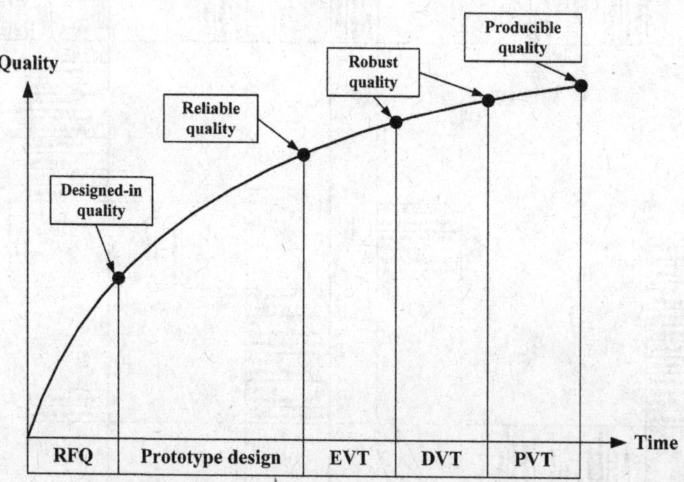

Figure 3.3. Quality progress from RFQ to PVT.

TABLE 3.2. EQFD Implementation Process (Shiu et al., 2007)

Business/product and Technology Development Planning

I. Business and Product Planning

1. Formulate a business strategy for maintaining current markets or developing new markets, and determine the required product portfolio.
2. Convert the important customer specifications (customer-proposed finished product quality characteristics and their specifications) of previously developed products into demanded qualities, and analyze the claims to make a demanded-quality deployment chart. Use scene deployment to survey the circumstantial issues in product applications to add potential demanded-quality items. **(1-I, 1-IV)**
3. Use the objective deployment structure of design-for-excellence (DFX) (comprising 14 DFX objectives) to ensure the completeness of demanded-quality deployment, and then perform quality positioning planning.

II. Technology Development Planning

4. Acquire the customer's advance product roadmap, and combine it with own company technology seeds deployment to formulate a technology development plan. **(1-II)**

Product Design

III. Request for Quotation (RFQ)

5. Obtain customer specifications to make a quality characteristics deployment chart, and enlarge this chart by adding the quality characteristics converted from the demanded qualities actively deployed by one's own company. **(1-I)**
6. Select a baseline product or reference model, and acquire the numerical descriptive measures (average and standard deviation) of the actual performance of its finished product characteristics.
7. Compare the customer specifications with the actual performance of the baseline product, and determine the target specifications and performance estimation (average and standard deviation) of the finished product characteristics with six-sigma design quality.
8. Assess the technical feasibility of six-sigma design quality based on the product-design technology deployment and design guidelines, and identify the design changes based on baseline and extract bottleneck-engineering (BNE). **(1′-II)**
9. Acquire the negative quality information about the baseline product to prevent the recurrence of quality problems resulting from product design. **(1-IV)**
10. Perform a competitive analysis of 4P/1S (product, price, place, promotion, and service) and customer target buying price to determine a target selling price, and thus identify the target product cost. **(1-III)**
11. Make a subsystems deployment chart, and compare it with that of the baseline to determine the target specifications and performance estimation (average and standard deviation) of the characteristics of subsystems and component units with six-sigma design quality. **(2-I)**

(*Continued*)

TABLE 3.2. (*Continued*)

12. Assess the technical and cost feasibility of six-sigma design quality of the subsystem and component unit characteristics, and identify the changed or important component units and BNE. **(2-II, 2′-III)**
13. Analyze the potential failure modes and their effects from the product design changes to prevent failures through early prediction, and update the relevant product-design guidelines. **(2-IV)**
14. Identify the "critical path" of the product development process, and draw up the necessary backup plans for critical activity delay.

IV. Prototype Design

15. Make a unit parts/materials deployment chart, and combine it with the subsystems deployment chart to identify the changed or important unit parts/materials and their target specifications. Assess the technical feasibility (or the supply feasibility of the vendor) and cost feasibility of the unit part/material specifications. **(2′-I, 2′-III)**
16. Build prototypes using different combinations of specific characteristic values (around the nominal value, upper limit or lower limit within the specification) of key components and parts to verify that the actual measured values of the prototype characteristics under all prototype-building scenarios conform to the expected specifications, and approximate the values calculated using the product-design guidelines.
17. Develop product test plans.
18. Manage design changes and conduct design reviews.

V. Engineering Verification Test (EVT)

19. Evaluate, and improve if necessary, the design quality of the quality characteristics critical to safety, reliability and lifetime.
20. Test product characteristics, analyze the test data and calculate the numerical descriptive measures (average and standard deviation) to verify product-design capability.
21. Based on the standard deviation of the product test data, select delivered samples with different test performance within the variation range to ensure the applicability of customer specifications.
22. Manage design changes and conduct design reviews.

Production Preparations

VI. Design Verification Test (DVT)

23. Make a process deployment chart, combine it with the subsystems deployment chart, and compare the combined chart with that of the baseline to determine the optimal specifications of process conditions (process characteristics and process recipes) for achieving the semi-finished product specifications. **(3-I, 3′-I)**
24. Assess the technical and cost feasibility of the specifications of process conditions based on the process-design technology deployment and design guidelines, and identify the changed or important process steps and BNE. Establish methods for checking process conditions. **(3-II, 3-III)**
25. Acquire the negative quality information about the baseline product to prevent the recurrence of quality problems resulting from process design. **(3′-IV)**

TABLE 3.2. (*Continued*)

26. Analyze the potential failure modes and their effects from the process design changes to prevent failures through early prediction, and update the relevant process-design guidelines. **(3-IV)**
27. Test product characteristics, analyze the test data and calculate the numerical descriptive measures (average and standard deviation) to verify product manufacturability.
28. Establish a control plan, and design a shop floor management system (comprising 10 subsystems) for its execution.
29. Convert the quality characteristics critical to safety, reliability and lifetime into the quality inspection system.
30. Manage design changes and conduct design reviews.

VII. Production Verification Test (PVT)

31. Evaluate, and improve if necessary, the adequacy of the design of the process recipes.
32. Evaluate, and improve if necessary, the appropriateness of the design of any shop floor management subsystem.
33. Test product characteristics, analyze the test data and calculate the percent defective rate to verify process capability.
34. Manage design changes and conduct design reviews.

Mass Production

VIII. Shop Floor Real-Time Management and Abnormality Management

35. Implement the shop floor management system and abnormality management to practice the control plan and continuously reduce the occurrence of shop floor problems.
36. Use the shop floor management experience to design the shop floor management system for future product development.

In the steps of Table 3.2, the codes in brackets represent the corresponding quality deployment charts shown in Figure 3.2 that need to be made at that step.

3.2 THIRTY-SIX STEPS OF THE EQFD IMPLEMENTATION PROCESS

In this section the contents of 36 steps in the EQFD implementation process are explained to clarify how to use NPD approach to establish EQFD system.

I. Business and Product Planning

1. Formulate a business strategy for maintaining current markets or developing new markets, and determine the required product portfolio.

The harmonization between product and market is the most important objective because NPD is regarded as a means to achieve the company's business purpose. The company uses valuable products to "exchange" the resources needed for survival and operation with customers, and further creates more added value through the accumulated resources in order to further pursue sustainable operation and growth.

The company must effectively use the concepts of product life cycle (PLC) and market evolution (Kotler, 2000) to develop the portfolio of current products or new products to protect current market share or develop a new market.

2. Convert the important customer specifications (customer-proposed finished product quality characteristics and their specifications) of previously developed products into demanded qualities, and analyze the claims to make a demanded-quality deployment chart. Use scene deployment to survey the circumstantial issues in product applications to add potential demanded-quality items. **(1-I, 1-IV)**

According to product knowledge, R&D personnel inversely convert important customer specifications of past products into demanded qualities, that is, toward quality characteristics of finished product and its specifications proposed by the customer, based on the adverse thoughts of "why the customer needs those quality characteristics and their specific specifications" and "while achieving those quality characteristics and their specific specifications, what requirements from the customer can we satisfy" to convert into demanded qualities (Ohfuji et al., 1994). These demanded qualities converted from the quality characteristics specification book are just like the requirements information acquired by asking the customer directly, which can be regarded as the requirements expressed by the customer (i.e., voice of the customer), and R&D personnel make the demanded-quality deployment chart according to these requirements.

The alternative way for the customer to express requirements is to propose claims that stand for "customer doesn't expect the emergence of some deficiency (negative quality)" so that claims analysis can increase the a demanded-quality items in a demanded-quality deployment chart and can enrich FTA contents used by R&D personnel to analyze fault symptoms of product in order to prevent the reoccurrence of claims by using both the "quality designing-in" approach of quality deployment and the "problem analysis" approach of FTA.

If to satisfy the customer's expressed requirements is the equivalent of "customer satisfaction," then to satisfy the customer's latent requirements is the equivalent of "customer delight" (Kano, 2002). Chan and Wu (2002) considered that QFD normally deals with satisfiers, not delighters, since the latter cannot be learned by directly asking the customers. Kano (1994) proposed the approach of "putting the focus on the customer's usage situation, not on product related issues" to effectively extract the customer's latent requirements. Thus, toward the deployment of demanded-quality, besides using the conversion of quality characteristics and its specifications and claims analysis to know well the demanded qualities expressed by the customer, we have to use scene deployment to survey circumstantial issues of product applications in order to increase latent demanded-quality items.

3. Use the objective deployment structure of design-for-excellence (DFX) (comprising 14 DFX objectives) to ensure the completeness of demanded-quality deployment, and then perform quality positioning planning.

As for the contents of demanded-quality deployment, they usually are the items related to function, features, and appearance—that is, for product quality, the three dimensions that are most easily associated. However, since a demanded-quality deployment chart is the only chart for QFD to be made in "world of customer," its completeness can significantly influence coverage and prioritization of QFD toward QA points for all deployments in "world of technology."

The objective deployment structure of DFX (X stands for excellence and completeness) is used to ensure the planning of all desirable dimensions of demanded quality. This book, based on the researches of Garvin (1987), Kano (1987), and Bralla (1996), proposes that an objective deployment structure contains 14 quality dimensions or DFX objectives for demanded quality:

- *Higher Functional Performance:* The higher functional level of the main operating characteristics of a product.
- *Physical Performance:* Product dimensions, volume, and weight.
- *User Friendliness:* The ease of use of a product.
- *Reliability/Durability:* Reliability indicates freedom from failure of a product during a period of time. Durabilify refers to product lifetime.
- *Maintainability/Serviceability:* The ease with which product availability can be restored after failure.
- *Effectiveness/Efficiency:* Effectiveness refers to the satisfactory combined effect of product reliability, maintainability, and performance in adapting certain environmental conditions. The term is also known as *dependability*. Efficiency refers to the lower life-cycle cost spent by a product.
- *Safety:* Freedom from injury and hazard from using a product.
- *Transportability:* Product design characteristics enabling a product to be carried, lifted, or hung.
- *Compatibility/Upgradability:* Compatibility describes the ease with which a product can be combined with other products for use. Upgradability indicates the ease with which a product can incorporate improved or additional features in the future.

- *Environmental Friendliness:* Freedom from environmental pollution and hazard of a product, including its manufacturing process and disposal.
- *Psychological Characteristics:* The product aesthetics perceived by the five senses of users, and the emotional characteristics (sensibility/*kansei*) that relate to user thinking, feeling, and identification regarding a product.
- *Short Time-to-Market:* The design characteristics enabling a product to shorten product time to market.
- *Manufacturability/Testability:* The ease of product manufacture and testing.
- *Low-Quantity Production:* Manufacturing flexibility enabling a product to be produced in small quantities.

The demanded qualities are the "true qualities" that the customer wants, hence the customer would never change the needs of all these quality dimensions and their deployed items, that is, while demanded-quality deployment is complete, it grasps the customer's all fundamental requirements about product quality. The directions for writing demanded-quality items are shown in Table 3.3.

A company doesn't need to pursue the excellence of all the quality dimensions; in fact, very few products can perform excellently in all the quality dimensions (Garvin, 1987), but the completeness and constancy is the foundation for the company to choose quality niches. The focus of demanded quality would vary from the customer's dynamic evaluation of importance or the company's quality positioning planning, in other words, the focus is determined by combining (a) the considerations of the importance evaluated in the customer's stance and (b) the product-market strategy, claims analysis, and Kano analysis in the company's stance, as shown in Figure 3.4.

II. Technology Development Planning

4. Acquire the customer's advance product roadmap, and combine it with own company technology seeds deployment to formulate a technology development plan. **(1-II)**

The customer's advance product roadmap describes the customer's development plan for future-generation products. If R&D personnel can get that product roadmap, it would help them confirm the availability of the company's existing technology seeds and define related technologies needed in the future; then formulate strategy and executive plan of advance technology development.

The flow chart of technology development deployment for advance product is shown in Figure 3.5.

TABLE 3.3. Directions for Writing Demanded Qualities

Item	Description
Focusing on product quality	• The "quality" deployed by QFD is originally referred to product quality so that there is so-called subsystems deployment, parts deployment, and process deployment. If we want to apply QFD to the management of service quality or business quality, then we have to implement the redesign of deployment contents and process.
The descriptions of demanded qualities should be specific enough so that it is easy to perceive which product they belong to.	• Since the demanded qualities at the lowest deployed level (generally, three levels) have to be directly converted into quality characteristics, the descriptions of them should be specific enough to be easily converted to the technical characteristics of that product. • For example, while deploying the demanded quality of a TV set, we have to make the demanded quality of the TV set at the lowest deployed level specific, which does not seem to be adaptable to all the electric appliances.
Using nontechnical language	• In "world of customer," the language used by the customer to describe the true qualities is different from the technical language used in "world of technology", by R&D personnel to describe the design quality achieved to satisfy the customer's requirements. • Take the picture quality of a TV set, for instance; the customer uses the customer language like "bright picture" and "the color is full and colorful," while R&D personnel use the technical language such as "luminance," "luminance uniformity," and "chromaticity."
One statement describes only one demanded-quality item	• One statement cannot include more than one demanded-quality item.
To describe ends not to specify means	• "Ends" (i.e., the true qualities wanted by the customer) is defined in "world of customer," while "means" is developed in "world of technology." The customer's needs for "true qualities" almost never change with time, but R&D personnel would change their "means used to achieve the ends" because of a different choosing decision or technological evolution. • Take the heat radiator of a personal computer (PC), for instance; "Hoping the PC can be equipped with a fan to be better heat radiated" is not describing the "ends," but specifying to set a "fan" as the "means of heat radiating." The true quality wanted by the customer should be "even the PC is under a long-time use; its temperature shouldn't be over heated or cause a crash." To satisfy this requirement, R&D personnel may find a more economic or effective way than a fan to achieve the "ends" of lowering the temperature of PC.

(*Continued*)

TABLE 3.3. (*Continued*)

Item	Description
Using the descriptive way of "under what circumstance there would be what benefits or wouldn't be what badness to the product"	• Considering at what time or under which condition the product would be noisy, bad heat radiating, etc. And, using the description of "the product wouldn't be noisy, bad heat radiating, etc., even at this time or under this condition" as demanded quality. In brief, using the descriptive way of "under what circumstance there would be what benefits or wouldn't be what badness to the product" to propose demanded-quality items. • For example, toward the "usage safety" of power supply, it should be described specifically as "any input voltage is available" or "even the input voltage is wrong, it won't burn down." Toward "better heat radiating" of PC should be described specifically as "even the PC is under a long-time use, its temperature shouldn't be over heated or cause a crash."
Considering the demanded-quality items under particular applied conditions	• Considering the demanded-quality items under particular applied and environmental conditions. • For example, to consider the applied conditions such as vibration, humidity, salt atmosphere, spray, electromagnetic radiation, etc.
Converting negative quality information into positive quality requirements	• Considering the quality deficiency items that ever happened to the product and converting that negative quality information into the description requiring "positive quality on the contrary". • For example, if a PC product has ever been complained by the customer about "easy to be overheated and crashed"; then we can, based on this problem, propose a demanded-quality item of "even if the PC is under a long-time use, its temperature wouldn't be over heated or cause a crash."
Considering complete quality dimensions	• Considering 14 dimensions of "design for excellence" (i.e., design for excellent quality), not only the items of function, features, and appearance which are easiest to be associated.
Considering all the customers'(end users, direct users, and internal customers) demanded quality toward the product	• Considering demanded-quality of end users (second-level external customers), direct users (first-level external customers), and internal customers (e.g., manufacturing factories) toward the product and arranging them according to customer groups. • Taking a printed circuit board (PCB) company, for instance; besides considering its director user's (system assembly plants) demanded quality, the company has to understand and consider, in end user's demanded quality of TV set and PC, the items related to PCB and to consider demanded-quality items of manufacturing factories toward "design for manufacturability" (DFM). R&D personnel should manage to satisfy these demanded qualities in "world of technology."

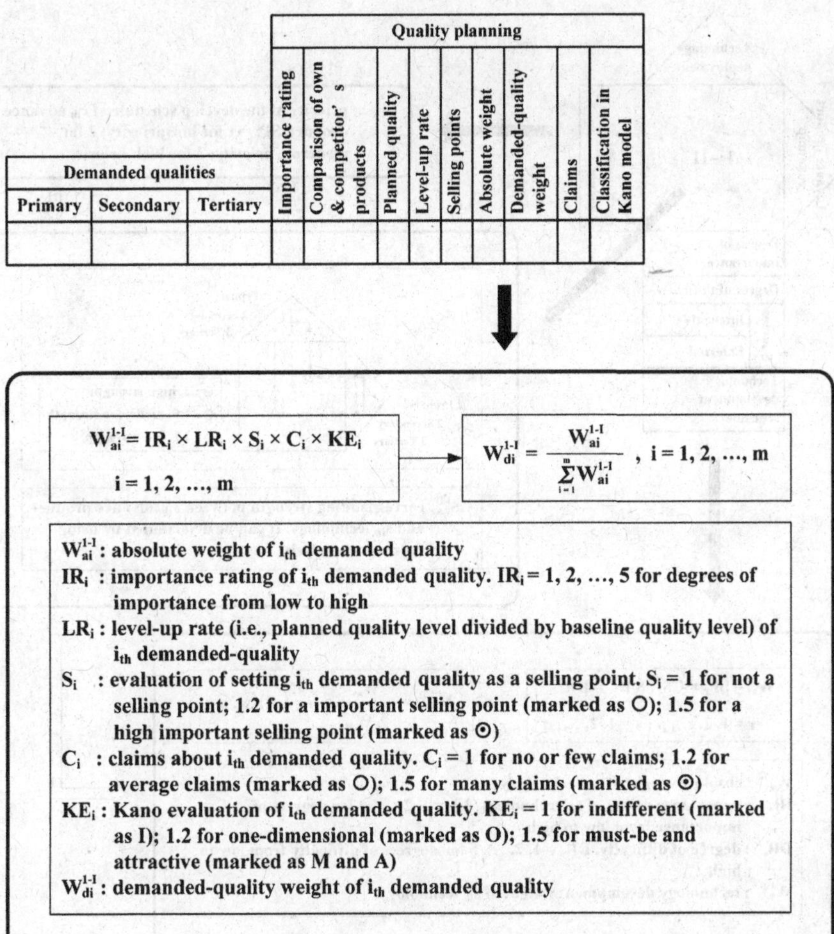

Figure 3.4. Flow chart of demanded-quality deployment.

III. Request for Quotation (RFQ)

5. Obtain customer specifications to make a quality characteristics deployment chart, and enlarge this chart by adding the quality characteristics converted from the demanded qualities actively deployed by one's own company. **(1-I)**

The contents of the customer specification book describes in detail the finished product items, units, and specifications needed by the customer; it's easy to convert them into the form of quality characteristics deployment chart. R&D personnel may increase the quality characteristics converted from the demanded quality of DFX deployment to expand the quality characteristics

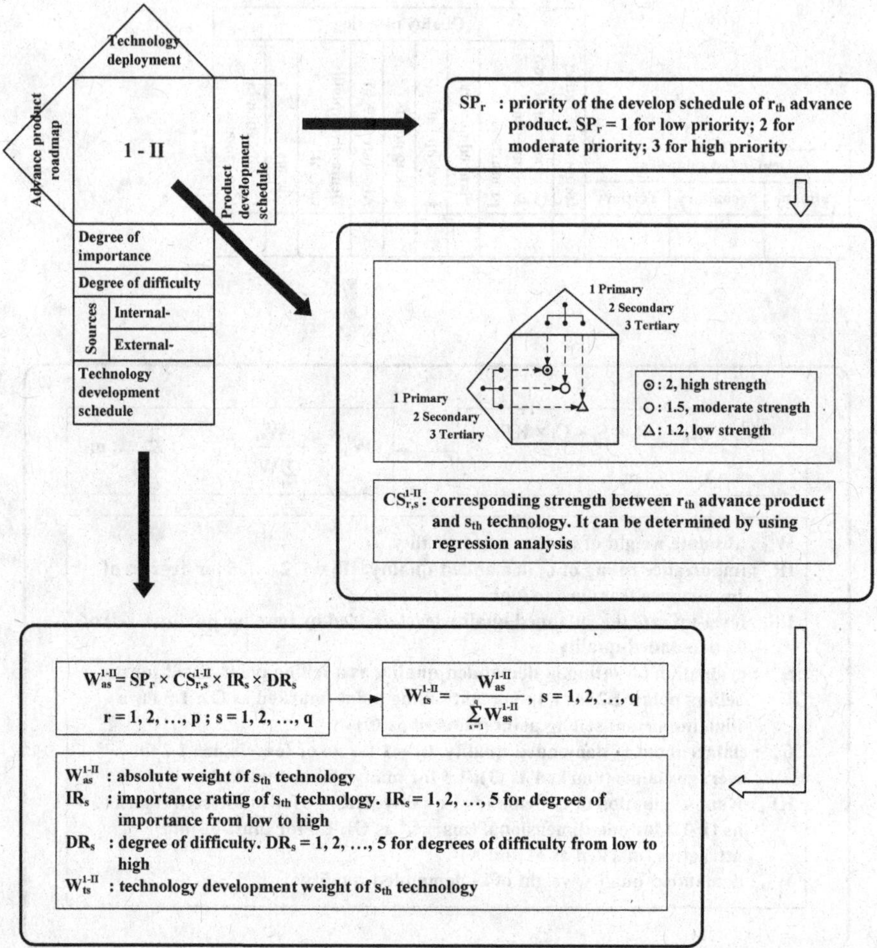

Figure 3.5. Flow chart of technology development deployment for advance product.

deployment chart, and they may combine a demanded-quality deployment chart to constitute a quality chart of two-dimensional matrix form.

The quality chart is the chart connecting "world of customer" and "world of technology" in quality deployment, which can convert customer language used by the customer to describe true qualities into technical language that can be comprehended by the company's R&D personnel. All the quality deployments following the quality chart are completed in "world of technology."

6. Select a baseline product or reference model, and acquire the numerical descriptive measures (average and standard deviation) of the actual performance of its finished product characteristics.

For a company with developing experiences, the specifications of a new product are usually not completely total new design requirements; one part is new, and another part is the similar design requirement of the company's existing developed products. R&D personnel can choose the products similar in specifications and performance to be baseline product or reference model to extend the past product developing experiences. The developing experiences of that baseline product exist in the visible form of product and process data. R&D personnel can use actual performance (average and standard deviation) of baseline finished product characteristic as the rough reference for the performance of new product.

7. Compare the customer specifications with the actual performance of the baseline product, and determine the target specifications and performance estimation (average and standard deviation) of the finished product characteristics with six-sigma design quality.

R&D personnel use the data of design specifications and actual performance of subsystems of baseline product to evaluate if the extension of design concept of baseline product can satisfy customer specifications of new product. Under the circumstance that baseline design concept and actual performance can satisfy customer specifications of new product, R&D personnel may go further to evaluate the following decisions:

- Target specifications or expected design specifications (i.e., expected nominal value and tolerance of finished product characteristics) of new product are beyond customer specifications to pass a message to the customer that the company has relative advantage against the competitors; or
- Toward the items with baseline actual performance that is beyond customer specifications, the company loses its design controls and process controls to reduce the cost.

Under the circumstance that baseline design concept and actual performance cannot fully satisfy the customer specifications of new product, R&D personnel must identify expected changing specification items and determine their design specifications to achieve the customer specifications of new product. R&D personnel may use actual performance (average and standard deviation) of baseline product, based on the formula of design guidelines of a new product, to calculate the expected design specification values. Without the formula for calculating design specification values, R&D personnel may, based on the data of design specifications and actual performance of subsystems of baseline product, conduct the experimental methods of "parameter design" and "tolerance design" and determine expected design specifications with six-sigma design quality (world-class quality level). The six-sigma quality refers to the following: Under the assumption that the average of the actual performance of product characteristics might shift 1.5 times the standard deviation away from target value, the width between target value and upper or lower specification limits is six times the standard deviation so that the defective rate can reach as low as 3.4 defects per million opportunities (DPMO).

The flow chart for finished product quality characteristics deployment implemented in steps 5–7 is shown in Figure 3.6. The flow chart for finished product design specifications determination in steps 6–7 is shown in Figure 3.7.

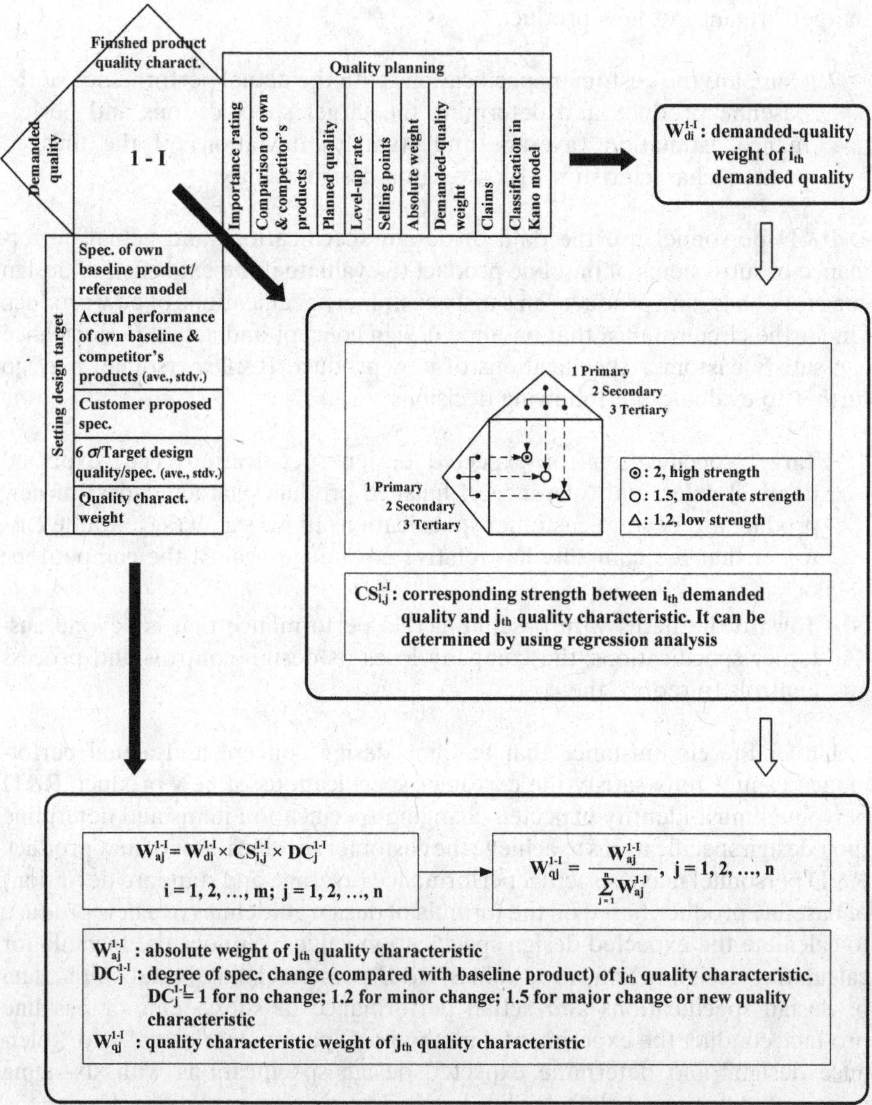

Figure 3.6. Flow chart of quality characteristics deployment.

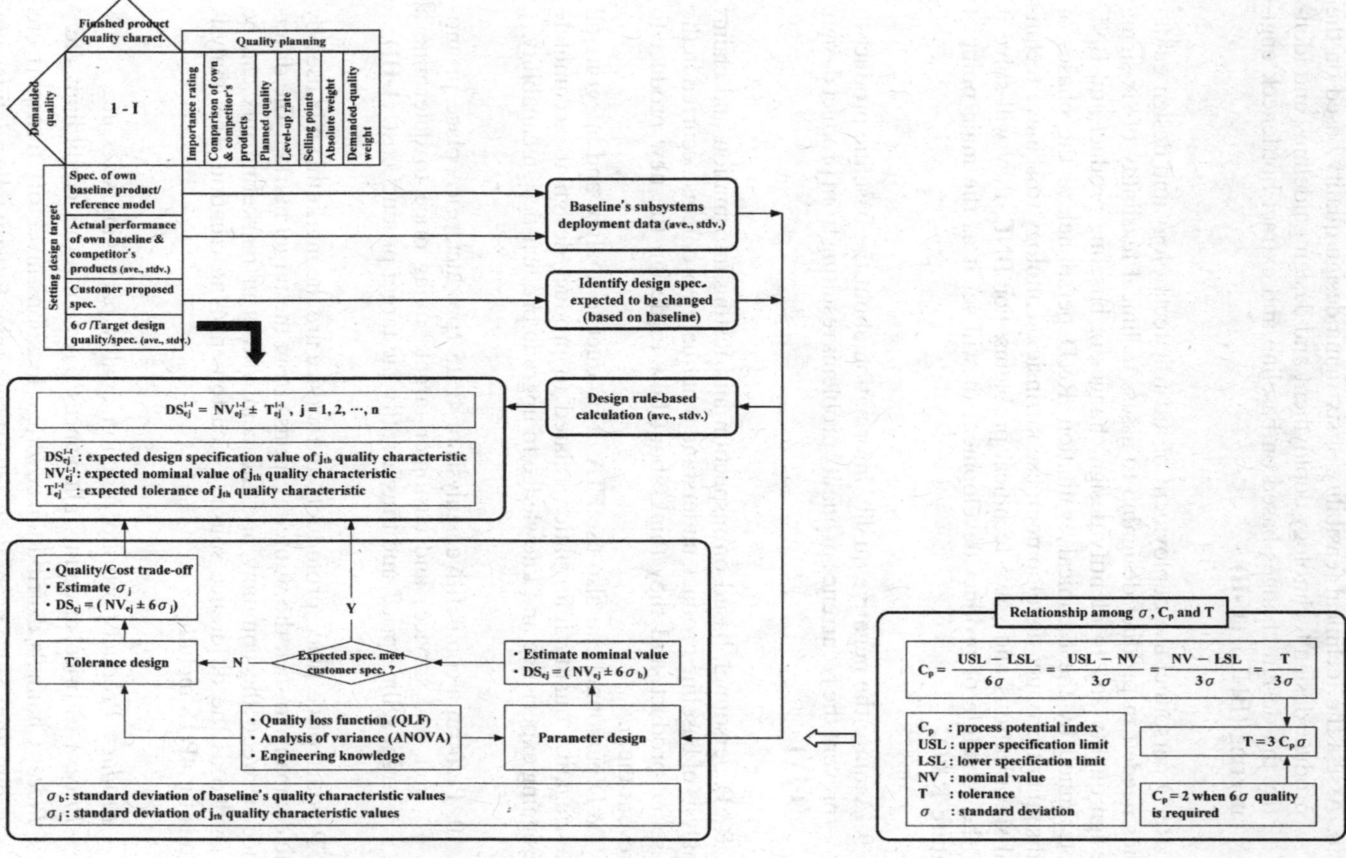

Figure 3.7. Flow chart of finished-product design specifications determination.

8. Assess the technical feasibility of six-sigma design quality based on the product-design technology deployment and design guidelines, and identify the design changes based on baseline and extract bottleneck engineering (BNE). **(1′-II)**

R&D personnel use deployment of intrinsic technology and design guidelines needed in product designing to assess technical feasibility of six-sigma design quality and to identify design changes of baseline product and BNE determined by the technical evaluation. R&D personnel, at the phases of subsequent product development, use intrinsic technology to achieve technical specifications and solve technical problems. For BNE, they will solve it in the schedule of product development or will set it as the main point of future R&D.

9. Acquire the negative quality information about the baseline product to prevent the recurrence of quality problems resulting from product design. **(1-IV)**

R&D personnel, based on inspection and testing information and claims analysis of baseline product, understand that quality problems occurred in the baseline product, and they think about how to design the new product to prevent them.

R&D personnel can also use FTA of product to analyze and integrate all the negative quality information related to product design to accumulate designing experience and knowledge to upgrade product-design capability.

10. Perform a competitive analysis of 4P/1S (product, price, place, promotion, and service) and customer target buying price to determine a target selling price, and thus identify the target product cost. **(1-III)**

Product, price, place, promotion, and service are the marketing mix used by the company to seek the wanted response from the target market. The difference between the company and the competitors in marketing mix can be regarded as the difference source of customer value creation. The analysis items for 4P/1S include:

- *Product:* Product portfolio, function, quality, packaging, sizes, etc.
- *Price:* List price, discounts, allowances, payment terms, credit limit, etc.
- *Place:* Channels, distribution coverage, locations, inventory, transport, etc.
- *Promotion:* Sales promotion, advertising, sales force, public relations, etc.
- *Service:* Ordering ease, delivery, installation, guarantee package, maintenance and repair quality, etc.

In the practice of fabrication supply, toward the same product but different fabrication suppliers, the customer may set a different target buying price. Thus, the company can evaluate if the company has a quality niche and relative advantage against the competitors by using (a) analysis of strength and weakness of 4P/1S mix between it and the competitors, (b) deviation analysis of target buying price for the customer toward a different company, and (c) the understanding of criteria for the customer to set price, in order to seek a dealing chance negotiating with the customer for a price better than that at present and then going further to decide the product's target profit and cost under this price.

With these analyses, even the company currently has no way to negotiate for a better price, but it can identify the issues such as the selling points and improving items that can strengthen future bargaining power.

11. Make a subsystems deployment chart, and compare it with that of the baseline to determine the target specifications and performance estimation (average and standard deviation) of the characteristics of subsystems and component units with six-sigma design quality. **(2-I)**

R&D personnel deploy subsystems and component units of a new product, and they set design specifications of finished product as the target, based on the formula of product-design guidelines or experimental methods of parameter design and tolerance design, to determine the expected design specifications of subsystems and component units with six-sigma design quality.

The flow chart of subsystems deployment is shown in Figure 3.8.

12. Assess the technical and cost feasibility of six-sigma design quality of the subsystem and component unit characteristics, and identify the changed or important component units and BNE. **(2-II, 2′-III)**

R&D personnel use intrinsic technology and design guidelines needed in product designing to assess the technical feasibility of six-sigma design quality of subsystem and component unit characteristics as well as to identify, after the technical evaluation, baseline design change items, important component units, and BNE. R&D personnel also have to assess the cost feasibility of component units for satisfying the target cost of subsystems (allocated based on target cost of finished product).

13. Analyze the potential failure modes and their effects from the product design changes to prevent failures through early prediction, and update the relevant product-design guidelines. **(2-IV)**

R&D personnel use FMEA of product design to analyze the potential impact of product design changes on quality, and prevent it through early

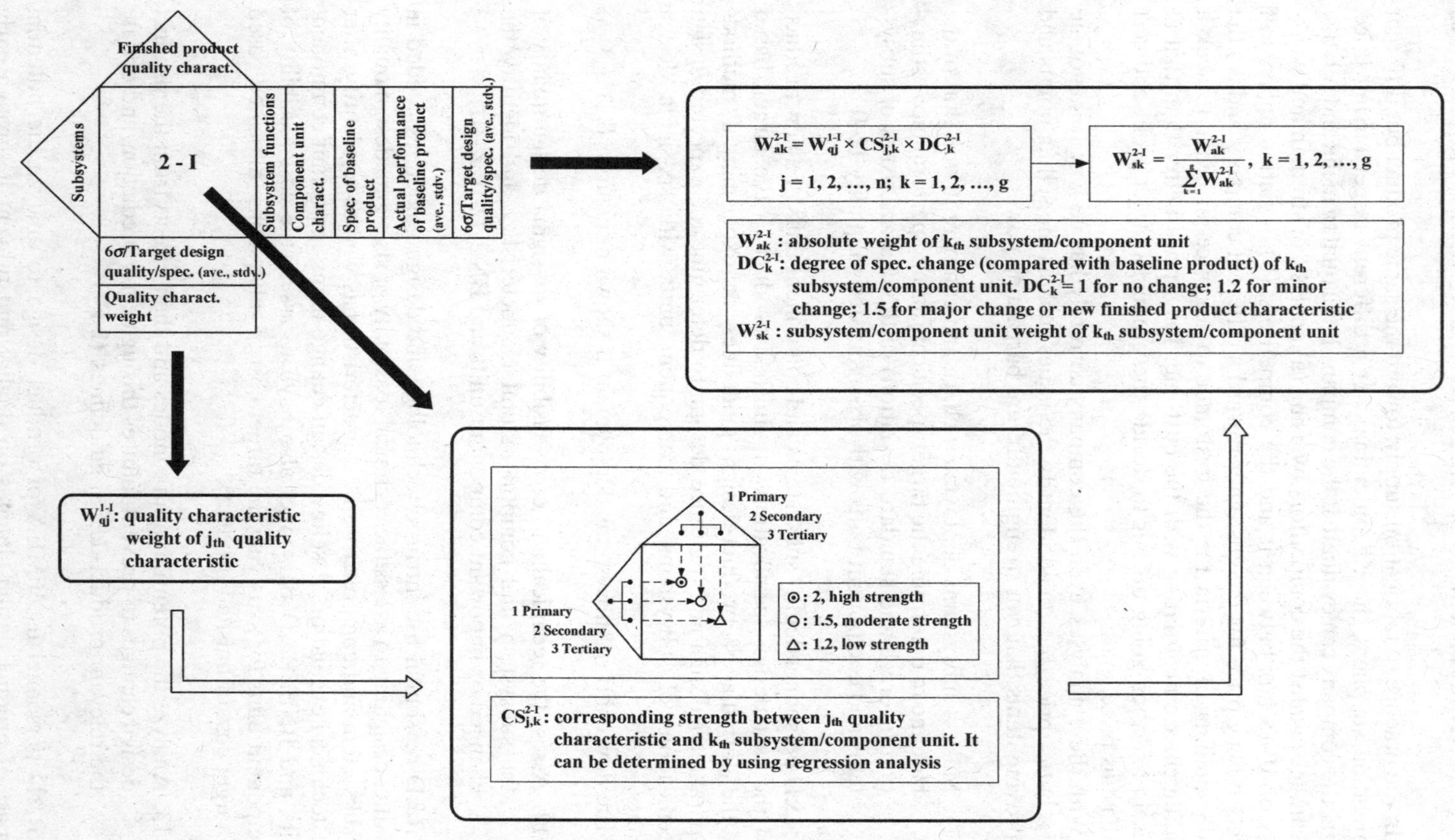

Figure 3.8. Flow chart of subsystems deployment.

prediction. The approach provided by FMEA is that "after the design, predict its potential design problem points and their effects to think earlier and prevent"; this is different from the debug approach, which suggests that after the design, use verification tests to find out design problem points.

While developing a new product, even the potential design problem points are predictable and preventable, and R&D personnel may have to choose between recommended actions taken and schedule limit and may directly send the risky product design to the next phase for verification tests. When the product cannot pass verification tests, R&D personnel can take the recommended actions whereby they know how to solve the design vulnerability "before the event" and go further to save time, instead of starting to analyze design vulnerability and thinking of actions "after the fact."

The FMEA of product design can be divided into design potential FMEA (DPFMEA) and design FMEA (DFMEA), which are respectively used, after design change, by R&D personnel to realize "predicting problems before potential failures happen" and "preventing the recurrence of failures after the first time of its happening." FMEA can ensure the prevention of failure modes and their effects before the event and after the fact, but DPFMEA is better than DFMEA in inspiring and upgrading R&D personnel's ability for designing and assessing the product and in ensuring the design quality.

According to FMEA information, R&D personnel may renew a product design guideline as the lesson learned of product design.

14. Identify the "critical path" of the product development process, and draw up the necessary backup plans for delaying critical activity delay.

The company has to identify, in the network of product development activities, the activity path spending longest time, that is, the critical path which can influence the total time of product development, in order to formulate project management plan.

Quality, cost, and delivery (QCD) are the three major objectives of product development management. With respect to quality and cost, we can ensure it with QFD deployment process; with respect to schedule, we have to focus on the management of the critical path, because it is not allowed for any activity in critical path to delay; otherwise, it will impact the total time of product development. Thus, the company has to draw up backup plans to not allow the delay of critical activities to happen, to ensure that product development can satisfy the demand of schedule.

IV. Prototype Design

15. Make a unit parts/materials deployment chart, and combine it with the subsystems deployment chart to identify the changed or important unit parts/materials and their target specifications. Assess the technical

feasibility (or the supply feasibility of the vendor) and cost feasibility of the unit part/material specifications. **(2′-I, 2′-III)**

According to information of subsystems deployment, R&D personnel identify the unit parts and materials that need design change or are important and identify their target specifications as well as assess the technical feasibility of unit part and material specifications and vendor's supply feasibility. R&D personnel also have to assess the cost feasibility of the cost of unit parts and materials for satisfying target cost of subsystems (allocated based on target cost of finished product).

The flow chart of unit parts and materials deployment is shown in Figure 3.9.

16. Build prototypes using different combinations of specific characteristic values (around the nominal value, upper limit, or lower limit within the specification) of key components and parts to verify that the actual measured values of the prototype characteristics under all prototype-building scenarios conform to the expected specifications, and approximate the values calculated using the product-design guidelines.

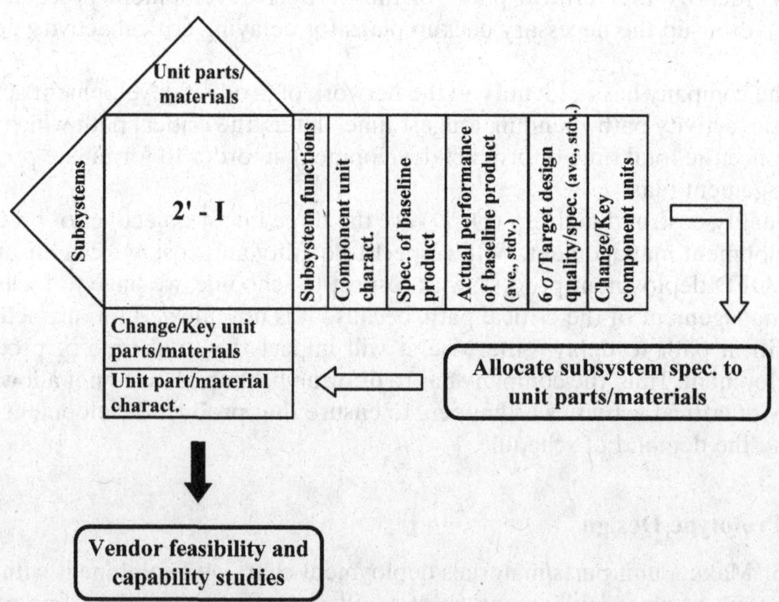

Figure 3.9. Flow chart of parts and materials deployment.

Prototype design involves building a product approximation with the quality characteristic whereby R&D personnel are interested in providing the prototype model for related experiments and tests.

A product is composed of many components and parts. The product has its specifications, and each component and part has its own allocated specification, but one needs to verify if a part conforming to specification values can compose a product conforming to specification values. At the phase of prototype design, R&D personnel have to ensure that when a characteristic value of important parts is the specification nominal value, for the composed prototype, the actual measured values of its characteristic also must be its specification nominal value and must be approximate to the calculated value based on design guideline so as to verify the design guideline. Likewise, when the characteristic values of important parts are around the upper limit or lower limit within the specification, for the prototype composed by R&D personnel, the actual measured values of its characteristic mush be around the upper limit or lower limit within the specification and must be approximate to the calculated value based on design guideline.

A prototype (product approximation) is composed of components and parts. The measured values of prototype characteristics may be different due to the effects of component-to-component variability or part-to-part variability. R&D personnel have to (a) ensure that, under the effects of component-to-component or part-to-part variability, the actual measured values of prototype characteristics can conform to an expected specification value and calculated value and (b) determine the gap between actual value and estimated value, and then think about how to eliminate that gap.

17. Develop product test plans.

The contents of various product test plans developed by R&D personnel must include testing items, testing equipment, testing methods, testing condition, and testing specifications, and they have to be consistent with those used by the customer in testing the company's product so that they can ensure that when failure of the product occurs, the company would find it first and wouldn't, because of the difference of test plan between the company and the customer, let the failed product be distributed out of the company and be discovered and condemned by the customer.

18. Manage design changes and conduct design reviews.

When R&D personnel are developing a new product, they are extending the past developing experiences about the baseline product and conducting management of "the changes of new product against baseline product in configuration" and "key QA points of new product," so R&D management can be regarded as "change management" with prioritization. The design changes of new product against baseline product would be identified by intrinsic technology and design guidelines, and key QA points of a new product may be determined by the QFD deployment process.

R&D personnel have to ensure the appropriate management of all design changes and their related items in the whole product development cycle and implement design reviews to fully assess design configuration to prevent quality deficiency.

V. Engineering Verification Test (EVT)

19. Evaluate, and improve if necessary, the design quality of the quality characteristics critical to safety, reliability, and lifetime.

R&D personnel have to evaluate design quality of quality characteristics critical to safety, reliability, and lifetime and to improve, if necessary.

20. Test product characteristics, analyze the test data, and calculate the numerical descriptive measures (average and standard deviation) to verify product-design capability.

R&D personnel shall, according to the contents of product test plans, implement various tests for product characteristics at EVT phase and analyze test data to verify if the design of new product can achieve expected design specifications, that is, to verify R&D personnel's product-design capability.

21. Based on the standard deviation of the product test data, select delivered samples with different test performance within the variation range to ensure the applicability of customer specifications.

Generally speaking, at this phase, R&D personnel are to deliver the so-called golden sample (i.e., the sample with product characteristic value that equals specification nominal values) to the customer to show that the company has the good capability to manufacture the product satisfying customer specifications and on target. However, when product characteristic value, because of inherent variability, does not always equal specification nominal values, it is necessary to confirm if the product would create a situation that the customer cannot apply the product, otherwise, the company would face the changes of customer specifications. In other words, even the product characteristic value satisfies customer specifications, but when the customer finds that the specifications cannot be applied in practice, he/she will still propose changes of specifications and ask the company to satisfy the new specifications.

Thus, R&D personnel should, based on standard deviation of product test data, select the samples (including golden sample) with different test performance in that variation range and deliver them to the customer and confirm earlier if the customer would find any application problem and propose specification change when he/she conducts an application test for the samples satisfying customer specifications. R&D personnel can use these selected delivered samples to confirm the applicability of customer specifications; and they can, at the earliest phase of product development, know if they have to implement design changes to satisfy the customer's new specifications. They don't need to wait until the late phases of product development and then decide if they need to implement design changes which may cause higher cost and possible schedule delay while the product is redesigned at this time.

22. Manage design changes and conduct design reviews.

R&D personnel have to ensure the appropriate management of all design changes and their related items in the whole product development cycle and implement design reviews to fully assess design configuration to prevent quality deficiency.

VI. Design Verification Test (DVT)

23. Make a process deployment chart, combine it with the subsystems deployment chart, and compare the combined chart with that of the baseline to determine the optimal specifications of process conditions (process characteristics and process recipes) for achieving the semi-finished product specifications. **(3-I, 3′-I)**

R&D personnel shall deploy the manufacturing process of the new product and set the design specifications of subsystems and component units (i.e., semi-finished product) as the aim to fulfill; based on process-design guidelines as well as experimental methods of parameter design and tolerance design, R&D personnel shall also determine the optimal specification values of process characteristics and the process recipes needed to achieve process characteristics specifications.

The flow chart of process deployment is shown in Figure 3.10. The flow chart of process-characteristic design specifications determination is shown in Figure 3.11.

24. Assess the technical and cost feasibility of the specifications of process conditions based on the process-design technology deployment and design guidelines, and identify the changed or important process steps and BNE. Establish methods for checking process conditions. **(3-II, 3-III)**

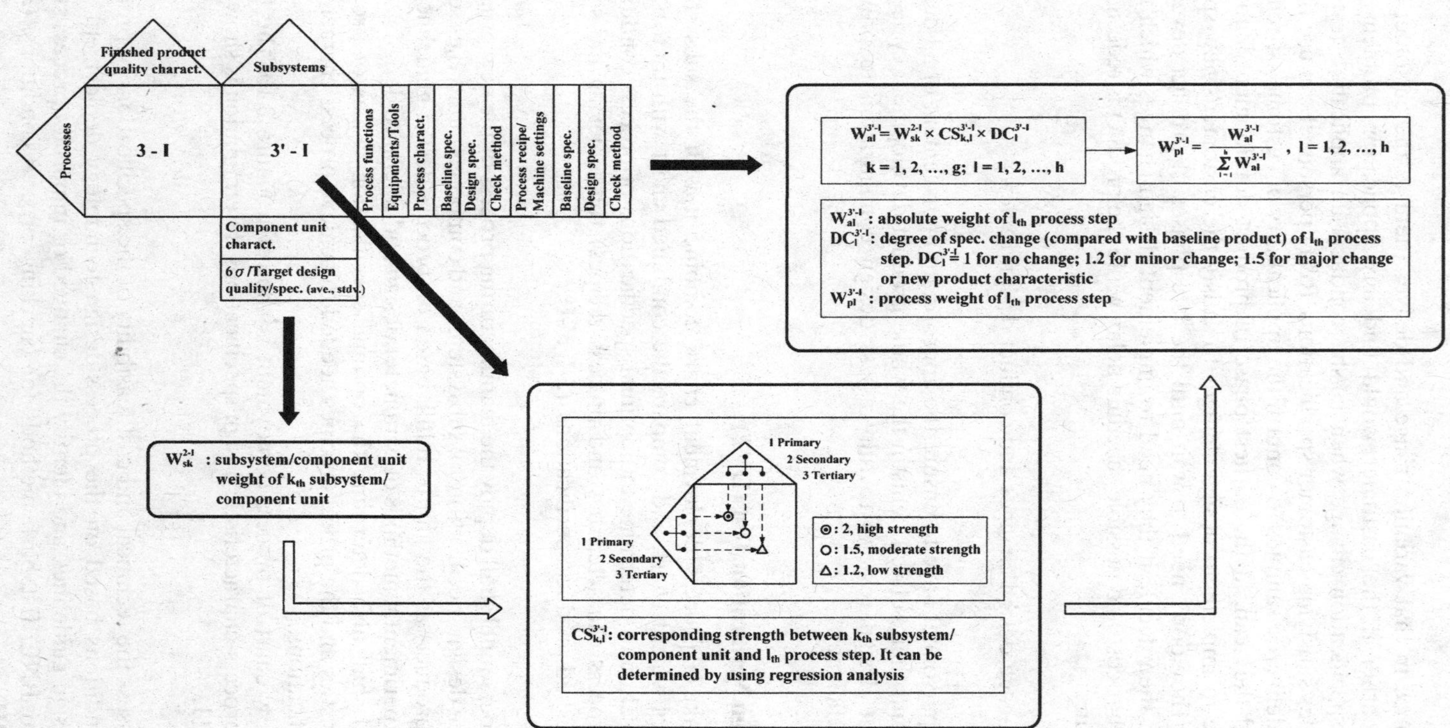

Figure 3.10. Flow chart of process deployment.

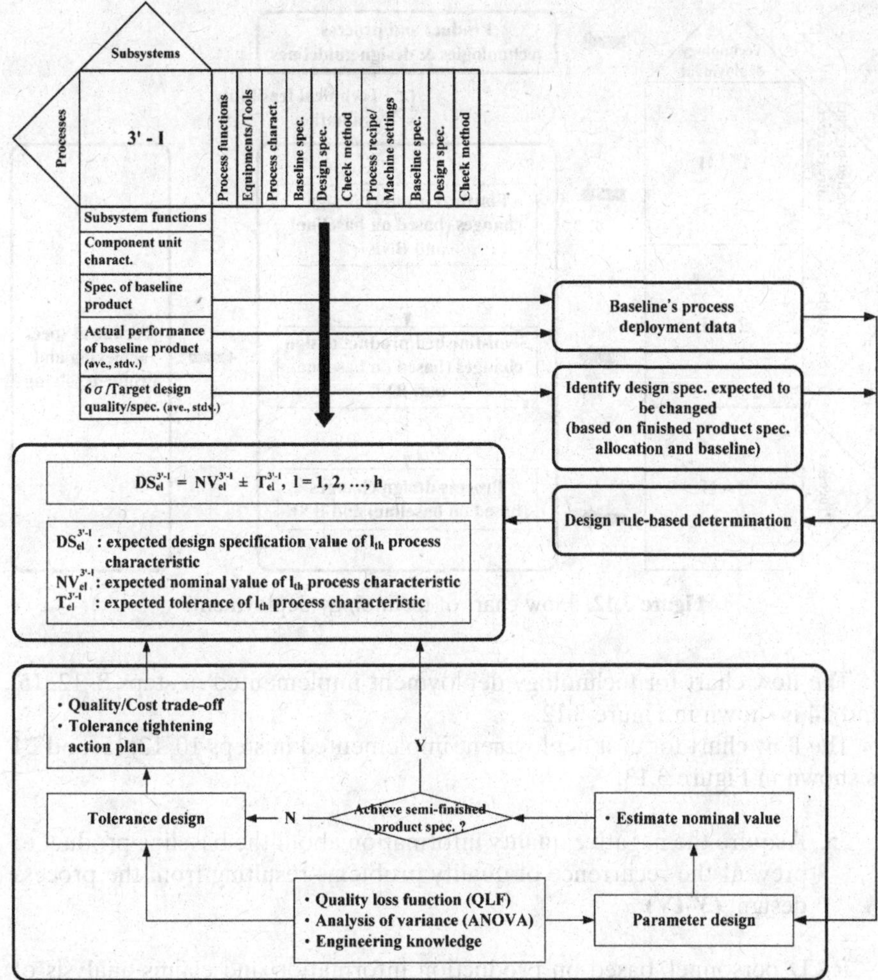

Figure 3.11. Flow chart of process-characteristic design specifications determination.

R&D personnel shall, using deployment of intrinsic technology and design guidelines needed in process designing, assess technical feasibility of the specifications of process conditions and identify design change items of baseline, important process steps, and BNE determined after the technical evaluation. R&D personnel also have to assess the cost feasibility of the cost items such as labor, equipment, tooling, and overhead for satisfying process target cost (allocated according to target cost of finished product).

In addition, R&D personnel also have to establish the check methods used to ensure no errors in process conditions.

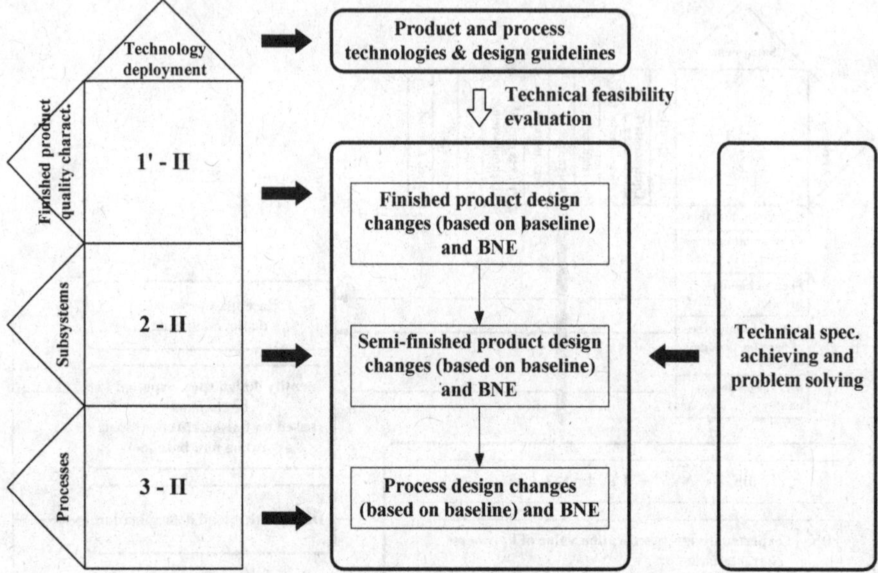

Figure 3.12. Flow chart of technology deployment.

The flow chart for technology deployment implemented in steps 8, 12, 15, and 24 is shown in Figure 3.12.

The flow chart for cost deployment implemented in steps 10, 12, 15, and 24 is shown in Figure 3.13.

25. Acquire the negative quality information about the baseline product to prevent the recurrence of quality problems resulting from the process design. **(3′-IV)**

R&D personnel, based on production information and claims analysis of baseline product, understand manufacturing quality problems occurring in the baseline product, and they think about how to design the process of new product to prevent future problems.

R&D personnel can also use the FTA of a machine to analyze and integrate all the negative quality information related to process design to accumulate designing experience and knowledge to upgrade process-design capability.

26. Analyze the potential failure modes and their effects from the process design changes to prevent failures through early prediction, and update the relevant process-design guidelines. **(3-IV)**

R&D personnel use FMEA of process design to analyze the potential impact of process design changes on quality, and they prevent it through early prediction.

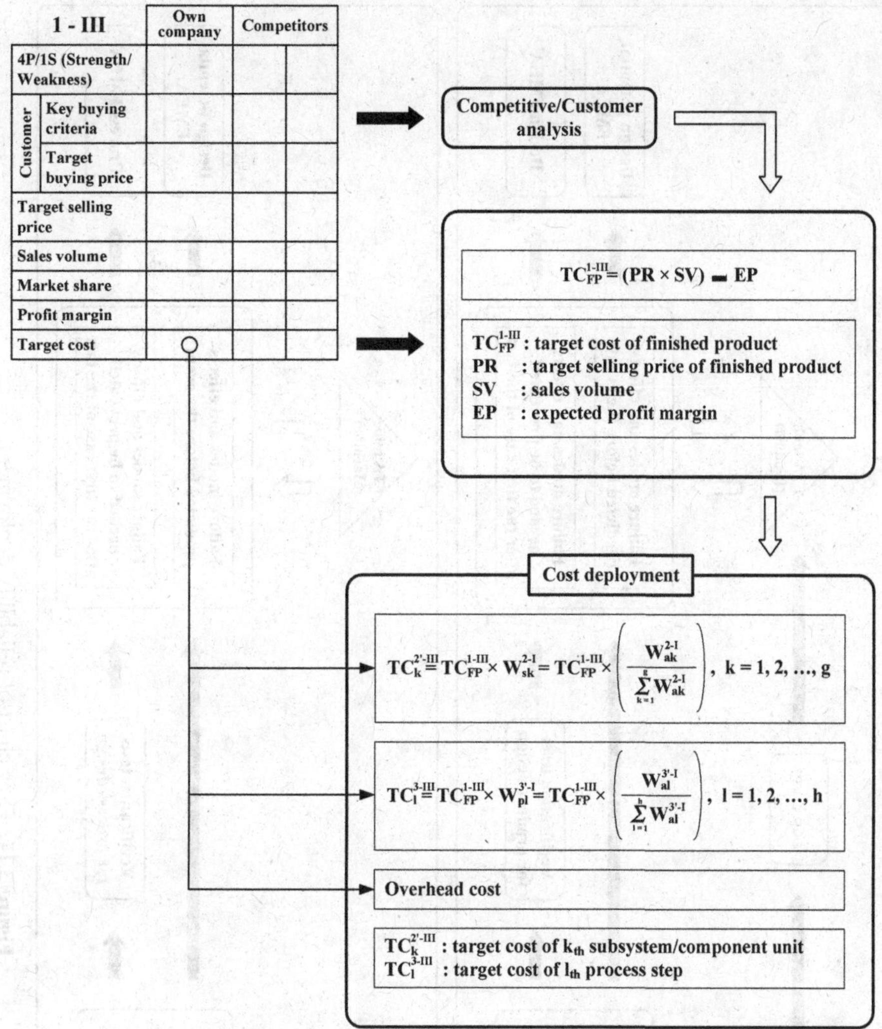

Figure 3.13. Flow chart of cost deployment.

The FMEA of process design can be divided into design potential FMEA (DPFMEA) and process FMEA (PFMEA), which are respectively used, after design change, by R&D personnel to realize (a) predicting problems before potential failures happen and (b) preventing the recurrence of failures after the first time of their happening.

According to FMEA information, R&D personnel may renew a process design guideline as the lesson learned of process design.

The flow chart for reliability deployment implemented in steps 9, 13, 25, and 26 is shown in Figure 3.14.

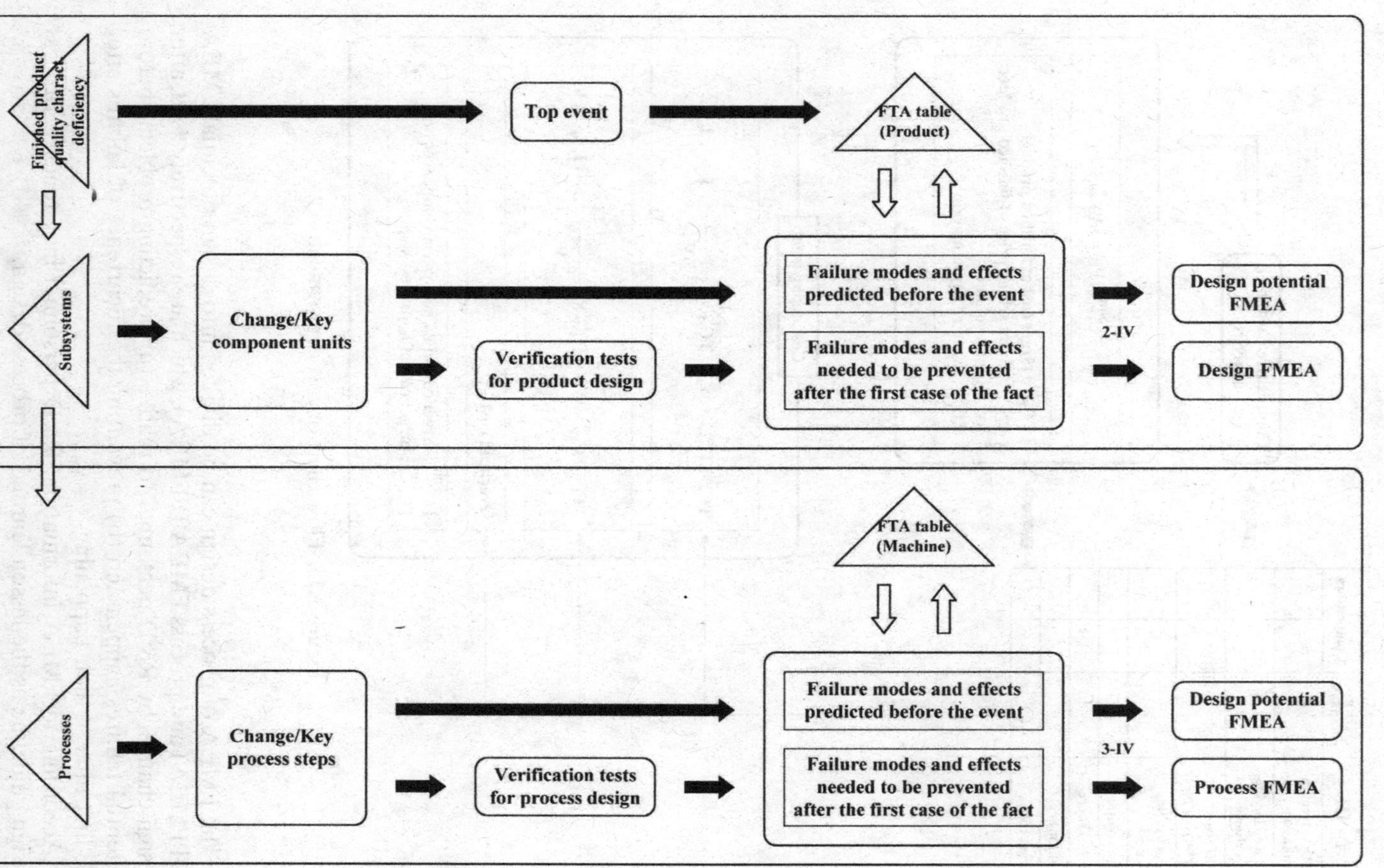

Figure 3.14. Flow chart of reliability deployment.

27. Test product characteristics, analyze the test data, and calculate the numerical descriptive measures (average and standard deviation) to verify product manufacturability.

R&D personnel shall, according to the contents of product test plans, implement various tests for product characteristics at DVT phase and analyze test data to verify if the design of new product design for manufacturability (DFM).

28. Establish a control plan, and design a shop floor management system (comprising 10 subsystems) for its execution.

R&D personnel shall establish a "control plan" to provide a structured approach for the design, selection, and implementation of value-added control methods for minimizing process and product variation. The control plan is the written summary of contents and standards of shop floor control so that its realization needs the implementation of various shop floor management activities to achieve control contents and standards. We integrate all the shop floor management activities into 10 management subsystems, and their objectives are described respectively as follows:

- *Vendor Quality Assurance System (*VQA*):* To ensure that defective components, parts, and materials are not approved, bought in, or slipped into the manufacturing site.
- *Preventive Maintenance System (*PM*):* To ensure that all equipment and tools are maintained in optimal condition to maintain the stability of the transfer relationship between process recipes and process characteristics, and that the deterioration and failures of equipment and tools can be deferred or reduced so as to avoid the manufacture of defective products.
- *Measurement Systems Analysis (*MSA*):* To ensure that all equipment, tools, and operators do not cause variation in the measurement of process characteristics and process recipes.
- *Pre-production Management System (*PPM*):* To ensure that all equipment, tools, and operators achieve preparedness for 5Ms/1E (man, machine, material, method, measurement, and environment) prior to production and are free of defect symptoms in preparations of processing and motions.
- *First-Article Inspection System (*FAI*):* To ensure that the measured values of the quality characteristics of the semi-finished products produced by each process workstation and the finished products tested by the final inspection station equal the nominal values of their specifications.
- *In-Process Quality Control System (*IPQC*):* To ensure that all equipment, tools, and operators are free of defect symptoms during processing and motion during production.

- *Statistical Process Control System (*SPC*):* To ensure that the measured values of the quality characteristics of the semi-finished products are in the state of statistical control.
- *Reliability/Safety Control System (*R/SC*):* To ensure proper process control and management over the quality characteristics critical to reliability and safety.
- *Quality Inspection System (*QIS*):* To ensure that finished product inspection and test plans are consistent between company and customers, and that the implementation of the inspection and test plans for the semi-finished products can promptly reflect quality deficiencies in the finished products.
- *Out-of-Control Action Plan (*OCAP*) and Abnormality Management System:* To ensure the quick elimination and prevention of the recurrence of abnormal problems.

R&D personnel have to develop forms, standard operating procedures (SOP), and operating instructions (OI) needed to execute these 10 management subsystems to ensure the achievement of contents and standards of control plan.

29. Convert the quality characteristics critical to safety, reliability, and lifetime into the quality inspection system.

R&D personnel shall convert the quality characteristics critical to safety, reliability, and lifetime into the quality inspection system to ensure that the company's inspection and test plans toward the quality characteristics of a finished product are in accordance with the customer; in the process, the inspection and test toward the quality characteristics of semi-finished product can reflect quality deficiency of finished product in time.

30. Manage design changes and conduct design reviews.

R&D personnel have to ensure the appropriate management of all design changes and their related items in the whole product development cycle and implement design reviews to fully assess design configuration to prevent quality deficiency.

VII. Production Verification Test (PVT)

31. Evaluate, and improve if necessary, the adequacy of the design of the process recipes.

R&D personnel shall evaluate the adequacy of the design of the process recipes to ensure the process recipes can achieve the needed process characteristic values and improve them, if necessary.

32. Evaluate, and improve if necessary, the appropriateness of the design of any shop floor management subsystem.

R&D personnel shall, according to performance of shop floor control, evaluate the appropriateness of design contents (including forms, SOP, and OI) of any shop floor management subsystem and improve them, if necessary, to upgrade performance of shop floor control.

33. Test product characteristics, analyze the test data, and calculate the percent defective rate to verify process capability.

R&D personnel shall, according to the contents of product test plans, implement various tests of product characteristics at PVT phase and analyze the test data and calculate the actual defective rate to verify if process design can achieve expected design specifications of semi-finished product, that is, to verify the process capability.

34. Manage design changes and conduct design reviews.

R&D personnel have to ensure the appropriate management of all design changes and their related items in the whole product development cycle and implement design reviews to fully assess design configuration to prevent quality deficiency.

VIII. Shop Floor Real-Time Management and Abnormality Management

35. Implement the shop floor management system and abnormality management to practice the control plan and continuously reduce the occurrence of shop floor problems.

The shop floor personnel shall execute shop floor management system (SFMS) and use real-time feedback system of shop floor data to reflect abnormal problems as well as conduct abnormality analysis and recurrence prevention to reduce occurrences of shop floor problem points, letting the shop floor return to its supposed performance.

36. Use the shop floor management experience to design the shop floor management system for future product development.

R&D personnel shall use the shop floor management experience as the foundation for designing shop floor management system of next-generation product development to continue upgrading their capabilities of problem prevention by prediction and variation control for managing a product when it enters the shop floor from the laboratory.

3.3 REINFORCEMENT OF EQFD FOR THE ORIGINAL QFD

The proposed EQFD characterizes eight key features between it and the original QFD as reinforcement of a renewed QFD. These features are described below (Shiu et al., 2007).

1. Design-for-Excellence (DFX) Objectives Deployment. The demanded qualities, in the original QFD, are collected from customer surveys by using the KJ method and expanded by adding the necessary items such as reliability and safety. This approach, however, cannot ensure the completeness of demanded-quality deployment; that is, demanded qualities are not added unless they was surveyed or considered. To address the problem, it is necessary to develop an objective deployment structure that can ensure planning of all desirable dimensions for demanded quality. In this book, they are called DFX, where X means excellence and completeness (Bralla, 1996).

We propose 14 dimensions or DFX objectives for demanded quality. These dimensions or X's include (1) higher functional performance, (2) physical performance, (3) user friendliness, (4) reliability/durability, (5) maintainability/serviceability, (6) effectiveness/efficiency, (7) safety, (8) transportability, (9) compatibility/upgradability, (10) environmental friendliness, (11) psychological characteristics, (12) short time-to-market, (13) manufacturability/testability, and (14) low-quantity production. Garvin (1987) argued that it was a challenge for the manager to use the quality dimension selected to engage in competition. Kano (2002) pointed out the necessity of giving consideration to two-dimensional recognition of quality to create attractive quality when quality was considered a competitive advantage. As a result, the deployment of DFX objectives can strengthen both the choice of quality niches and the quality planning of creating attractive quality for the original QFD.

2. Six-Sigma Specification Tolerance Design. For fabrication suppliers with sufficient development experience, contract specifications rarely have new content. In most cases, a new requirement is added, while the other requirements existed in past development experiences. Consequently, a usual design approach is to (1) use the specifications of a similar product that has been developed in the past as a baseline, (2) extract the items that are different in the newly specifications, and (3) attempt to compensate for differences in specification values by changing the reference product's design concept. For example, some specification target of a new product is over 80 units whereas the reference specification has required 70, assuming that the other specifications of the new and old are identical. The previously described approach of concept design makes use of partial design changes to the reference configuration to compensate for the difference of 10 units. If it is assumed that the reference product's actual performance has already reached 85 units on average, with a standard deviation of 2 units, the aforementioned design changes would be meaningless. In this case, whether a design change should be made or not is decided by considering our intent: (a) reducing cost by leveling down our good performance to just conform to the design requirements or (b) taking advantage of the excess of the design requirements to create selling points.

Therefore, as far as an evaluation or determination of a new product's specifications is concerned, the previously attained performance of similar products developed in the past should be assessed in order to connect with intrinsic technology sufficiently and to decide how to make design changes in a more strategic manner. Using the statistical tolerance design method, the orientation can be embodied as an orientation of design for Six Sigma (DFSS) for finished product characteristics and semi-finished product characteristics. Consequently, the method of QFD combining with the 6σ design is an effort as follows: In quality characteristics deployment and subsystems deployment, using the information of the actual performance of the baseline product, predict separately the actual performance of the characteristics of finished products and semi-finished products; then, based on the aforementioned, decide the specifications for their 6σ design; finally, achieve the target specifications by detailed product and process designing.

3. Process Characteristics Design. Generally, a product's process design refers to the design of a process recipe to decide the optimal settings for process parameters, in order to meet the product characteristic requirements for the semi-finished product during the process. A great number of studies that explore how to combine QFD with design of experiments (DOE) discuss how to decide the optimal settings for main process parameters in process deployment without increasing costs. Figure 3.15 shows that the objective for DOE is to decide the optimal temperature setting S_A that can obtain the expected output energy P_0.

Even with machines of the same model, however, various differences may also exist, such as machine's age, maintenance state, initial temperature during

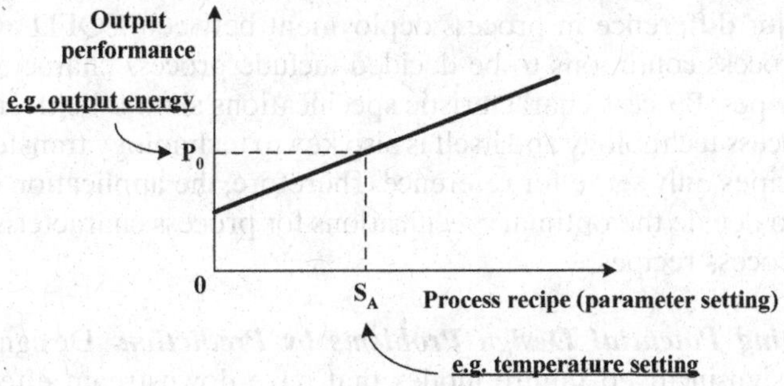

Figure 3.15. Expected output performance and process recipe.

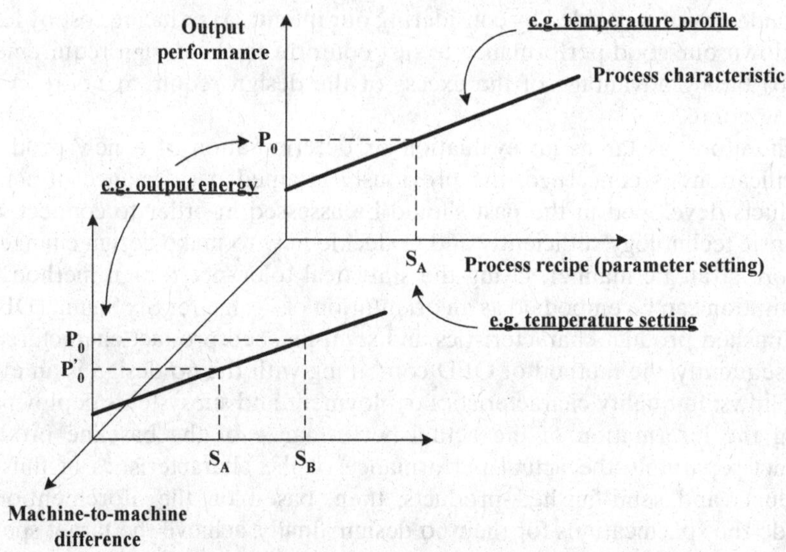

Figure 3.16. Process characteristic and process recipe.

operation, or speed of temperature rising. Therefore, same process recipes would not be able to ensure same output energy. Output energy can only be directly affected by the profile of temperature variations. As a result, only when the optimal range of a temperature profile is decided and controlled can the response value of output energy be ensured. At this point, to ensure the same output energy, different parameter settings may be needed, as shown in Figure 3.16. When the optimal temperature setting S_A is transferred to another machine, the output energy becomes P_0' rather than the expected P_0. To reach P_0 relies on monitoring the temperature profile; and, by setting the temperature to S_B, the profile is adjusted to the optimal range that can reach P_0. In this book, "process characteristics" are used to describe the process control item, "temperature profile," which is different from "temperature settings" referred to as "process recipes" or "process parameter setting."

The major difference in process deployment between EQFD and QFD is that the process conditions to be decided include process characteristics and process recipes. Process characteristic specifications should be deemed as the core of process technology and itself is also key to technology transfer, whereas process recipes only serve for reference. Therefore, the application of DOE in EQFD is to decide the optimal specifications for process characteristics rather than for process recipes.

4. Preventing Potential Design Problems by Prediction. Design problems involve design-incurred failure modes that have downstream effects of failures. During the design and development of a product, the "failure" concept surfaces when the developed object has become specific, so from the object per se the potential failure modes can be predicted. FMEA can effectively prevent the occurrence of design problems.

The reliability deployment of the original QFD carries out an FMEA on key parts. And the prediction of a parts failure mode combines with FTA. It deploys the failure modes of similar products that have been developed so that when new parts are designed, the recurrence of the failure mode can be effectively prevented. Despite this, the recurrence prevention means in FMEA may be treated as an inseparable part of design work. That is, when the designer designs new parts, both "design" itself and "giving consideration to past problem points and preventing them in design" cannot be separated from each other.

Two important differences in reliability deployment between EQFD and QFD are as follows: To redefine the timing of FMEA by laying stress on its functions of prediction and prevention; and to implement the FMEA on the process design which the original QFD emphasizes less, in order to separately deal with "design" itself and "after design, predicting the potential problem points and developing a strategy to prevent them" for product and process.

In Figure 3.2, charts 2-IV and 3-IV are product design FMEA and process design FMEA, respectively. They were implemented with design changes to the baseline product for the product or the process. In this case, FMEA was used after a design change rather than when "one thinks how to design and consider if a design change is needed." The purpose for the FMEA means is to make an early prediction and prevent the potential impact of the design change on quality. As a result of this, in subsystems deployment and unit parts/materials deployment and process deployment, key items also include design-change items.

For differences between product design FMEA and process design FMEA, refer to Table 3.4.

5. Technology Development Deployment for Advance Products. Renewal of the technology deployment part of the original QFD aims to strengthen thinking and formulating of an advance technology strategy. Acquiring the customer's advance product roadmap helps to make sure of the availability of existing technology seeds for the advance product and to define various related technologies for the future, in order to harmonize product planning and technology deployment.

In Figure 3.2, matrix 1-II offers an analytical mechanism for formulating an advance technology strategy, while technical details that can embody the requirements and characteristics of existing new products are deployed using matrices 1′-II, 2-II, and 3-II.

6. Competitive Analysis on Quality Positioning and Market Pricing. Maybe the issue that draws the most attention in NPD is a trade-off between quality assurance and cost reduction. In the practice of fabrication supply, the

TABLE 3.4. Product Design FMEA vs. Process Design FMEA

Item	Product Design	Process Design
Application areas	System, subsystems, components	Equipment, tooling, measurement instruments, operation procedures, setup, environment (5Ms/1E)
Failure mode	Potential weakness in product design, or occurred product deficiencies	Potential weakness in process design, occurred noncompliance to process requirements or manufacturing deficiencies
Recommended actions	• Technical solutions • Product-design guidelines updating	• Technical solutions • Process-design guidelines updating • Process controls
Used format	• Design potential FMEA (after product design changed, and before potential failures occur) • Design FMEA (after product design changed, and after failures occurred)	• Design potential FMEA (after process design changed, and before potential failures occur) • Process FMEA (after process design changed, and after failures occurred)

customer may have different target buying prices in mind for the same products from different fabrication suppliers. On the other hand, the fabricator must set a target cost that can (a) maintain his target profit when his price is restricted and (b) complete product development at the target cost. Therefore, the customer's price-setting evaluation criteria and the fabricator's bargaining power both decide whether the fabricator can acquire a "price premium" for his product.

The fabricator can adopt an active position by using the SW (strength and weakness) analytical mechanism of 4P/1S (product, price, place, promotion, and service) mixes between competitors to define key issues—such as selling points and items to be improved—that can strengthen his bargaining power for the future.

A similar analytical method is quality deployment for market pricing (QD_m) proposed by Mochimoto (1996, 1997). Its purpose is to ensure competitive quality design and price design for product development.

In respect of renewal of the cost deployment part of the original QFD, the most important feature is the mindset transformation from passive cost deployment to active price positioning.

7. QA Function Deployment System Development. This book develops a QA function deployment system—which comprises a "design engineering management system" (DEMS) and a "shop floor management system" (SFMS)—as the deployment structure for narrowly defined QFD. All NPD's relevant QA activities can be deployed based upon DEMS and SFMS.

DEMS aims to (a) plan and design the optimal products that meet customer requirements most and (b) design the optimal processes for manufacturing products. The system is constituted by 11 subsystems, which are described below:

- *Sales Quality Assurance System (*SQA*):* To ensure that the demanded-quality and market information is fully grasped, as well as to develop an effective value (quality and price) proposition.
- *Vendor Quality Assurance System (*VQA*):* To ensure that unqualified, in terms of feasibility and capability, vendors of components, parts, materials, tools and equipment are not chosen or approved.
- *Design for Excellence (*DFX*):* To ensure the successful planning, characterization, and development of all desirable quality dimensions.
- *Product and Process Data Management System (*PPDM*):* To ensure the availability of all the important qualitative and quantitative product and process data, from product planning to product disposal after usage, for every product.
- *Configuration Management System (*CM*):* To ensure that the technical feasibility of the specifications of the finished and semi-finished product characteristics—including the development of the design concept, the analysis of the design changes of the product and process based on identified configuration, and the impact of those design changes on quality and cost—are fully evaluated.
- *Process Deployment System (*PDS*):* To ensure that the target quality—characteristics and specifications of the finished product—can be communicated to the manufacturing stage by deploying the characteristics and specifications of the semi-finished products and process conditions in manufacturing process flows.
- *Design Optimization System (*DO*):* To ensure that the specifications (nominal value and tolerance) of semi-finished product characteristics and process conditions are optimized so as to minimize variability of product and process performance.
- *Failure Mode and Effects Analysis (*FMEA*):* To ensure early identification of potential problems in product design and process design, as well as prevent their impact on quality.
- *Design Reviews (*DR*):* To ensure full evaluation and verification of the design concepts of product and process in every phase of product development to prevent quality deficiency.
- *Management of Technology (*MOT*):* To ensure effective development, accumulation, and transfer of product-related and process-related technologies.

- *Project Management System (*PJM*):* To ensure that the dynamic allocation of management resources in simultaneously managing project scope, time, cost, risk, and quality can meet the project requirements.

The original QFD ensures mass-production quality by bringing operating standards into practice and through problem analysis. This book renews the shop floor deployment part of the original QFD by developing SFMS to achieve proactive management for mass-production QA. The design of SFMS serves to never buy in, slip in, manufacture, or pass on defective products. SFMS is constituted by 10 subsystems that were described in Section 3.2, VI, 28: (1) vendor quality assurance system (VQA), (2) preventive maintenance system (PM), (3) measurement systems analysis (MSA), (4) pre-production management system (PPM), (5) first-article inspection system (FAI), (6) in-process quality control system (IPQC), (7) statistical process control system (SPC), (8) reliability/safety control system (R/SC), (9) quality inspection system (QIS), and (10) out-of-control action plan (OCAP) and abnormality management system.

The operating standard emphasized by the original QFD refers to the standard operating procedure for equipment, tooling, and operators. The definition for the operating standard of the proposed EQFD also includes operation instructions on how to execute 11 design-engineering management subsystems and 10 shop floor management subsystems.

EQFD is composed of a quality deployment system and a QA function deployment system. Quality deployment system refers to the information network formed by deployment charts of quality, technology, cost, and reliability, whereas QA function deployment system refers to the activity network formed by QA job functions deployed by 11 subsystems of DEMS and 10 subsystems of SFMS. Figures 3.17 and 3.18 show how DEMS and SFMS relate to the NPD process, as well as how the QA function deployment system supports the quality deployment system, respectively. This figure also shows the importance of intrinsic technology and information technology (IT) toward EQFD: EQFD is a management technology managing QA of NPD, and it has to be combined with various intrinsic technologies needed in a developing product to realize the product. The usage of IT may enhance the efficiency of information exchange at implementing EQFD and help establish a knowledge management system to further achieve Rdb-QFD.

8. Implementation Process Development by Using the NPD Approach. QFD's history shows that the past development breakthroughs of quality deployment came from the creation of some "forms" (e.g., process assurance items tables, quality charts); so, generally, quality deployment models are presented by using matrices and forms to express systematic deployment processes as well as their interrelationships, as shown by the quality deployment system proposed by Mizuno and Akao in 1978. Such an expression form of "seeing a whole on one page" that can enhance system thinking capability has also been emulated by researchers afterwards.

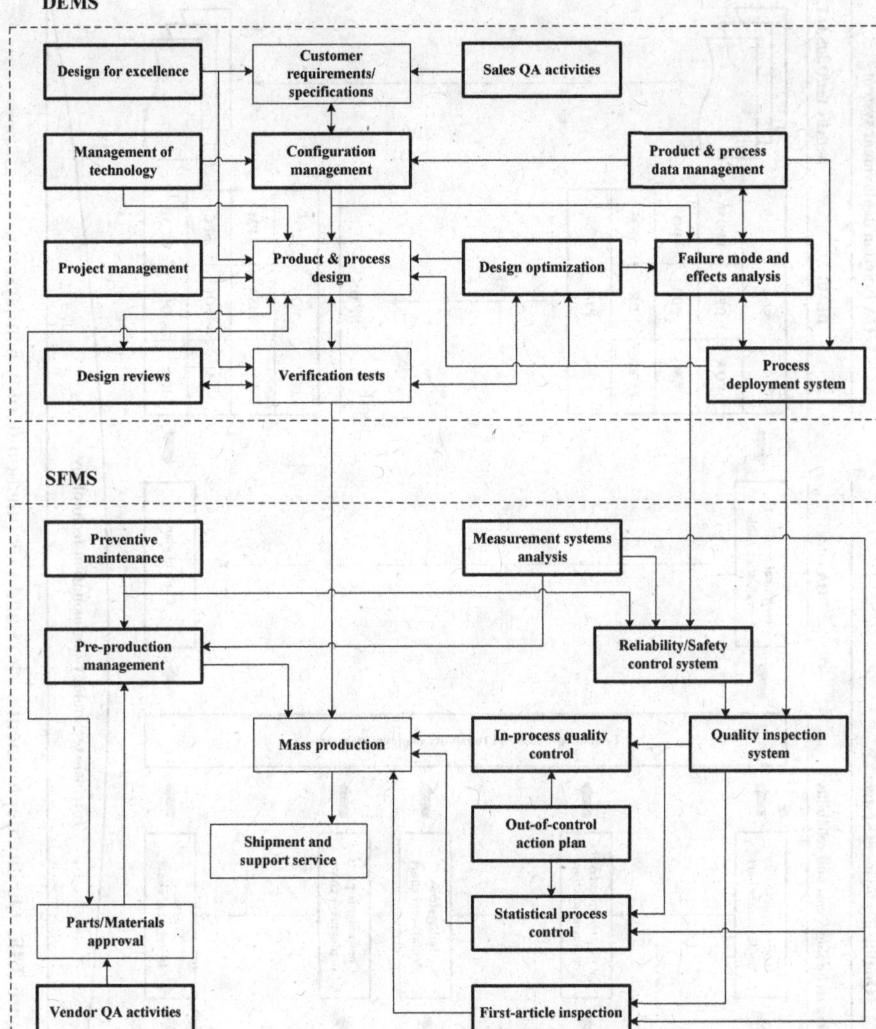

Figure 3.17. DEMS and SFMS.

Although QFD refers to the combination of quality deployment and narrowly defined QFD, as its integration model is not easy to be specific in implementation, the current QFD contents are almost considered as quality deployment while neglecting the narrowly defined QFD. The quality deployment model constituted by matrices and forms has become a common QFD form. Figure 3.19 shows the information flows of EQFD's quality deployment. For its differences with the original quality deployment system (Mizuno and Akao, 1978), refer to Table 3.5.

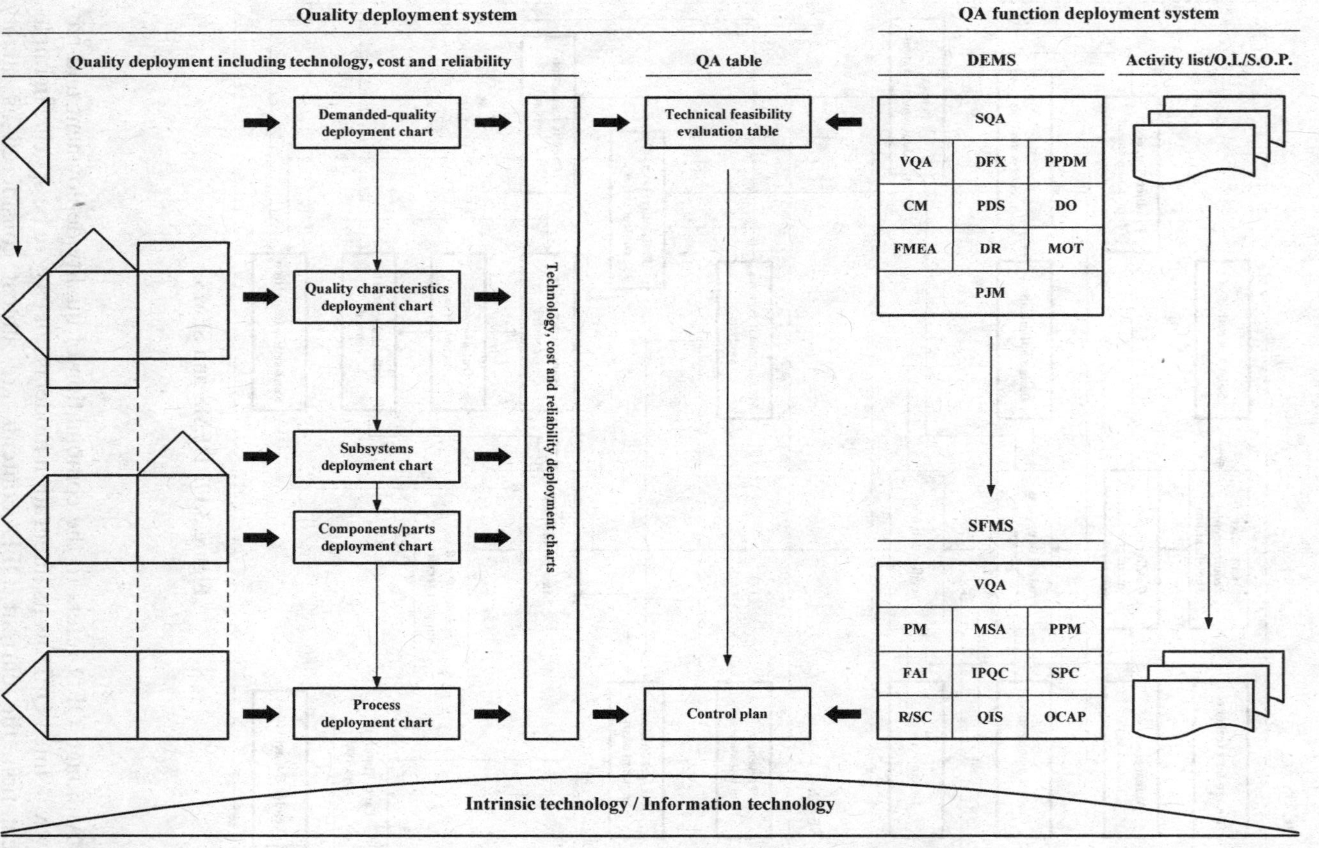

Figure 3.18. Quality deployment system and QA function deployment system.

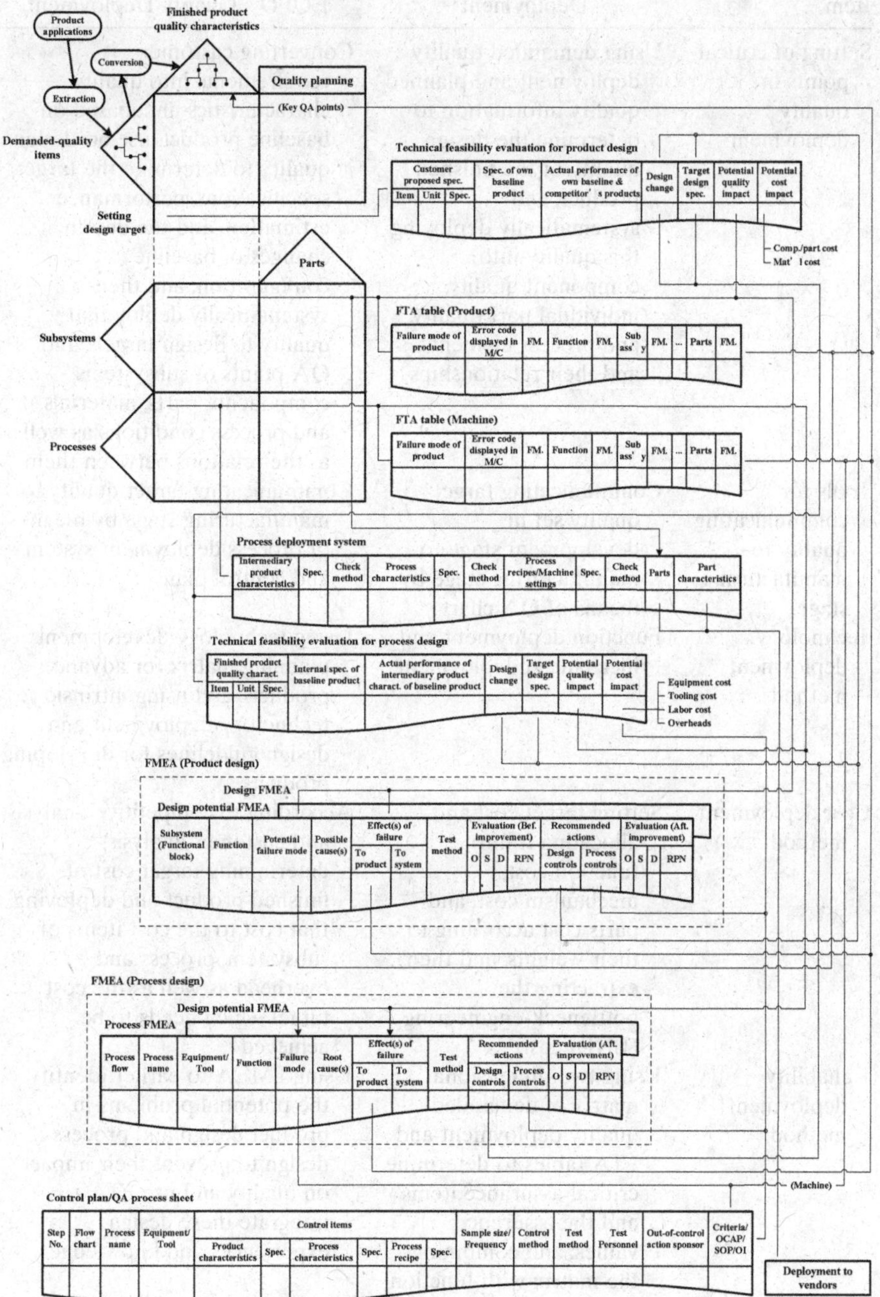

Figure 3.19. Information flows of EQFD's quality deployment (Jiang et al., 2007c).

TABLE 3.5. Comparison of Quality Deployments in Terms of Information Flows

Item	Original Quality Deployment	EQFD's Quality Deployment
Setting of critical points in quality deployment	Using demanded-quality deployment and planned quality information to determine the design quality of the finished product; and systematically deploying this quality into component quality, individual part quality, and process elements, and their relationships	Converting customer requirements into quality characteristics and, based on baseline product, use 6σ design quality to determine the target specifications, performance estimation, and the design changes of baseline configuration; and then systematically deploy that quality to design targets and QA points of subsystems, components, parts, materials, and process conditions as well as the relations between them
Tools for communicating quality to manufacturing stage	Communicating target quality set in development stage to manufacturing stage by means of QA chart	Communicating target quality to manufacturing stage by means of process deployment system and control plan
Technology deployment method	Function deployment and mechanism deployment	Using technology development planning matrix for advance products, and using intrinsic technology deployment and design guidelines for developing products
Cost deployment method	Setting target cost and allocating it into function cost, mechanism cost, and parts cost according to their weights, and then extracting the bottleneck-engineering (BNE) cost	According to competitive analysis and customer analysis, determining target cost of finished product and deploying that cost to the cost items of subsystem, process and overhead as well as the cost target which needs to be achieved
Reliability deployment method	Using two-dimensional matrix of demanded-quality deployment and FTA tables to determine critical assurance items and the assurance values; and combining the matrix with function deployment, quality characteristics deployment, and component and parts deployment to implement FMEA of key parts	Using FMEA to earlier identify the potential problems in product design and process design to prevent their impact on quality and use FTA to integrate these design experiences and knowledge

Broadly defined, QFD refers to how detailed matrices and forms are made and linked in different phases of the NPD cycle. The proposed EQFD specifies an integral model for simultaneously executing QFD and NPD activities, as shown in Table 3.2 and Figure 3.20.

The EQFD system is composed of a quality deployment system and a QA function deployment system, and its overall deployment structure is shown in Figure 3.21. In this figure, codes I to IV refer to the comprehensive quality deployment including technology, cost, and reliability, while code V is QA function deployment. Since "people make quality," QA function deployment ensures, through deploying various QA job functions, the achievement of design targets defined by quality deployment and major QA points.

Toward the design processes of the finished product, the semi-finished product, and the process in the NPD cycle, the relevant personnel may ensure the quality of the execution of the processes through benchmarking, management audit and diagnosis, and QA job functions deployed by the 11 subsystems of DEMS, as shown in the charts of 1-V, 1′-V, and 2-V in Figure 3.21. In addition, toward the shop floor management for the new product at mass production phase, the relevant personnel may, based on the design of control plan and QA job functions deployed by the 10 subsystems of SFMS, minimize process and product variation, as shown in chart 3-V in Figure 3.21.

The additional benefits of EQFD can be described as follows:

1. An EQFD system can be regarded as an integrated mechanism that can provide the organization, while facing various new or evolved R&D philosophies and methodologies, such as IPPD, DFX, DFSS, or BOS, with an architecture to analyze the positioning and usefulness of these methodologies, or help the organization, while introducing these methodologies, implement management in a more integral and "internalized" way.
2. An EQFD system provides a living architecture that can retain NPD experiences and knowledge in a systematic manner and that can be easily applied to future similar designs as well as to the common frame of standardization and documentation among different plants.
3. Reinforcing NPD's QA management before production start-up, as well as supporting the company's TQM in the establishments of quality information network and NPD operating system.
4. Just like the "value chain" proposed by Porter (1985), which is a fundamental tool for diagnosing and finding ways to enhance corporate competitive advantage, the overall deployment structure of the EQFD system can provide an analytical framework to diagnose and seek to improve R&D's development management (DM) level.

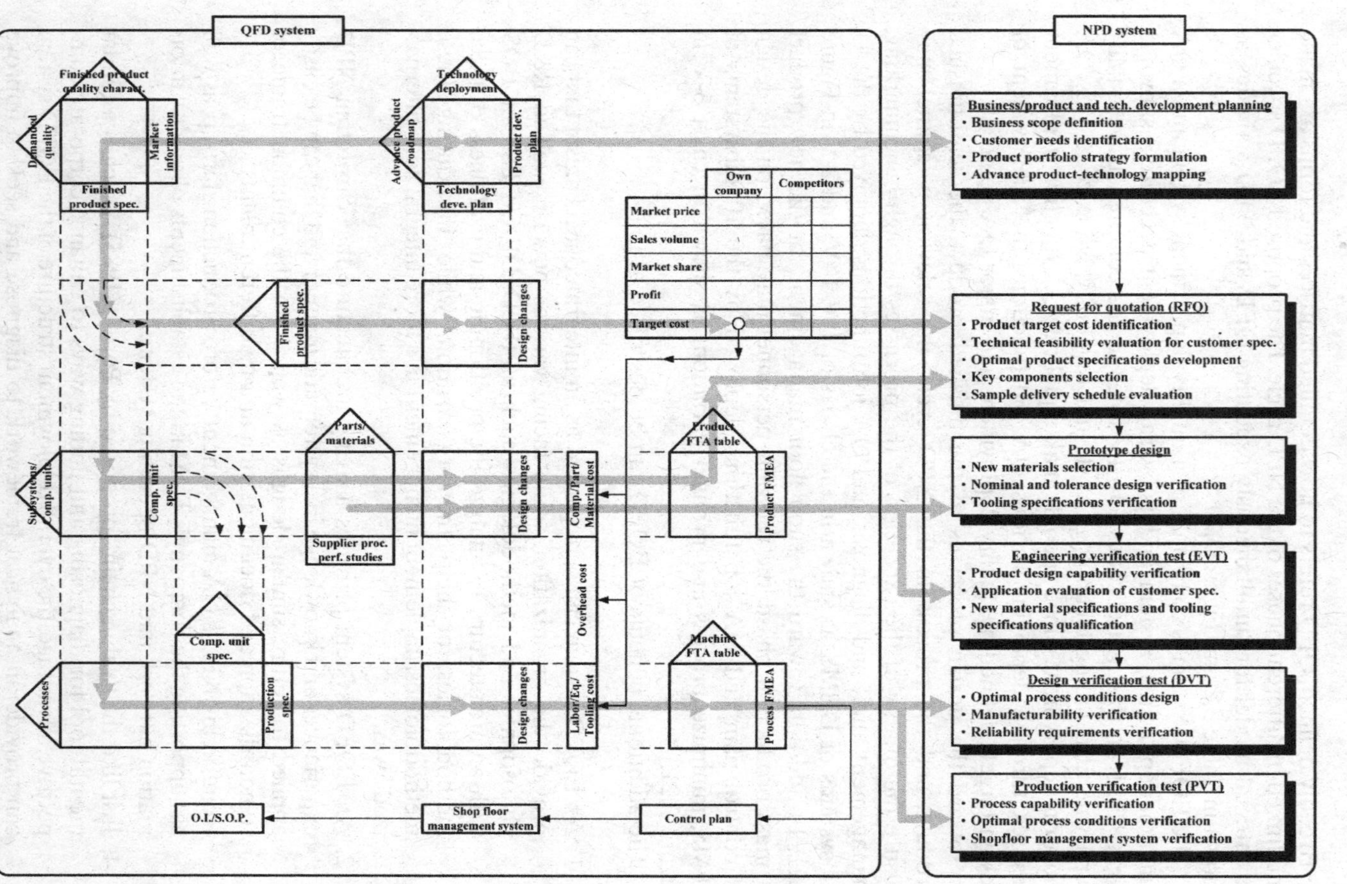

Figure 3.20. Conceptual model of implementing QFD by the NPD approach.

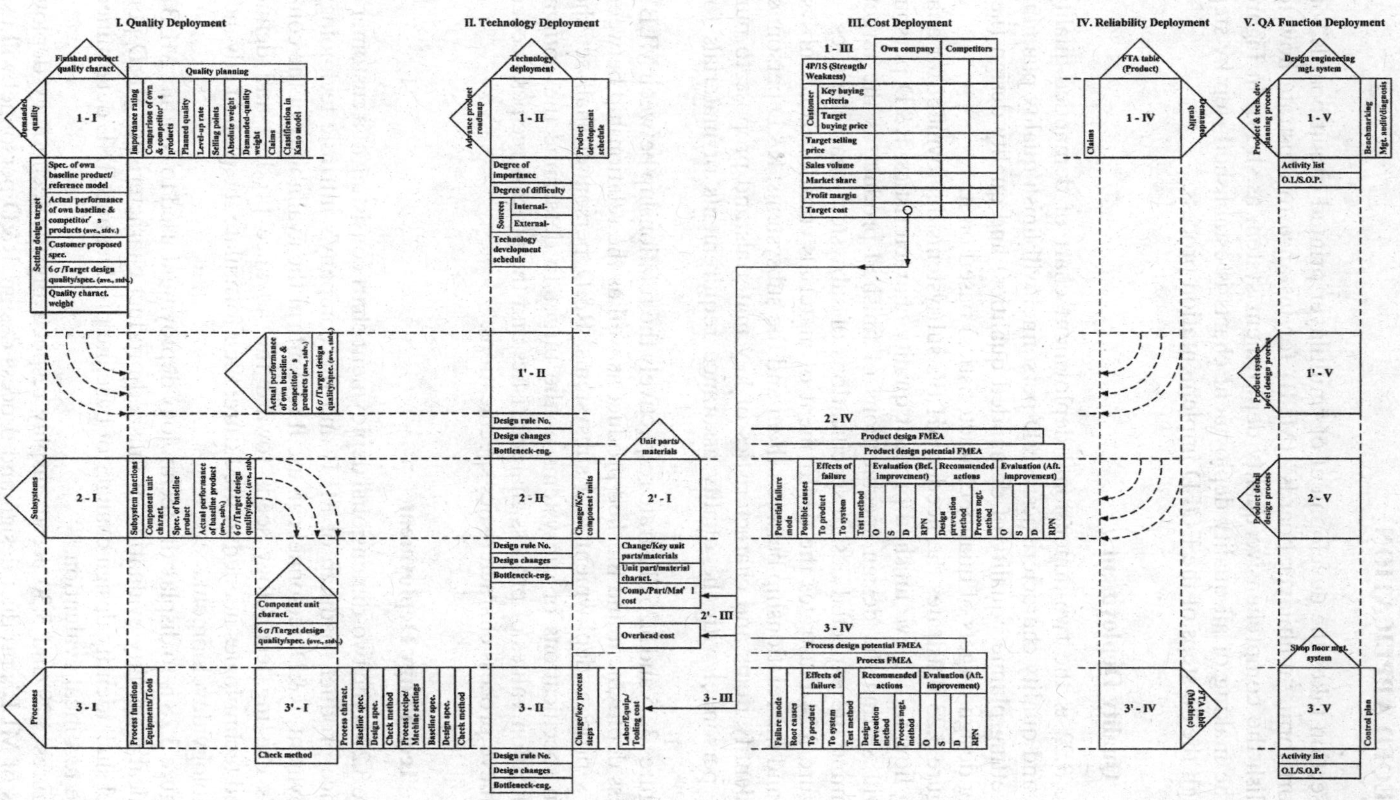

Figure 3.21. Overall deployment structure of the EQFD system.

3.4 EQFD APPLICATION

This section takes the development of "multilayer printed circuit boards" or, in short form, "multilayer boards" (MLB) for instance to illustrate how to establish the comprehensive quality deployment system as shown in Figure 3.2. The making of all quality deployment charts is established step by step through the 36 steps of the EQFD implementation process.

3.4.1 Quality Deployment

Figure 3.22 is the two-dimensional deployment chart of demanded-quality items and quality characteristics; R&D personnel will finish quality planning (i.e., setting planned quality of demanded qualities) and quality design (i.e., setting design targets of quality characteristics) based on it.

Figure 3.23 illustrates the needed MLB's subsystems deployment for analyzing how to achieve finished product quality characteristics; R&D personnel will, setting the design specifications of finished product as the target, determine the expected design specifications of subsystem characteristics.

Figure 3.24 illustrates the deployment of materials constituted of subsystems and the relationship between them and the subsystems; R&D personnel can identify items and characteristics of key materials and propose to purchase personnel with the quality assurance requirements of materials for vendors.

Figures 3.25 and 3.26 illustrate respectively the relationship between MLB's process deployment and finished product as well as the relationship between MLB's process deployment and subsystems. R&D personnel shall set the design specifications of subsystems as the target to determine the optimal specification values of process characteristics and the process recipes needed to achieve process characteristics specifications.

3.4.2 Technology Deployment

Figure 3.27 is the two-dimensional deployment chart consisting of a customer's advance product roadmap of MLB and the company's intrinsic technology deployment. R&D personnel can use it to confirm the availability of the company's existing technology seeds toward the advance product and define related technologies needed in the future; then formulate the plan of advance technology development.

Figure 3.28 is to display the technology deployment used to conduct MLB's technical feasibility evaluation of finished product characteristics. R&D personnel shall identify design changes of baseline product and BNE determined by the technical evaluation.

Figures 3.29 and 3.30 are to display respectively the technology deployments of MLB's product design and process design. R&D personnel will use them to assess the technical feasibility of the specifications of subsystem

Demanded quality deployment / Finished product quality characteristics deployment	1st level	. . .	Overall board thickness		Holes					Conductors and pads			Solder mask		Electrical characteristics			. . .	Quality planning										
																				Competitive analysis									
1st level	2nd level	. . .	Board thickness	. . .	PTH diameters	Plated hole wall thickness	Voids	Hole solderability	Hole location accuracy	Conductor widths	Conductor spacing	Pad diameters	Soldermask thickness	Exposed traces	Continuity	Isolation	. . .	. . .	Importance rating	Own company	Competitor A	Competitor B	Planned quality	Level-up rate	Selling points	Absolute weight	Demanded-quality weight	Claims	Classification in Kano model
Functional performance	Providing an electrical interconnect for electronic components				△	○	◎	△	△	◎	◎	◎	△	△	◎				5	5	5	5	5	1.00		7.50			M
	Controlling impedance		△		△	◎	○	△	△	△	△	○	△	△	○				5	5	5	4	5	1.00		6.00			O
	Dissipating heat				◎	◎	○	△	△	△	△	○	△	△					4	4	4	4	5	1.25	○	7.20			O
	. . .																												
Physical performance	Providing a mechanical support for electronic components		○		◎	△	△	◎	◎	△	△	◎	△	○					5	5	5	5	5	1.00		7.50			M
	Low volume		◎				△			△	△	△							5	5	5	4	5	1.00	◎	11.25			A
	Light weight		◎				△			△	△	△							5	4	4	4	5	1.25	◎	14.06			A
Reliability/ Durability	Protecting circuit		△		△	△	△	△	△	◎	△	△	◎			○													
	. . .																												
Effectiveness/ Efficiency	Offering uniformity of electrical characteristics from assembly to assembly					△	△	△	○			○	△																
	. . .												△																
Safety	Providing a safe working environment for electronic components		△													○													
. . .	. . .																												
Setting design target	Spec. of own baseline product		2.20±8%		2.50±0.055	>0.038	Void area less than 10%	No more than 10% de-wetting of solder	0.28±0.12																				
	Actual performance of own baseline product		(ave., stdv.) (2.22, 0.05)		(2.52, 0.018)	(0.040, 0.001)	Less than 8%	Less than 10%	(0.28, 0.03)																				
	Actual performance of competitor's product		(2.26, 0.05)		(2.52, 0.023)	(0.040, 0.002)	Less than 8%	Less than 10%	(0.27, 0.06)																				
	Customer proposed spec.		2.25±10%		2.50±0.065	>0.035	Less than 10%	No more than 10%	0.28±0.12																				
	Target design spec.		2.25±8%		2.50±0.065	0.038	Less than 10%	Less than 10%	0.28±0.10																				
	Quality characteristic weight																												

Figure 3.22. Quality chart 1-I.

1st level	2nd level	...	Overall board thickness		Holes					Conductors and pads			Solder mask		Electrical characteristics			...
Subsystems deployment \ **Finished product quality characteristics deployment**		...	Board thickness	...	PTH diameters	Plated hole wall thickness	Voids	Hole solderability	Hole location accuracy	Conductor widths	Conductor spacing	Pad diameters	Soldermask thickness	Exposed traces	Continuity	Isolation	...	...
Mechanical configuration	Core material		◎															
	Plated hole				△	△			△						◎	◎		
	Trace														◎	◎		
	Solder mask												◎	○				
	...																	
Electrical configuration	...																	
	...																	
Setting design target	**Spec. of own baseline product**		2.20±8%		2.50±0.055	>0.038	Void area less than 10%	No more than 10% de-wetting of solder	0.28±0.12									
	Actual performance of own baseline product		(ave., stdv.) (2.22, 0.05)		(2.52, 0.018)	(0.040, 0.001)	Less than 8%	Less than 10%	(0.28, 0.03)									
	Actual performance of competitor's product		(2.26, 0.05)		(2.52, 0.023)	(0.040, 0.002)	Less than 8%	Less than 10%	(0.27, 0.06)									
	Customer proposed spec.		2.25±10%		2.50±0.065	>0.035	Less than 10%	No more than 10%	0.28±0.12									
	Target design spec.		2.25±8%		2.50±0.065	0.038	Less than 10%	Less than 10%	0.28±0.10									
	Quality characteristic weight																	

Subsystem functions	Component unit characteristics	Spec. of baseline product	Actual performance of baseline product	Target design spec.
	Thermal gradient (Tg)	220220220		
	...	...		
	Copper plating thickness	>0.028	(ave., stdv.) (0.032, 0.002)	0.032
	Tin/Lead solder thickness	>0.0038	(0.0044, 0.0002)	0.0044
	Copper foil thickness	...	...	
	Dielectric thickness	...	...	

Figure 3.23. Subsystems deployment chart 2-I.

Subsystems deployment / Materials deployment		1st level: • • •	Etching solution		Plating solution			Soldering material			• • •						
1st level	2nd level	2nd level: • • •	Ferric chloride	Ammonium persulphate	Acetic acid	Potassium iodide	Sulfuric acid	Solder	Flux	• • •	• • •	Subsystem functions	Component unit characteristics	Spec. of baseline product	Actual performance of baseline product	Target design spec.	Change/Key component units
Mechanical configuration	Core material												Thermal gradient (Tg)	220	220	220	• • •
													• • •	• • •	• • •	• • •	
	Plated hole				◎	◎	◎						Copper plating thickness	>0.028	(ave., stdv.) (0.032, 0.002)	0.032	• • •
													Tin/Lead solder thickness	>0.0038	(0.0044, 0.0002)	0.0044	• • •
													Copper foil thickness	• • •	• • •	• • •	• • •
													Dielectric thickness	• • •	• • •	• • •	Dielectric type
	Trace																
	Solder mask							◎									
	• • •																
Electrical configuration	• • •																
	• • •																
Change/Key unit parts/materials																	
Unit part/material characteristics																	
Component/part/material cost																	

Figure 3.24. Materials deployment chart 2′-I.

Process deployment \ Finished product quality characteristics deployment			1st level: . . .	Overall board thickness		Holes					Conductors and pads			Solder mask		Electrical characteristics			. . .	Process functions	Equipments/Tools
1st level	2nd level	3rd level	2nd level: . . .	Board thickness	. . .	PTH diameters	Plated hole wall thickness	Voids	Hole solderability	Hole location accuracy	Conductor widths	Conductor spacing	Pad diameters	Soldermask thickness	Exposed traces	Continuity	Isolation	. . .	. . .		
Inner layer	. . .	. . .																			
	Etching	Developing																			Developing machine
		Etching																			Etching machine
		Stripping																			Stripping machine
	MLB lamination	Lamination		○																	Lamination press
		Post treatment		○																	X-ray target drilling equipment Routing machine Target furnish equipment
	Drilling			△		◎	○	○	○	◎											Drilling machine
Outer layer	Plated through hole (PTH)	Preliminary treatment						△	○												Scrubbing machine
		Desmear						△	△												. . .
		E-Less Cu					△	△	◎												. . .
	Panel plating			△		△	△				△	△	△								Horizontal plating line
	Etching	Developing									◎	◎	◎								Developing machine
		Etching									◎	◎	◎								Etching machine
		Stripping																			Stripping machine
	LPI solder mask	Preliminary treatment												△	△						Scrubbing machine
		Solder mask coating												○	△						Spray coater
		. . .																			
	. . .	. . .																			
Setting design target	Spec. of own baseline product			2.20±8%		2.50±0.055	>0.038	Void area less than 10%	No more than 10% de-wetting of solder	0.28±0.12											
	Actual performance of own baseline product			(ave., stdv.) (2.22, 0.05)		(2.52, 0.018)	(0.040, 0.001)	Less than 8%	Less than 10%	(0.28, 0.03)											
	Actual performance of competitor's product			(2.26, 0.05)		(2.52, 0.023)	(0.040, 0.002)	Less than 8%	Less than 10%	(0.27, 0.06)											
	Customer proposed spec.			2.25±10%		2.50±0.065	>0.035	Less than 10%	No more than 10%	0.28±0.12											
	Target design spec.			2.25±8%		2.50±0.065	0.038	Less than 10%	Less than 10%	0.28±0.10											
	Quality characteristic weight																				

Figure 3.25. Process deployment chart 3-I.

Process deployment: 1st level	2nd level	3rd level	Mechanical configuration: Core material	Plated hole	Solder mask	Electrical configuration	Process functions	Equipments/Tools	Process characteristics	Baseline spec.	Design spec.	Check method	Process recipe/ Machine settings	Baseline spec.	Design spec.	Check method
Inner layer	. . .	. . .														
	Etching	Developing						Developing machine	. . .				. . .			
		Etching						Etching machine	Temperature profile	46±10	50±8		Temperature	48	52	
									. . .				. . .			
		Stripping						Stripping machine	. . .				. . .			
	MLB lamination	Lamination						Lamination press	. . .				. . .			
		Post treatment						X-ray target drilling eq. Routing machine	. . .				. . .			
	Drilling		○	△				Drilling machine	. . .				Drill speed	22000	26000	
													Feed rate	40	50	
													. . .			
Outer layer	Plated through hole (PTH)	Preliminary treatment		○				Scrubbing machine	. . .				. . .			
		Desmear		○				. . .	. . .				. . .			
		E-Less Cu		◎				. . .	. . .				. . .			
	Panel plating			○				Horizontal plating line	Temperature profile	30±10	30±10		Temperature	28	28	
									. . .				Concentration	32%	33%	
													. . .			
	Etching	Developing						Developing machine	. . .				. . .			
		Etching						Etching machine	. . .				. . .			
		Stripping						Stripping machine	. . .				. . .			
	LPI solder mask	Preliminary treatment			○			Scrubbing machine	Scrubbing pressure				. . .			
		Solder mask coating			◎			Spray coater	Viscosity				. . .			
		. . .														
	. . .	. . .														

Subsystem functions					
Component unit characteristics	Thermal gradient (Tg)	Copper plating thickness	Tin/Lead solder thickness	Copper foil thickness	Dielectric thickness
Spec. of baseline product	220	>0.028	>0.0038	. . .	. . .
Actual performance of baseline product	220	(ave., stdv.) (0.032, 0.002)	(0.0044, 0.0002)	. . .	. . .
Target design spec.	220	0.032	0.0044	. . .	. . .
Check method	Thermo-mechanical analysis	Micro section sample coupons shall be taken from three locations diagonally across the board, and then ...	. . .	. . .	. . .

Figure 3.26. Process deployment chart 3′-I.

Technology deployment / Advance product roadmap												
1st level	. . .	Mechanical design		Electrical design		Electrical design			. . .	Product development schedule		
2nd level	. . .	High layer count		Fine line		High circuit density			. . .			
3rd level / Attribute	. . .	Multilayer pressing capability	. . .	Etch factor	Throwing power	Microvia	Stack via	. . .	. . .	2013	2014	2015
. . .												
Min. trace (Starting copper)				◎						0.0060 (0.5 oz)	0.0050 (0.5 oz)	0.0040 (0.5 oz)
										0.0070 (1 oz)	0.0060 (1 oz)	0.0050 (1 oz)
										. . .	. . .	. . .
Trace tolerance (per oz)				◎						0.0016	0.0015	0.0014
Min. space (Starting copper)				◎						0.0060 (0.5 oz)	0.0050 (0.5 oz)	0.0040 (0.5 oz)
										0.0070 (1 oz)	0.0060 (1 oz)	0.0050 (1 oz)
										. . .	. . .	. . .
Plating min.					◎		△					
Finished drill aspect					○	◎						
Min. drill					◎	○	○					
Drill tolerance PLTH					◎	○	○					
Body size		○					△					
Board thickness		◎					△					
. . .												
Degree of importance		5		5	5	5	5					
Degree of difficulty		4		5	5	4	4					
Sources – Internal – Mother company												
Sources – Internal – R&D dept.		○				○						
Sources – Internal – Laboratory				○	○							
Sources – External – Company A							○					
Sources – External – Company B												
Technology development schedule – 2013												
Technology development schedule – 2014												
Technology development schedule – 2015												

Figure 3.27. Advance technology deployment chart 1-II.

Finished product quality characteristics deployment / Technology deployment		1st level: Mechanical design		1st level: Electrical design			1st level: Functional design	1st level: Environmental design						
1st level	2nd level	Layer stack-up design	. . .	Dielectric material selection	Film design	Trace width and pad diameters design (to account for etch factor and etch-back)	. . .	. . .	Spec. of own baseline product	Actual performance of own baseline product	Actual performance of competitor's product	Customer proposed spec.	Target design spec.	Quality characteristic weight
. . .	. . .													
Overall board thickness	Board thickness	◎		△					2.20±8%	(ave., stdv.) (2.22, 0.05)	(2.26, 0.05)	2.25±10%	2.25±8%	
	. . .													
Holes	PTH diameters			△		◎			2.50±0.055	(2.52, 0.018)	(2.52, 0.023)	2.50±0.065	2.50±0.065	
	Plated hole wall thickness					◎			>0.038	(0.040, 0.001)	(0.040, 0.002)	>0.035	0.038	
	Voids			△					Void area less than 10%	Less than 8%	Less than 8%	Less than 10%	Less than 10%	
	Hole solderability			△					No more than 10% de-wetting of solder	Less than 10%	Less than 10%	No more than 10%	Less than 10%	
	Hole location accuracy								0.28±0.12	(0.28, 0.03)	(0.27, 0.06)	0.28±0.12	0.28±0.10	
Conductors and pads	Conductor widths				◎	◎								
	Conductor spacing				◎	◎								
	Pad diameters					◎								
Solder mask	Soldermask thickness													
	Exposed traces					◎								
Electrical characteristics	Continuity			△	◎	○								
	Isolation			△	◎									
	. . .													
. . .	. . .													
Design rule No.														
Design changes		Layer increase from 10 to 20 layers				• Width reduction • Aspect ratio reduction								
Bottleneck-engineering						Etch factor								

Figure 3.28. Technology deployment chart 1′-II.

Subsystems deployment / Technology deployment		Mechanical design (1st level)			Electrical design (1st level)				Functional design	Environmental design						
1st level	**2nd level**	Layer stack-up design		• • •	Dielectric material selection	Film design	Trace width and pad diameters design (to account for etch factor and etch-back)		• • •	• • •	**Subsystem functions**	**Component unit characteristics**	**Spec. of baseline product**	**Actual performance of baseline product**	**Target design spec.**	**Change/Key component units**
Mechanical configuration	Core material	◎			◎							Thermal gradient (Tg)	220	220	220	• • •
												• • •	• • •	• • •	• • •	
	Plated hole						◎					Copper plating thickness	>0.028	(ave., stdv.) (0.032, 0.002)	0.032	• • •
												Tin/Lead solder thickness	>0.0038	(0.0044, 0.0002)	0.0044	• • •
												Copper foil thickness	• • •	• • •	• • •	• • •
												Dielectric thickness	• • •	• • •	• • •	Dielectric type
	Trace						◎									
	Solder mask					◎										
	• • •															
Electrical configuration	• • •															
	• • •															
Design rule No.																
Design changes		Layer increase from 10 to 20 layers					• Width reduction • Aspect ratio reduction									
Bottleneck-engineering							Etch factor									

Figure 3.29. Product design technology deployment chart 2-II.

1st level	2nd level	3rd level	. . .	Etching technique determination	Equipment selection	Drilling machine selection	Drill path design	Drill entry and back-up material selection	Plating technique determination	Equipment selection	. . .	. . .	Process functions	Equipments/Tools	Process characteristics	Baseline spec.	Design spec.	Check method	Process recipe/Machine settings	Baseline spec.	Design spec.	Check method	Change/Key process steps	Labor/Equipment/Tooling cost	Overhead cost
Process deployment / Technology deployment (1st level)			. . .	Etching process design		Drilling process design			Plating process design			. . .													
Inner layer	. . .	. . .																							
	Etching	Developing												Developing machine	. . .				. . .						
		Etching		◎	◎									Etching machine	Temperature profile	46±10	50±8		Temperature	48	52				
															. . .				. . .						
		Stripping												Stripping machine	. . .				. . .						
	MLB lamination	Lamination												Lamination press	. . .				. . .						
		Post treatment												X-ray target drilling eq. Routing machine Target furnish eq.	. . .				. . .						
	Drilling					◎	○	○						Drilling machine	. . .				Drill speed	22000	26000				
																			Feed rate	40	50				
																			. . .						
Outer layer	Plated through hole (PTH)	Preliminary treatment												Scrubbing machine	. . .				. . .						
		Desmear												. . .	. . .				. . .						
		E-Less Cu												. . .	. . .				. . .						
	Panel plating								◎	◎				Horizontal plating line	Temperature profile	30±10	30±10		Temperature	28	28				
															. . .				Concentration	32%	33%				
																			. . .						
	Etching	Developing												Developing machine	. . .				. . .						
		Etching												Etching machine	. . .				. . .						
		Stripping												Stripping machine	. . .				. . .						
	LPI solder mask	Preliminary treatment												Scrubbing machine	Scrubbing pressure				. . .						
		Solder mask coating												Spray coater	Viscosity				. . .						
		. . .																							
	. . .	. . .																							
		Design rule No.																							
		Design changes				Change drill bit type	Shorten drill path	Use aluminium composite																	
		Bottleneck-engineering																							

Figure 3.30. Process design technology deployment chart 3-II.

		Own company		Competitors		
				A		B
4P/1S (Strength/Weakness)	Product	Strength: • Stable product quality • High process	Weakness: • Vendor management	Strength: • Wide product line • Short cycle time	Weakness: • Unstable quality level	
	Price	Strength: • Scale economics	• • •	Strength: • Low material cost	• • •	
	Place	Strength: • High capacity • Low logistics cost	• • •	Strength: • Low inventory level	• • •	
	Promotion	Strength: • Strong sales force	• • •	Strength: • Low minimal order quantity	• • •	
	Service	Strength: • High delivery reliability	• • •	Strength: • Good technical	• • •	
Customer	Key buying criteria	• Product quality • Process technology development capability • Price				
	Target buying price					
Target selling price		• • •				
Sales volume		16, 392, 982 square feet		9, 985, 117 sqft.		
Market share		12.08%		9.32%		
Profit margin		25%				
Target cost		• • •				

Figure 3.31. Cost deployment chart 1-III.

characteristics and process conditions, and they will identify design changes of baseline product, important items, and BNE determined by the technical evaluation.

3.4.3 Cost Deployment

Figure 3.31 is the cost deployment chart of finished MLB. It can be used to evaluate if the company has a quality niche and relative advantage against the competitors, in order to seek a dealing chance negotiating with the customer for a better price than the present or the competitors and go further to decide product's target profit and cost under this price.

R&D personnel have to allocate the target cost of subsystems and processes according to the target cost of finished product as well as to assess the cost feasibility of cost items such as material, labor, equipment, tooling, and overhead for satisfying the target cost of subsystems and processes.

3.4.4 Reliability Deployment

Figure 3.32 illustrates the relationship between MLB's defect symptoms and demanded-quality items. Claims analysis is conducted to extract negative quality information so that sales personnel and R&D personnel can take it

Demanded quality deployment	FTA table (Product)	Top event								Quality planning										
											Competitive analysis									
1st level	2nd level	. . .	Discontinuity	Lineresistance	Hole plating open		Finish failure	Incorrect impedance	. . .	Importance rating	Own company	Competitor A	Competitor B	Planned quality	Level-up rate	Selling points	Absolute weight	Demanded-quality weight	Claims	Classification in Kano model
Functional performance	Providing an electrical interconnect for electronic components		◎	○	◎		○	◎		5	5	5	5	5	1.00		7.50			M
	Controlling impedance		◎	○	◎			◎		5	5	5	4	5	1.00		6.00			O
	Dissipating heat		○		○			○		4	4	4	4	5	1.25	○	8.64		○	O
	. . .																			
Physical performance	Providing a mechanical support for electronic components									5	5	5	5	5	1.00		7.50			M
	Low volume									5	5	5	4	5	1.00	◎	11.25			A
	Light weight									5	4	4	4	5	1.25	◎	14.06			A
Reliability/ Durability	Protecting circuit						○													
	. . .																			
Effectiveness/ Efficiency	Offering uniformity of electrical characteristics from assembly to assembly						○													
	. . .																			
Safety	Providing a safe working environment for electronic components																			
. . .	. . .																			

Figure 3.32. Reliability deployment chart 1-IV.

into consideration while they are conducting quality planning of the finished product.

Figures 3.33 and 3.34 are FMEA tables for R&D personnel to analyze the potential impact of product design changes on quality and to prevent it at the earliest possible time. The FMEA of product design can be divided into DPFMEA and DFMEA, which are respectively used, after design change, by R&D personnel (a) to predict problems before potential failures happen and (b) to prevent the recurrence of failures after the first time of their happening.

Figure 3.35 illustrates the relationship between the defect symptoms of MLB and the defect symptoms of working machines, in order to provide a basis for R&D personnel to analyze process failure modes and their effects.

Figures 3.36 and 3.37 are FMEA tables for R&D personnel to analyze the potential impact of process design changes on quality and to prevent it at the earliest possible time.

3.4.5 Shop Floor Management

Figure 3.38 is to display the written summary of contents and standards of shop floor control, that is, control plan, to provide as the control standards that

Subsystem/Material	Function	Potential failure mode	Possible causes	Effects of failure		Test method	Evaluation (Bef. improvement)				Recommended actions		Evaluation (Aft. improvement)			
				To product	To system		O	S	D	RPN	Design prevention method	Process management method	O	S	D	RPN
. . .																
Plated hole	Provide internal pad connection in multilayer board	Conducting track permeated by chemical solution on hole wall	Spacing between plated hole and conducting track too small	Circuit short	System failure	Magnifier	7	8	4	224	• Diameter of drilled hole before plating shall greater than diameter of plated hole about 0.2 mm • Spacing between edge of drilled hole and conducting track shall greater than 0.5 mm		5	8	3	120
. . .																

Figure 3.33. Product design potential FMEA 2-IV.

Subsystem/Material	Function	Failure mode	Root causes	Effects of failure		Test method	Recommended actions		Evaluation (Aft. improvement)			
				To product	To system		Design prevention method	Process management method	O	S	D	RPN
. . .												
Outer layer	Extend plane to deal with a large number of interconnections and cross-over	Film image washed away by chemical solution	Track width too small and adhesion bad	Circuit discontinuity	System failure		Film image compensation		4	6	4	96
. . .												

Figure 3.34. Product design FMEA 2-IV.

FTA table (Machine)	FTA table (Product) — 1st level	. . .	Discontinuity				Lineresistance					Hole plating open	Incorrect impedance		. . .
	2nd level	. . .	Non-etched copper	Undercut	Outgrowth	. . .	Non-etched copper	Undercut	Voids	Plating nodules	. . .	Drilling defects	Laminating defects	. . .	. . .
Error code	**Symptom**														
. . .	. . .														
. . .	Etching speed too high									△					
. . .	Etching speed too slow		△	△	△		○	○		△					
. . .	Laminating pressure sensor damage/disconnect												◎		
. . .	Laminating pressure not stable												◎		
. . .	Laminating pressure too high												△		
. . .	Drilling overspeed											△			
. . .	Drill bit breakage											◎			
. . .	Drilling program missing											◎			
. . .	Plating current density too high								△	△					
. . .	Plating solution overflow								△	△					
. . .	. . .														

Figure 3.35. Reliability deployment chart 3′-IV.

have to be achieved for shop floor management to minimize process and product variation.

3.4.6 Summary

Figure 3.39 illustrates the quality deployment diagram constructed by all the deployment charts shown in Figures 3.22 to 3.38 and use a plated hole or a plated through hole (PTH) for instance to illustrate how that item and related items create linkage among the deployments of quality, technology, cost, and reliability and how it is helpful to achieve quality assurance of MLB.

All these quality deployment charts are established through the development process of MLB along with implementation steps of EQFD. If the interrelated information about market and technology becomes transparent and are linked this way, from the viewpoint of applying EQFD in an organization, "deployment chart management" is a key issue for managing profitable knowledge and making decision-making rules to be seen. Hence, the organizations implement EQFD need a mechanism to operate the management of deployment charts (shown in Figure 3.40).

Process flow	Process name	Equipment/ Mechanism/Tooling	Function	Potential failure mode	Possible causes	Effects of failure		Test method	Evaluation (Bef. improvement)				Recommended actions		Evaluation (Aft. improvement)			
						To product	To system		O	S	D	RPN	Design prevention method	Process management method	O	S	D	RPN
. . .																		
☐↓	Lamination	Lamination Press	Press copper-clad layers and insulate material to form a solid board	Conducting track sliver	Conducting track excessive undercut	Circuit discontinuity	System failure	Visual check	6	7	4	168	Minimal width of conducting track shall greater than 0.3 mm	• Use fast-working etchant and exercise exact control on the etching time • Adjust laminating stress	4	7	4	112
. . .																		

Figure 3.36. Process design potential FMEA 3-IV.

Process flow	Process name	Equipment/ Mechanism/Tooling	Function	Failure mode	Root causes	Effects of failure		Test method	Recommended actions		Evaluation (Aft. improvement)			
						To product	To system		Design prevention method	Process management method	O	S	D	RPN
. . .														
☐↓	Plating	Horizontal Plating Line	Protect copper tracks, via holes and PTH from oxidation	Plating nodules or outgassing	Co-deposition of organic additives during plating of tin-lead	Surface defect	Lineresistance	Magnifier		Batch carbon treatment	4	6	4	96
. . .														

Figure 3.37. Process FMEA 3-IV.

Step No.	Flow chart	Process name	Equipment/Tool	Control items						Sample size/ Frequency	Control method	Test method	Test personnel	Out-of-control action sponsor	Criteria/ OCAP/ SOP/OI
				Product characteristics	Spec.	Process characteristics	Spec.	Process recipe	Spec.						
		. . .	. . .												
7		Etching	Etching machine	Etching rRate						Once/ P/N		Visual check			
						Temperature profile					Temp. monitor				
								Temperature				Thermometer			
								. . .				. . .			
8		Stripping	Stripping machine	Line width						2 Pnls/3Hrs	Xbar-R chart	Magnifier			
				Line spacing						2 Pnls/3Hrs	Xbar-R chart	Magnifier			
						Temperature profile					Temp. monitor				
								Temperature				Thermometer			
								. . .				. . .			
18		Drilling	Drilling machine	Hole location						1st Pnl/Lot	X-MR chart	Drilling sample			
				Hole roughness						1st Pnl/Lot	X-MR chart	Drilling sample			
								Drill speed				. . .			
								Feed rate				. . .			
								. . .				. . .			
21		Plating	Horizontal plating line	Back light						2 Pnls/2 Hrs	X-MR chart				
				Copper thickness						3 Pnls,Once/Shift	Trend chart				
						Temperature profile					Temp. monitor				
								Temperature				Thermometer			
								Concentration				Content analysis			
								. . .				. . .			
30		Preliminary treatment	Scrubbing machine	Appearance						3Pnls/Per 150 Pnls		Visual check			
						Scrubbing pressure					Pressure gauge				
								. . .				. . .			
31		Solder mask coating	Spray coater	Appearance						10 Pnls/Week		Visual check			
				Thickness						10 Pnls/Week	Trend chart	Micro section			
						Viscosity					. . .				
								. . .				. . .			
		. . .	. . .												

Figure 3.38. Control plan.

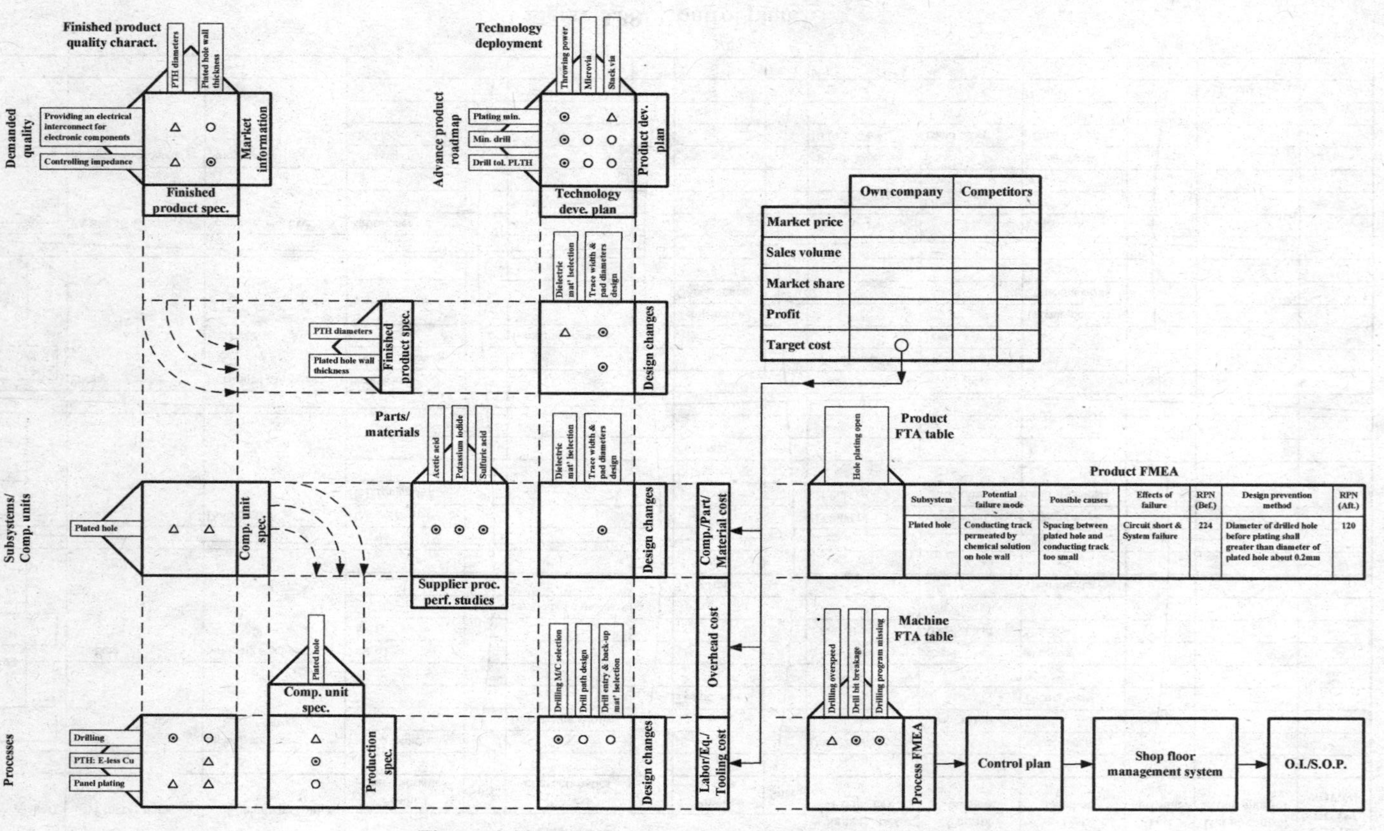

Figure 3.39. Quality deployment diagram of MLB.

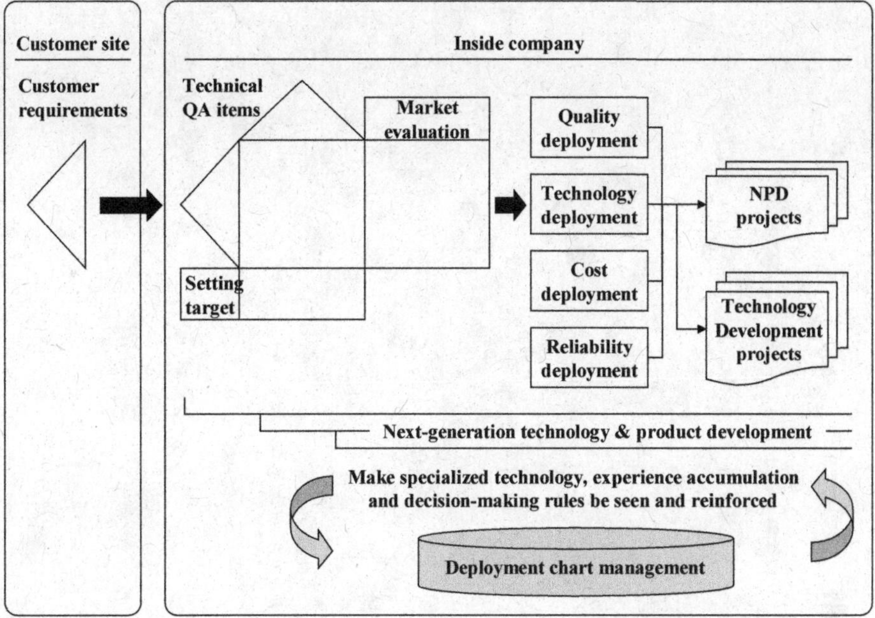

Figure 3.40. Managing deployment charts for corporate memory.

PART II
Optimizing Design for Functionality

4 R&D Paradigm

4.1 R&D STRATEGY AS PREDICTION AND PREVENTION

When a company competes with its competitors in the market, it's just like a war. In a real war, we have to, outside the battlefield, formulate "strategy" as the long-term and overall operational policy, and we must use "tactics" to plan practical actions in order to win the "fighting" inside the battlefield. Likewise, when a company competes with its competitors in the market through products, it also has to fight, exercise tactics, and develop long-term strategy. Though it seems that the company competes with the competitors in terms of products, in fact, it's a competition of technology development capability (in other words, "technology-fighting"); specifically, it means the capability of speedily developing the technology with low cost and high quality. The company attempts to design the market-competitive new products through technology leadership; so "technology," like tactics, is the solutions developed for specific new products and customer requirements. Facing keen competition, dynamic or unceasing created customer requirements, and short product life cycle caused by such circumstances, a company must pursue technology leadership continuously to deal with the competition of technology capability comprehensively and for long term. Therefore, we also have to develop "technological strategy." The concept described above is shown in Figure 4.1. This "technological strategy" signifies the so-called "generic technology," which is a little different from the ordinarily recognized "technological strategy" in meaning:

1. The ordinarily recognized "technological strategy" is a technological blueprint used to achieve the company's business objectives or advanced planning for technological development direction.
2. Generic technology is the universal framework used to comprehensively rationalize and efficientize the technology development in the company's various product fields (including totally new technological fields).

Quality Strategy for Research and Development, First Edition. Ming-Li Shiu, Jui-Chin Jiang, and Mao-Hsiung Tu.

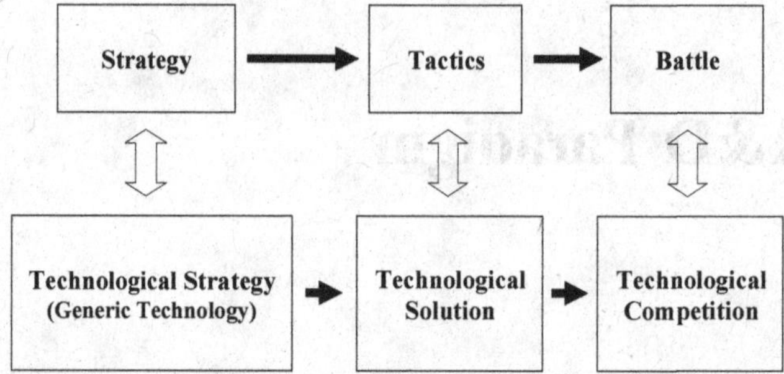

Figure 4.1. Strategy, tactics, and battle in terms of technology development.

Because the latter one has the universality for diverse fields, it is regarded as the strategy of technology development itself. In R&D, one of managers' most important work is to choose, introduce, or develop good technology development strategy to reinforce technology development itself. As mentioned above, "strategy" is different from short-term-focused "tactics" that emphasize something specific and deal with contingencies; instead it is systems thinking and framework which are long-term, universal, and institutionalized. Regarding "unknown items," strategy is utilized to provide an effective thinking direction and a framework of dealing with them. For R&D, unknown items mean the functional performance of product under various market usage conditions. Since this is related to technological quality, "technological quality prediction" is the core of the strategic thinking mentioned here. When technological quality can be effectively predicted at the laboratory research stage, we can take appropriate actions for unknown items (the functional performance of product under various unknown conditions) as early as possible to be able to make improvements before problem occurs. Unlike the tactics that take according actions and solve problems for known items, the so-called "strategy" must be utilized to effectively deal with the prediction and early improvement for unknown items.

For unknown items, the most commonly seen strategy is to adopt the methods such as reliability tests and life tests; however, technological quality is still not easy to be assured so that, after sales, product quality problems often arise. Consequently, the company has no choice but falls into the "firefighting" of technology development, product design, and manufacturing. A good R&D strategy should be able to guide, for long term, the technology development in various product fields to, at the possible lowest cost, effectively "resist" various unknown usage conditions (sources of variability) in the market and further reform the current paradigm, enhance technological capability, and win technology fighting in long term. Figure 4.2 shows the difference between technology and technology development strategy.

Part II of this book focuses on the universal methodology for rationalized and efficient technology development (including totally new areas). More

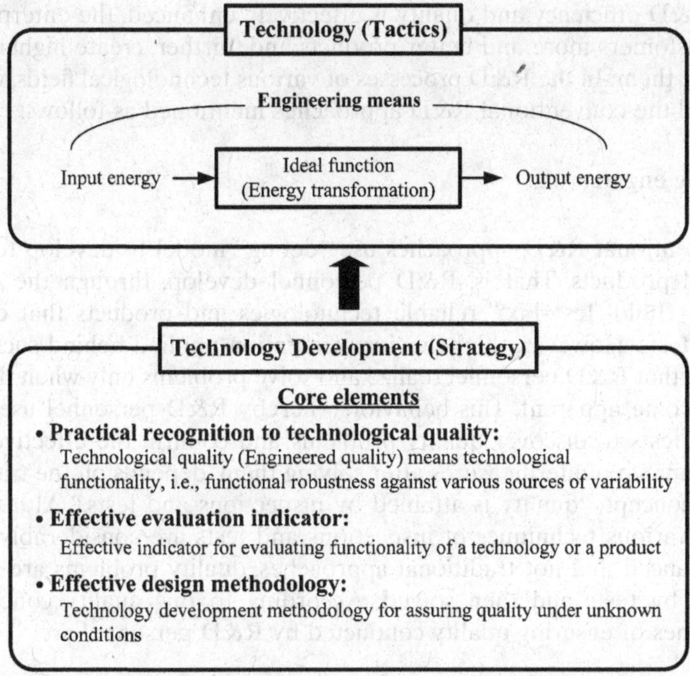

Figure 4.2. Technology and technology development.

specifically, this part of this book is to explore how to apply leading-edge quality methodology to enable R&D personnel to, under the possible lowest costs (such as using low-cost/loose-tolerance/high-variability materials or component parts), predict and early improve the functional performance of technologies and products to be maintained in a stable state (also known as "robustness") under various sources of variability without trial runs, tests, and inspections. This can be regarded as the powerful quality strategy for a company to face the severe commercial challenge, and R&D managers have to assume the leadership responsibility to comprehensively reform their conventional R&D paradigm as early as possible.

4.2 CONVENTIONAL APPROACH TO R&D

Nowadays, although the development of new technologies and new products changes with each passing day, the execution approach urgently needs some new development to upgrade R&D's efficiency, quality, and productivity. If the life cycle and time interval of launch of new technologies and new products are short while the efficiency of R&D performs poorly, it shows that the current R&D approach hides improvement opportunities for breakthrough.

Once R&D efficiency and quality is effectively enhanced, the enterprise can bring customers more and better products and, further, create higher added-value for them. In the R&D processes of various technological fields, we often observed the conventional R&D approaches mentioned as follows:

1. Debug engineering.

Conventional R&D approaches use "debug" model to develop technologies and products. That is, R&D personnel develop, through the cycle of "Design–Build–Test–Fix," reliable technologies and products that can pass various tests. However, the behavior orientation implied behind such debug model is that R&D personnel realize and solve problems only when the problems become apparent. This behavior, whereby R&D personnel use inspections or tests to discover quality problems and confirm the effectiveness of improvement countermeasures after solving them, depends on the traditional quality concept, "quality is attained by inspections and tests." Although the current various techniques of inspections and tests are considerably precise and advanced and not traditional approaches, quality problems are still discovered by tests and then solved. According to this quality concept, the approaches of ensuring quality conducted by R&D personnel are:

- Using more precise inspection and test equipment;
- Increasing the coverage ratio of test items; or
- Increasing test samples to enhance problem detection rate or test reliability.

Besides, while implementing the evaluations and tests of quality and reliability, we need to set various testing conditions, and, under such conditions, confirm the performance of quality and reliability using many samples for a long period of time. Even so, it's still not easy for us to ensure the reliability of the product under various customer usage conditions; thus the reliability tests are not efficient and effective enough.

In product development process, another activity used to evaluate and validate product design is "design reviews." Many technical discussions at design reviews, as well as the review opinions proposed by managers or senior designers, revolve around some aspects such as "Will it be better if we adopt some certain design?" and "Will this kind of problems be solved if we adopt some certain design?" However, such discussion may be a "guidance based on experiences" and also may be a "guess based on experiences" because we can know how the product performance becomes by adopting design changes only after we complete the evaluation. Many good design ideas may be, based on senior designers' development experiences, evaluated as "hard to achieve or technically infeasible," so we have to change the design concept or terminate the development. Nevertheless, if we use other evaluation strategies instead of depending on empirical rules, would we obtain a different evaluation result to change our decisions from "changing design concept" or "terminating development" to "continuing development"? If possible, then we need an evaluation technology that is more efficient and effective than empirical rules. The difference of evaluation technologies will influence the decision risk and opportunity cost.

2. Measure quality to get quality.

In order to ensure quality, R&D personnel develop various quality characteristics and measure them. According to the measured values of quality characteristics, R&D personnel can know the effects of product design on various quality characteristics and, through it, identify the quality characteristics needed to be improved (to be heightened or to be lowered) and further modify design to improve quality. Such improvement efforts usually need empirical rules, and the response analysis of each design-change trial. R&D personnel choose the better design change through observing the changes of quality characteristic values.

Although measuring multiple quality characteristics can make the quality problem apparent and be helpful to find out the improvement items, the frequently seen situation is that the improvement of a quality characteristic may cause the deterioration of anther quality characteristic; thus when R&D personnel want to simultaneously improve multiple quality characteristics, they usually need to make trade-off decisions. In other words, they need to keep a balance state between the improvement of certain characteristic values and the deterioration of other characteristic values, so as to make the design comprehensively better than that before improvement. Nevertheless, such approach of "quality design" cannot improve quality fundamentally.

3. Hit the target first, and then control variability.

While exploring the new product design process, we can point out frequently seen design approaches as follows:

a. For the design of new products, R&D personnel usually don't implement "zero-based designing," but determine a reference model to modify its design instead. Since the target specifications of a new product and reference model are different, R&D personnel need to modify the design of the reference model to make its response value achieve target performance. R&D personnel usually use the experimental method of experience-based trial and error or "one-factor-at-a-time" (change one factor at a time to examine the factorial effects) to conduct design modifications of reference model. At this time, the experience of senior designers is the key to efficiency of conducting effective design modifications.

b. The effects of design modification upon response values would reflect on "mean value" and "variability." Generally speaking, senior designers know very well about whether a certain design parameter influences response value; thus they can quickly judge which parameters need to be studied to modify the design and achieve target performance. However, what the senior designers really understand is whether certain design parameters have effects on the mean value of response values. They cannot know the effects of each parameter on variability until completing reliability tests. If the reliability testing results of new designs don't meet the requirements, R&D personnel have to implement design changes. As mentioned above, since R&D personnel fail to clarify which parameter can be used to reduce variability and not to affect mean value, they might have to conduct back-and-forth changes, tests, and modifications of the design; at this time, they are falling into the "debug" cycle (Design–Build–Test–Fix) until mean value and variability of new design are all in the acceptable range.

c. Since R&D personnel use various tests (such as reliability test, life test, etc.) to verify the new design, when that design fails to pass these determined test conditions, it means that the design has to be changed. At this time, R&D personnel utilize the "cause-and-effect relationship between design parameters and their effects" to propose the countermeasures that can cope with various test conditions. More specifically, R&D personnel focus on the test failures of a product under certain conditions and manage to find out the problem points and causes of failing to pass the tests. Furthermore, R&D personnel use the cause-and-effect relationship to tune design parameter values so that the new design can pass the same test conditions. If the modified design can pass the tests and has no other test failures, it means that R&D personnel can complete the new design. In this way, R&D personnel, according to the response analysis of a product under test conditions, modify the design to pass all the determined test conditions; however, as for those conditions other than test conditions and various unknown customer usage conditions, since R&D personnel have no ways to conduct response analyses one by one to tune design based on them, such approach cannot ensure that product performance will be good under these unknown conditions, and R&D personnel can only conduct trials and design changes again and again according to customers' feedbacks or complaints about product performance.

d. Although R&D personnel know whether design parameters have effects on product performance, they cannot identify an individual parameter's contribution toward the performance and individual effects of each parameter on mean tuning and variability reduction. Therefore, R&D personnel analyze and tune response value until that design shoots the target value. Once target value is achieved, design work stops. For the situation that needs to control over-scaled variability, R&D personnel tend to choose higher-class components and parts (increasing cost) to control variability within an acceptable range.

In summary, what is mentioned above is a structural model of "seeking to hit design target first and then control variability" that R&D personnel used in designing products, and consequently the R&D process includes a lot of debug operations.

In addition, R&D personnel usually don't use quality design to ensure quality. In fact, the activities ensuring quality in practice such as "modifying a design to pass test conditions according to response analysis" as well as "ensuring a design conformance to various design guidelines and checking items" are all operational basic works which, in strict definition, are not quality design activity. The real quality design is to efficiently achieve high quality at low cost—that is, to achieve target performance and control variability without debug approach and increasing cost—and further make product performance robust against determined test conditions and various unknown conditions.

4. The quality indicators used are too "downstream" (which are the lagging indicators for detecting occurred problems and thus tend to be used at downstream stages), thus they are useless at upstream stages.

The indicators generally used to measure quality are conformance-to-specification, reliability test data, process capability (such as process potential index, C_p, and process performance index, C_{pk}), failure rate, and so on. Observing these indicators, we can know that they are used to detect occurred problems so their utility is to detect problems after some quality problems happened. Though these indicators are helpful to us for knowing the occurrence of problem after some quality problems occurred, its disadvantage is that the evaluation about quality is too late (after the occurrence of quality problems). Therefore, they are lagging indicators and are often used at downstream stages of product development (large-scale manufacturing stage and customer usage stage).

Since, at the R&D stage, R&D personnel usually cannot find proper quality leading indicators for measuring quality and design capability, in practice they can only use downstream indicators (conformance-to-specification, reliability test data, process capability, failure rate) as substitutes. However, these indicators seem useless at the R&D stage where we need to predict and prevent the occurrence of quality problems. Moreover, it's not that R&D personnel lack design ideas, but they usually would choose a design concept, which is considered the most effective and feasible, to implement development and, until making sure that the design idea failed, they would not conduct the trials, tests, or reviews of the next design idea. Hence, it's very important for R&D personnel to effectively predict quality before the occurrence of problems. In other words, helping R&D personnel know design failure as early as possible (to detect poor design concepts as early as possible) is the key to enhance R&D efficiency. The lagging indicators mentioned above have no way to effectively predict quality before the occurrence of problems.

4.3 R&D PARADIGM SHIFT

In order to let R&D personnel avoid the disadvantages of the above-mentioned conventional approach and, with a more efficient and effective model, execute R&D works, R&D managers have to promote R&D paradigm shift as well as play the leading role. Table 4.1 describes the conventional R&D approach, improvement opportunities, and the key issues needed to be dealt with in changing the paradigm. To deal with these key issues needs a method that is more effective than the current model, and this method is what this book's Part II focuses on.

The operational model of R&D designers is limited by the management system designed by managers; thus managers have the biggest responsibility and influence over the development of new paradigm. Therefore, R&D managers must know whether the conventional approach described in Table 4.1 is the model currently adopted by the R&D department and are conscious of the necessity and leadership responsibility of changing this model. This book provides the effective methods for facilitating managers to revolute the R&D paradigm. This revolutionary strategy is suitable for various technology fields and can realize the following advantages toward technology itself:

1. *Precedency (Technology Readiness):* Technology can be evaluated with more effective indicators so that technology readiness can be achieved without product design targets at the product planning stage. Also, it ensures that there are no technological problems at product design and its downstream stages.
2. *Universality (Technology Flexibility):* Generic technologies can be identified and optimized, and they widely apply to the whole product family or next-generation products. If the technology has universality, then R&D personnel don't need to conduct technology optimization for each new product, and consequently the efficiency of R&D itself is effectively enhanced.
3. *Reproducibility (Technology Stability):* A product's quality in market usage can be effectively predicted at the laboratory research stage; thus R&D personnel make improvements for unknown items before quality problems occur. This method enables the technological achievements obtained in laboratory to be reproduced in large-scale manufacturing or various customer usage conditions.

TABLE 4.1. R&D's Structural Improvement Opportunities

Current Model	Improvement Opportunity Analysis		Key Issues
Debug engineering	Use debug model ("Design–Build–Test–Fix" cycle)	• The behavioral orientation of R&D personnel concealed in the debug model is that they are aware of, deal with, and solve problems as the problems become apparent; and, the quality concepts it concealed are "quality is gotten by inspections and tests" and "zero defects." Although the inspections and tests mentioned here are not the conventional approach, but use very precise and advanced test methods, quality problems are still found through tests first and then solved. (In R&D, the approaches of using more precise test equipment, higher test coverage, and more test samples are usually utilized to detect design problems or enhance the reliability of verification tests.)	• Apply effective design methods to achieve "quality is gotten by design." Inspections and tests are just used for confirmation.

(*Continued*)

TABLE 4.1. (*Continued*)

Current Model	Improvement Opportunity Analysis	Key Issues
Lack effective and efficient quality (functionality, reliability) evaluation methods/ techniques	• Lack the robustness assessment for design itself or materials used so that it may cause over-restricted or wrong design guidance and lead to ineffective quality assurance. • R&D personnel may not know how to choose materials/components or determine their specifications because of not understanding the characteristics of materials/components. • Reliability estimation/tests need to set various test conditions and acquire many samples and over a long period of time. Even so, it's not easy to ensure product's reliability under various customer usage conditions, so as to make reliability estimation not efficient and effective enough. • Investments on more precise and advanced test equipment for reliability. • Since lacking methods that can effectively ensure the reproducibility of laboratory research conclusion at downstream stages, it's necessary to make the customer willing to try field tests.	• Use functionality evaluation technology to efficiently conduct the quality evaluation of design itself or materials. • Use the materials/components of the lowest cost as possible to conduct experiments, and achieve high quality by such materials/components and set the range of variability as specification. • Use efficient (less samples and frequencies) and effective (high robustness and reliability) reliability estimation/testing methods. • Investing on the development of measurement capability for input value and output value of ideal function. • Use effective approach to confirm the reproducibility (if there exists interactions among factors).

Insufficient design reviews	• Many technical discussions at design reviews revolve around some aspects like "Will it be better if we adopt some certain design?" as well as "Will this kind of problem be solved if we adopt some certain design?" However, such discussion may be a "guidance based on experiences" and also may be a "guess based on experiences" because we can know how the product performance becomes by adopting design changes only after we complete the evaluation. • Many good design ideas may be evaluated as "hard to achieve or technically infeasible" and determined to "change the design concept" or "terminate the development" (feasibility evaluation based on technology development experiences). Nevertheless, is it possible to use other evaluation strategies to obtain different evaluation results to change our decisions from "changing design concept" or "terminating development" to "continuing development"? If possible, the effectiveness of evaluation technology can influence the decision risk and opportunity cost.	• At the research stage, use the evaluation technology of "considering, at a time, various possible design concepts in design space and efficiently evaluate their feasibility and quality levels." • Use the assessment method distinct from empirical rules to conduct the evaluation of technological feasibility.

(*Continued*)

TABLE 4.1. (*Continued*)

Current Model	Improvement Opportunity Analysis		Key Issues
Measure quality to get quality	Measure multiple quality characteristics and, according to them, identify the items needed to be enhanced and improve quality	• The improvement of one quality characteristic may cause the deterioration of another quality characteristic; hence it usually needs to make trade-off while simultaneously improving multiple quality characteristics. Consequently, fundamental quality improvement cannot happen.	• Do not use the approach of measuring quality to get quality.
Hit the design target first and then control variability	Utilize "the cause-and-effect relationship among design parameters and their effects" to propose the countermeasures dealing with various test/ usage conditions	• For the design of new model, R&D personnel usually use the methods of changing one factor at a time (one-factor-at-a-time) or experience-oriented trial and error to modify the design of reference model. • R&D personnel only know whether design parameters have effects on the result (response value), and the effects means the effects on mean value. The effects of design parameters on variability are known by reliability tests. If the result of test cannot conform to requirements, R&D personnel conduct design changes; but since they have no way to clarify which parameter can be used to reduce variability and not to influence mean value, they get involved in the debug cycle (Design–Test–Modify).	• Not to find out the causes of creating variability, which is not easy to be clarified; meanwhile, not to use the "cause-and-effect relationship among design parameters and their effects" to propose the countermeasures that can deal with test conditions. Instead, to study how to make product performance can be robust against various sources of variability (including unknown conditions other than test conditions) through the setting of parameter levels, so as to make quality problems not happen. • Use effective experimental strategies and design procedures to efficiently implement design work and grasp cost reduction opportunities.

<table>
<tr><td></td><td>• Even if we know whether design parameters have effects on response value, the contribution of each parameter to the result and the individual effect of parameters on mean tuning and variability reduction cannot be identified.
• Analyze and tune response value until the design hits the target value under the set extreme conditions.
• Once target value is achieved, design work stops, and the cost reduction opportunity which is possibly brought by improving variability first would be lost.</td><td>• Concretely grasp the individual effect of various parameters on mean tuning and variability reduction to quickly and effectively reduce variability or tune the response value to the target value.</td></tr>
<tr><td>While designing a product, quality is not ensured by quality design</td><td>• To conduct tuning according to response analysis and to ensure the design complying with design guidance/checklists are operational/basic works, not quality design.
• Rely on tolerance design approach (improve quality through cost-increasing approach such as changing the materials/components with higher grade) to control variability.</td><td>• Apply quality design method to reinforce design work toward ensuring product quality.
• Use low-cost approach (use the low-grade materials/components with wide tolerance and high variability) to achieve high quality (small variability).</td></tr>
</table>

(*Continued*)

TABLE 4.1. (*Continued*)

Current Model	Improvement Opportunity Analysis		Key Issues
The quality indicators used are too "downstream" (which are the lagging indicators for detecting occurred problems and thus tends to be used at downstream stages); thus they are useless at upstream stages	Use the lagging indicators such as conformance to specification, reliability testing data, process capability indices (C_p/C_{pk} or P_p/P_{pk}), and failure rate to measure quality or to assess engineers' design capability	• At R&D stage, since there is no proper quality leading indicators to measure quality or assess engineers' design capability, the downstream indicators (conformance to specification, reliability testing data, process capability, failure rate, etc.) are substituted for them. • Design engineers usually conduct the reviews for the next design concept after making sure current design ideas/ concepts are failed. Hence, detecting poor design concepts as early as possible is the key to enhance R&D efficiency.	• Use more upstream quality leading indicator to effectively evaluate design capability and predict quality. • Utilize rational and accurate method for concept evaluation to detect poor design concepts as early as possible, so as to enhance R&D efficiency.

5 Functionality Evaluation

5.1 ENERGY TRANSFORMATION AND TECHNOLOGY DEVELOPMENT

Technology and product can be regarded as man-made engineered systems. The existence of them is to provide certain specific functions. R&D personnel utilize laws of physics to study how develop a man-made engineered system to achieve the function demanded or required by customers. One engineered system has input and output, and input is transformed to be output by the laws of physics. Such input/output transformation according to the laws of physics is a transformation of energy form; in other words, the development of technology and product (man-made engineered systems) is a process of seeking the optimization of input/output energy transformation. Each engineered system has its own "ideal function," that is, in energy transformation process, the perfect input/output transformation without losing any energy. It is the technological quality that R&D personnel want to ensure. While speaking of quality, we can classify quality into two types as follows (Taguchi et al., 2000):

1. *Customer-Driven Quality:* What customer wants (e.g., function, appearance, color, etc.). The way we improve customer-driven quality is to fulfill the customer's expressed and latent requirements.
2. *Engineered Quality:* Freedom from what customer does not want (e.g., noise, vibrations, failures, pollution, etc.). The way we improve engineered quality is to reduce the variability around ideal function caused by various sources of variability.

Part II of this book focuses on the latter—that is, to explore how to use leading-edge quality methodology to early predict and improve the functional performance of technology and product under various sources of variability to make it approach as close as possible to the ideal function; and further, to eliminate or reduce what customers do not want. The ideal function of an engineered system is shown in Figure 5.1.

Quality Strategy for Research and Development, First Edition. Ming-Li Shiu, Jui-Chin Jiang, and Mao-Hsiung Tu.

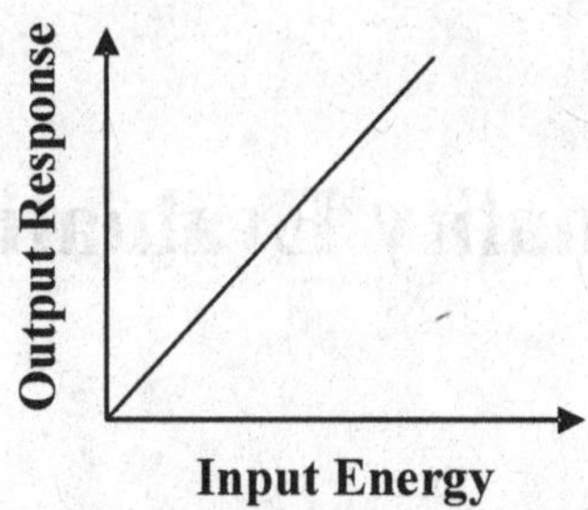

Figure 5.1. Ideal function of an engineered system.

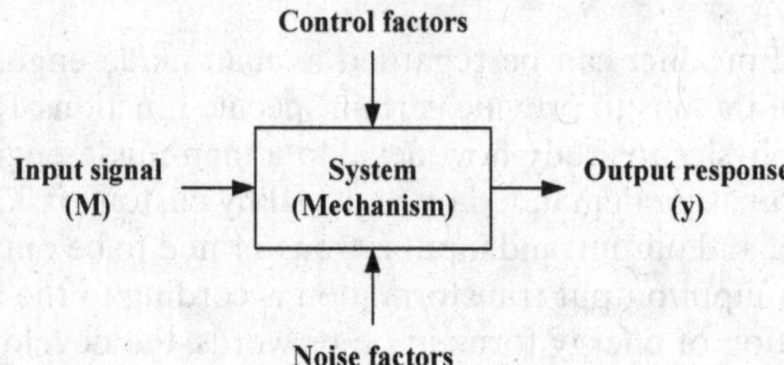

Figure 5.2. Parameter diagram (P-diagram).

"Input/output relationship" or "ideal function" sounds like a regression equation or functional relationship in mathematics; however, in this book, such a relationship in nature, doesn't mean mathematical relationship or cause-and-effect relationship, but instead means the "intention" conceived by R&D personnel for a man-made engineered system. That is, we expect, by clearly defining and optimizing this relationship, to make the system realize that intention stably and sufficiently. The composition of every engineered system can be generalized to be three factors and output response, as shown in Figure 5.2. In that parameter diagram (P-diagram), input signal (input energy) is represented by M and output response by y; thus the input/output relationship can be expressed with the function as $y = \beta M$.

The meanings of these factors and output response are interpreted as follows:

1. *Response:* The output value of an engineered system. This output value represents the functional performance of that system.
2. *Signal Factor:* The factor that can be used to change output value or obtain the wanted response value. Designers or users can utilize the proportional relationship of ideal function between signal and response value to change output response value by different input signal freely.
   a. Designers can create the wanted changes on response value according to different signal levels (settings) to achieve customers' different

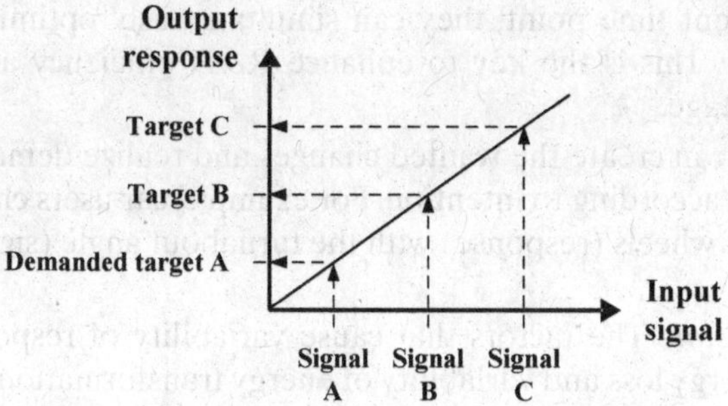

Figure 5.3. Using signals to generate target responses.

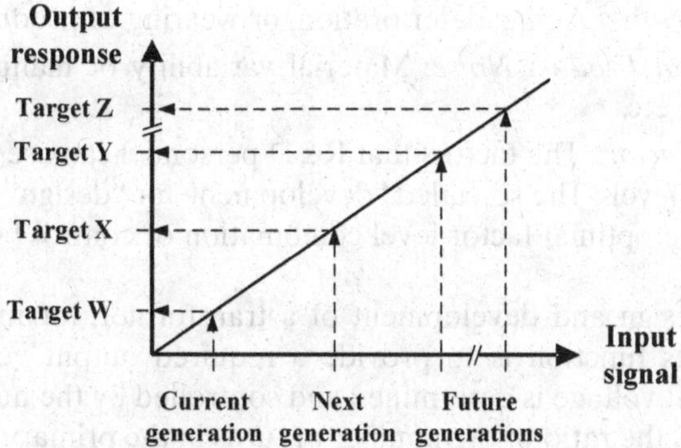

Figure 5.4. Input/output relationship using technology roadmapping.

demanded response values (different target values may derive from customers' different product sizes or product applications), as shown in Figure 5.3. Besides, different signal factor levels can be regarded as the technology levels of different generations; R&D personnel can study early how to achieve the response values (functional performance) which must be realized by next-generation technology according to technology and product roadmap. For example, regarding technology development of a semiconductor, designers can set "line width" as a signal factor as well as set current line width and the line width according to a future technology roadmap as the different levels of a signal factor; input/output relationship between input signal and output response is shown in Figure 5.4. When that input/output relationship reaches optimization, R&D personnel can realize the technology of "no matter how the line width of current and future level is, its corresponding ideal output responses all can be achieved." When R&D personnel implement the optimization of existed technology at a current time point, they can simultaneously optimize future technology. This is the key to enhance R&D efficiency and competitive advantage.

b. Users can create the wanted changes and realize demanded response values according to intention. For example, car users change the direction of wheels (response) with the turnabout angle (signal) of steering wheel.

3. *Noise Factor:* The factors that cause variability of response value (i.e., cause energy loss and variability of energy transformation). These factors are usually uncontrollable or their control cost is very high. Noise factor can be classified into three types:
   a. *Outer Noise:* Environmental conditions such as temperature, humidity, pressure, dust, wind, etc.
   b. *Inner Noise:* Aging, deterioration, or wearing of product itself.
   c. *Between Product Noise:* Material variability or manufacturing variability, etc.
4. *Control Factor:* The factors that R&D personnel can use freely to determine the levels. The so-called "development" or "design" is the activities to find the optimal factor-level combination of control factors.

Take the design and development of a transformer, for instance; we can identify that its function is to provide a required output voltage, and this required output voltage is determined and controlled by the number of "turns ratio" (which is the ratio of the number of turns in the primary side of a transformer to the number of turns in the secondary side). We can express a system of transformer by P-diagram, as shown in Figure 5.5.

Technology development means the process in which R&D personnel enable the technology to achieve ideal function by determining the optimal

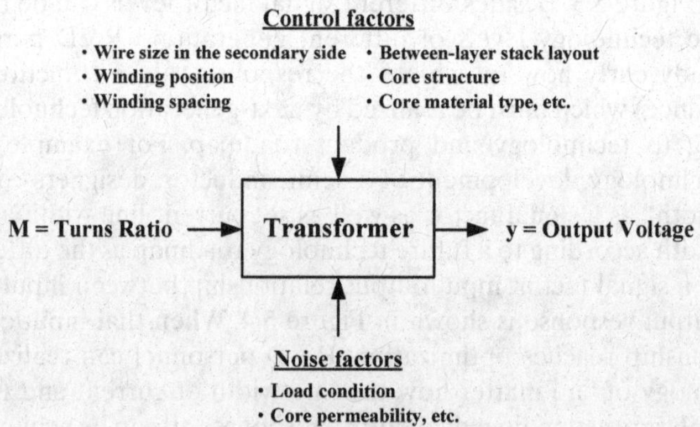

Figure 5.5. P-diagram of a transformer.

factor-level combination of controllable factors. While the proportional relationship of energy transformation between signal factor and output response can maintain minimum variability under the influence of noise factors, it can be regarded as approaching or achieving ideal function. At this time, R&D personnel consider that the technology is successfully developed.

5.2 EVALUATION OF TECHNOLOGY

All technologies can be regarded as the means of controlling energy transformation. With that energy transformation, some specific function can be provided. Nevertheless, there are various sources of variability in the real world to cause the loss and variability of that energy transformation, and consequently the function cannot fully activate and perform, as shown in Figure 5.6.

Based on conservation of energy, such lost energy is used to create nonintentional function and further to produce various defect phenomena, such as noise, vibration, failure, and so on. In other words, the better the technological quality, the stronger the product is to resist various sources of variability (unknown usage conditions); the worse the technological quality, the weaker the product is to resist various unknown conditions to make functional performance have large variability. Therefore, good technological quality means that technology has robustness (as shown in Figure 5.7). Robustness is defined as the state of performance where the technology is insensitive to factors causing variability at the lowest possible cost.

As mentioned above, the existence of technology or product is to provide a certain specific function; further more, the functionality of technology or product means that the functional performance of technology or product achieves the state of robustness (small functional variability). The so-called technology development involves use of the approaches such as small-scale experiment, test pieces, and simulation to predict the functionality of technology under various customer usage conditions. Most contents of R&D work

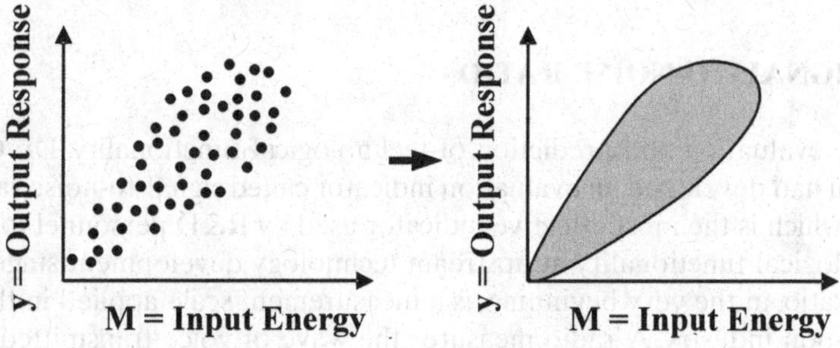

Figure 5.6. Functional variability in energy transformation.

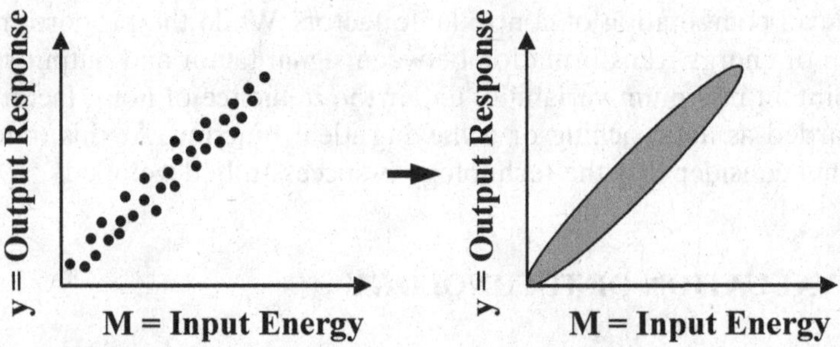

Figure 5.7. Functional robustness in energy transformation.

are focused on researching how to reduce the functional variability of technology and product caused by various sources of variability, thus R&D (research & development) work can be regarded as robustness development (RD) of technology and product. Therefore, how to evaluate technological quality is a very important issue for R&D work. The efficient and effective indicators and methods for evaluation is helpful to predict and improve in advance the functionality (the functional robustness of resisting various sources of variability) of technology, so as to enhance R&D productivity with a breakthrough. The cause of poor R&D productivity is usually not that R&D personnel lack design ideas, but they would choose a design concept that they think is the most feasible to develop with and would not conduct trials, tests, and reviews of the next design idea until they make sure that the design idea has failed. The most frequently seen evaluation approach of technology and design is to adopt various tests such as reliability tests and life tests; R&D personnel, based on the test results, study how to change the design to pass the tests according to the arisen problems, and they consequently get involved in the cycle of "Design–Build–Test–Fix" (debug cycle). Judging from this, if R&D personnel can quickly and effectively predict the success or failure of a design before the problems arise, the productivity of R&D itself can be improved with a breakthrough.

5.3 SIGNAL-TO-NOISE RATIO

For the evaluation and prediction of technological functionality, Dr. Genichi Taguchi had developed an evaluation indicator called signal-to-noise ratio (SN ratio), which is the most effective indicator used by R&D personnel to predict technological functionality at upstream technology development stages.

SN ratio, in the very beginning, is a measurement scale applied in the communication industry. A radio measures the wave of voice transmitted from a broadcasting station and converts the wave into sound. The larger the voice sent (i.e., input signal), the larger the voice received (i.e., output). Actually, the magnitude of the voice is mixed with the audible noise in the space, making voices unclear. Ideally, a good radio catches the voice sent and is not affected by the influence of noise (Taguchi et al., 2005). With this concept, Dr. Taguchi regarded "a factor used to adjust the output" as "voice sent" and regarded "product performance" as "voice received" and also regarded "factors causing variability" (either in the manufacturing environment or user's environment) as "noise." Therefore, the concept of SN ratio is expressed as the equation below:

$$SN = \frac{\text{Power of the voice sent that is catched}}{\text{Power of the noise that is heard}}$$

The stronger the magnitude of signal or the weaker the noise, the clearer the voice received; that is, the larger the SN ratio, the better the technological quality. If we look at this with the perspective of energy transformation, then the stronger the input energy used to create intended function or the lower the energy loss used to create nonintentional function (various defective symptoms), the higher the SN ratio (i.e., quality). In Sections 5.3.1 and 5.3.2, we will further explain the SN ratio used in technology evaluation and its forms.

5.3.1 Dynamic SN Ratio

As mentioned above, technology is energy-related. It can be regarded as a man-made engineered system of controlling energy transformation. R&D personnel intend to make that energy transformation approach as close as possible to ideal function; however, the influence of various sources of variability causes input/output energy transformation to present variability near ideal function. The core of R&D work is to study how to enable the input/output relationship to achieve its ideal function, and not to be influenced by various sources of variability. Therefore, when the above-mentioned SN ratio concept is applied to technology evaluation, we regard energy strength used to create intended function as "signal," and we regard the lost energy used to produce various nonintentional defect phenomena as "noise"; the ratio between the two represents the functionality of technology. At this time, SN ratio can be expressed as

$$SN = \frac{\text{Energy transformed into intended output}}{\text{Energy transformed into unintended output}}$$

In other words,

$$SN = \frac{\text{Useful output}}{\text{Harmful output}}$$

Since useful output is created according to the strength of input signal while harmful output is produced by noises, the above equation can be regarded as

$$\text{SN} = \frac{\text{Work done by Signal}}{\text{Work done by Noises}}$$

In $y = \beta M$ (the relationship of energy transformation shown in Figure 5.7), since β is the amount of y changed by one-unit changes of M, β (the slope of input/output relationship, also called the efficiency of energy transformation) can be used to quantify the effects of signal factor; and the variability of data points around the proportional equation of $y = \beta M$ (like the gray area in Figures 5.7) is regarded as work done by noises. Hence,

$$\text{SN} = \frac{\text{Linear slope between Signal }(M)\text{ and Response }(y)}{\text{Variability around the slope}} = \frac{\text{Efficiency}}{\text{Variability}}$$

Since using the "square" of data value to express variation is a common approach, we use β^2 to quantify the effects of signal factor. Likewise, we also can use σ^2 to quantify the effects of noise factor. Thus, the above equation can be rewritten as

$$\text{SN} = \frac{\beta^2}{\sigma^2} \tag{5.1}$$

In what we mentioned above, we used the most understandable approach to explain the SN ratio. In fact, at the beginning, the SN ratio was proposed in studying the optimization of measurement equipment; the development of its concept is described as follows:

1. For a measurement equipment, we expect that the value of a certain object measured through that equipment equals its true value (unknown); thus when its ideal function is described as $y = \beta M$, y represents the object's measured value and M represents the object's true value. In other words, the object's true value is regarded as input value; we know the object's true value according to the output value of the measurement equipment. Ideally, there is no error between input and output values.
2. Since $y = \beta M$, M can be expressed as

$$M = \frac{y}{\beta} \tag{5.2}$$

3. There are many sources of variability in the real world to prevent the ideal function from being achieved. For a measurement equipment, if

$e(M)$ is used to symbolize the measurement errors caused by various noises, the actual function of that equipment is

$$y = \beta M + e(M)$$

4. Since the actual situation is $y = \beta M + e(M)$, M can be expressed more precisely as

$$M = \frac{y - e(M)}{\beta} \tag{5.3}$$

5. When we use Equation (5.3) instead of Equation (5.2) to estimate M, the difference between the two represents the deviation between actual estimated value and ideal estimated value (true value):

$$M - M = \frac{-e(M)}{\beta}$$

6. If we have n entries of data, then we can use the mean sum of squares to conduct overall estimation on the deviation expressed by the above equation. That is;

$$\frac{\sum_{i=1}^{n} [-e(M)/\beta]^2}{n-1} \tag{5.4}$$

7. The variance (σ^2) between actual function and ideal function can be expressed as

$$\sigma^2 = \frac{\sum_{i=1}^{n} [-e(M)]^2}{n-1} \tag{5.5}$$

8. If we replace a part of Equation (5.4) with Equation (5.5), then we can rewrite the deviation between actual estimated value and ideal estimated value (true value) of M from Equation (5.4) as

$$\frac{\sigma^2}{\beta^2} \tag{5.6}$$

9. Since we want to enable Equation (5.6) to be transformed to the form of "the larger the value, the better it is," then we transform Equation (5.6) to be the form of Equation (5.1). Moreover, we take logarithm for Equation (5.1) and multiply it by 10. The reason why we take the logarithm is that the value obtained under that scale will be within the interval of $(-\infty, +\infty)$ and wouldn't have limit values. Multiplying by 10 is

a frequently seen approach. Therefore, when we analyze the data, the following form of the SN ratio is adopted and the larger the value, the better it is:

$$10\log\frac{\beta^2}{\sigma^2}$$

From what was mentioned above, we know certain object's true value according to the output value of measurement equipment. However, that true value is actually "unknown"; the output value of measurement equipment is just a "possible" true value. Because true value is unknown, we can't know the deviation between measured value and true value. Nevertheless, the nature and strength of the SN ratio are that even though the true value is unknown, we can still use it to estimate the deviation. Although the deviation here represents the error between ideal measurement capability (which is capable of measuring true value) of measurement equipment and actual measurement capability, the SN ratio can be applied to various technology fields and can be used to evaluate the deviation between ideal or true state and actual state of a certain object's functionality. If the deviation (i.e., the degree deviated from ideal function) mentioned above which is hard to be estimated and even quantified can be evaluated by the SN ratio, many problems can be early predicted (the deviation from ideality can be detected) and prevented (to adopt the countermeasures for approaching ideality); and further, we can improve the efficiency and productivity of R&D itself obviously.

When having data information, we can use unbiased estimates to calculate the SN ratio; the related formula can be found in Section 5.4.2. In addition, the SN ratio is highly relevant to three elements, as shown in Figure 5.8.

1. *Linearity:* The linear degree of input/output relationship. From the viewpoint of technology, our expectation of linearity is high, because it means that there is a good proportional relationship between input and output. If the linearity is not good, it would be hard to predict the correspondence between input and output and output performance.
2. *Variability:* The variability of input/output energy transformation near ideal function. The lower the variability, the closer the input/output relationship to ideal function and the less the energy lost to create nonintentional function. Simultaneously, the occurrences of various defect symptoms are reduced.
3. *Sensitivity:* The efficiency of input/output energy transformation, that is, the change amount of response value (output) when we change signal value (input) for one unit. The higher the sensitivity, the better the efficiency of the input/output energy transformation. However, under some circumstances, a higher sensitivity may not be the better one; it must be tuned to a specific target value. The most frequently seen sensitivity

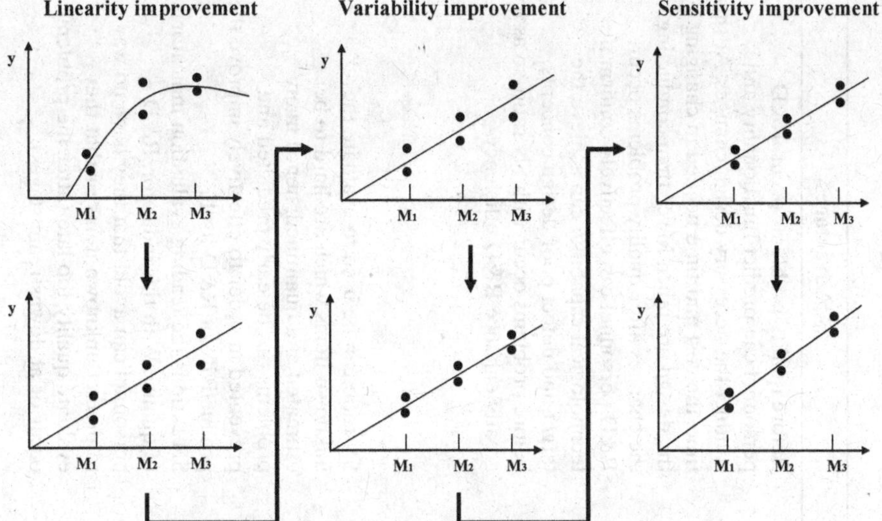

Figure 5.8. Three relevant elements of the SN ratio.

target is 1, under such situation we have to maintain the input value equal to the output value in input/output energy transformation.

Once the SN ratio is improved, this means that we heighten linearity and lower variability simultaneously; that is because, in the formula of the SN ratio, σ^2 includes the nonlinearity between M and y as well as the variability caused by noises. β^2 is related to the tuning of sensitivity.

Since the SN ratio is used to evaluate functionality (what's right and what we want), not to detect or verify the occurrence of quality problems (what's wrong), it can be used to predict functional robustness (i.e., quality) before quality problems occur. The needed cost and time for predicting technological or product performance before quality problems occur and for conducting the necessary design changes are much lower than conducting changes after quality problems occur. In summary, the features and advantages of the SN ratio can be described as shown in Table 5.1.

5.3.2 Static SN Ratio

The SN ratio formula mentioned above is used for dynamic response. In this section, we introduce the SN ratio used for static response. The main difference between dynamic response and static response is whether the target value of response is known or a fixed value. For example, technology development is executed before product planning; at this time, the specific design target or customer specification of product is unknown, and even the application fields

TABLE 5.1. Features and Advantages of SN Ratio

Feature	Description	Advantages
Early evaluate functionality	• SN ratio is used to evaluate functionality (what's right and ideal), not to detect or verify the occurrence of quality problems (what's wrong); thus it can be used, before quality problems occur, to predict the functional robustness of product in market.	• Before quality problems occur, R&D personnel can predict functionality and conduct the necessary design changes. At this time, the cost and time needed in changing the design are much lower than conducting the changes after quality problems occur.
Early confirm technological feasibility and detect poor design concepts	• The cause of poor R&D productivity is not that R&D personnel lack design ideas, but they would choose a design concept, which is considered the most feasible, to implement development and, while making sure that the design idea failed, they would conduct the trials, tests, and reviews of the next design idea. SN ratio can make R&D personnel know the insufficiency of technological capability or design failure as soon as possible to take appropriate early actions.	• R&D personnel can effectively confirm if technological capability can achieve the target and detect poor design concepts before problems occur with SN ratio, so as to greatly enhance R&D efficiency.
The display of evaluation capability for "unknown items"	• The nature of SN ratio is an evaluation index for "unknown items." It is used to evaluate the degree of certain object's function deviates from ideality, and can be applied to various technology fields. • At the technology development stage before product planning, even though the specific design targets or customer specification of product is unknown, and even the application fields of technology is unknown, we still can use SN ratio to evaluate the capability of technology to dynamically achieve multiple target values. • By means of the deviation degree from ideality to evaluate quality is better than the approaches using various determined test conditions to conduct evaluation.	• SN ratio can be used to evaluate the unknown items which are hard to be estimated and quantified; hence many problems can be early predicted and prevented in order to effectively improve the productivity of R&D itself. • SN ratio is the leading evaluation indicator of quality. With this indicator, R&D personnel can avoid that they have no way to predict unknown items or that they evaluate quality too late (after the problems occur or at downstream stages).

Use few samples and the data don't need to satisfy the assumptions related to probability distributions	• While calculating SN ratio, the data don't need to satisfy the assumptions related to probability distributions, and only a few samples are needed, thus it is more useful and efficient than the statistics (such as process capability, reliability testing data, etc.) which needs to know the distributions and more samples.	• In R&D practice, the number of samples used to predict quality is usually very few and its distribution shape must be assumed. In such situation, using SN ratio is better than using other quality evaluation indicators and it can provide better prediction performance.
Comprehensively evaluate the linearity and variability of input/output relationship of technology	• When we improve SN ratio, we simultaneously enhance linearity and reduce variability.	• SN ratio is a single comprehensive evaluation indicator for linearity and variability.
Compare and benchmark the technology itself	• For comparing the design configurations between ours and competitors' or material quality among suppliers, SN ratio can be used to analyze their differences in technological functionality.	• R&D personnel can sufficiently evaluate and compare technological functionality only by one indicator, SN ratio.

of technology are also unknown. R&D personnel research and optimize an energy transformation or function, hence the target is dynamic. However, in product development, a specific design target or customer specification is known and R&D personnel try to achieve that static target (fixed value). To sum up:

1. *Dynamic Case:*
   - Response target value is unknown or dynamic (e.g., the technology development before product planning).
   - Researching and optimizing an energy transformation or function (the target is multiple values).
2. *Static Case:*
   - Response target value is known or customer specification is known (e.g., product development).
   - Making a certain quality characteristic value achieve target value (fixed value).

In the dynamic case we analyze data with dynamic SN ratio, while in static case we calculate static SN ratio. In fact, the analysis of the dynamic SN ratio can be regarded as the analyses for a series of static Nominal-the-Best (NTB) characteristics, as shown in Figure 5.9. In the dynamic case, when we want to optimize an entire energy transformation or function, we need to set signal factor as the input energy to conduct the dynamic optimization of input/output relationship. However, regarding the static optimization of achieving the specific target in energy transformation is achieved, we don't need signal factors to research the functionality of the entire functional range.

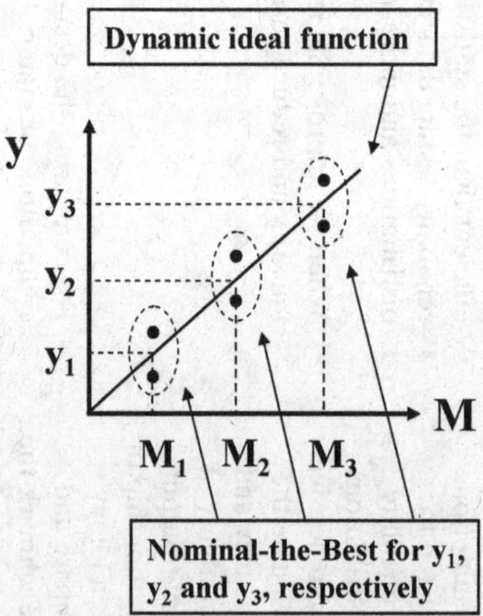

Figure 5.9. Nominal-the-Best responses and dynamic functionality.

Just like the concept of dynamic SN ratio, the static SN ratio is expressed as

$$\mathrm{SN} = \frac{\text{Useful output}}{\text{Harmful output}} = \frac{\overline{y}^2}{\sigma^2}$$

When we have *n* entries of data and use SN ratio to analyze them, the static SN ratio is calculated with Equations (5.7), (5.8) and (5.9); and the larger its value, the better it is (the smaller the variability is):

$$\overline{y} = \frac{\sum_{i=1}^{n} y_i}{n} = \frac{y_1 + y_2 + \cdots + y_n}{n} \tag{5.7}$$

$$\sigma_{n-1}^2 = \frac{\sum_{i=1}^{n} (y_i - \overline{y})^2}{n-1} = \frac{(y_1 - \overline{y})^2 + (y_2 - \overline{y})^2 + \cdots + (y_n - \overline{y})^2}{n-1} \tag{5.8}$$

$$\mathrm{SN} = 10\log\frac{\overline{y}^2}{\sigma_{n-1}^2} \tag{5.9}$$

In Equation (5.9), the reason why we take logarithm is that the scale doesn't have limit values and multiplying by 10 is the common way.

In practice, we suggest using the dynamic SN ratio to conduct functionality evaluation so that we can evaluate the entire energy transformation or function at one time and don't need to individually evaluate if the specific target in energy transformation is achieved.

5.4 COMPARATIVE ASSESSMENT OF FUNCTIONALITY

Once functionality can be evaluated, we can use its evaluation approach to implement the comparison and benchmarking among technologies themselves. Since the SN ratio is the single indicator used to evaluate the functional robustness of technology or product against various sources of variability (environmental conditions such as temperature, humidity, and pressure; aging, deterioration and wearing of product itself; material variability or manufacturing variability), the advantage of using the SN ratio is that we can sufficiently predict technological quality with only that indicator.

5.4.1 Conventional Evaluation Indicators

In practice, R&D personnel usually use the following evaluation indicators to implement the comparison and benchmarking of quality performances:

1. *Defective Rate:* Calculating the rate of the measured values of quality characteristics exceeds its specification limits. The higher the defective rate, the worse the product manufacturability or process quality.
2. *Process Capability:* Calculating the capability of quality characteristic values to satisfy specifications under the state of statistical control (stable state). Process capability value is expected to be 1.33 or even over 1.67; if it is smaller than 1.33, it urgently needs to be improved. The process capability value of the so-called quality level of "Six Sigma" is high, with values of up to 2.
3. *Data of Reliability Test and Other Technical Tests:* In order to implement reliability engineering, R&D personnel need to collect and analyze failure data as well as fit and establish a statistical model. Based on this model, R&D personnel can conduct the trial prediction of reliability to verify the model's effectiveness according to its performance. Monte Carlo simulation is another commonly used method for evaluating variability and reliability. It uses simulators to generate plenty of simulation samples based on the probability distribution of the product's performance assumed by R&D personnel and the distribution features, and R&D personnel estimate the product's variability and reliability according to the measured values. Besides, some companies conduct the calculation of the risk priority number (RPN) of defect symptoms according to the severity, occurrence, and detection. The higher the RPN, the more the defect symptom needs to be highlighted and solved.
4. *Hypothesis Testing for Mean Value and Variance:* Using the statistical method of hypothesis testing to conduct the testing of whether mean value or variance equals a specific value. If that hypothesis is not rejected, then the mean or variance (substitute indicators for quality) equals that specific value (quality level); if that hypothesis is rejected, then the mean or variance doesn't equal that specific value. We also can conduct the testing for larger or smaller than a specific value.

However, the above-mentioned indicators cannot assess the robustness of technology or product under various sources of variability. Moreover, because the concept of functionality is to maximize what's right and what we want, which focuses on ideal function and not quality problems, R&D personnel can, at the most upstream stage before quality problems occur, finish functionality evaluation. On the contrary, defective rate, process capability, or reliability testing focuses on "what's wrong and correcting error" so R&D personnel can only know, after quality problems occur, the results of defective rate, process capability, or reliability testing. In the view of another angle, when R&D personnel need to use the lagging indicators such as defective rate, process capability, or reliability testing data to evaluate quality performance, it doesn't mean that they have confidence regarding the quality assured by these approaches, but means that they have difficulty regarding how to effectively

TABLE 5.2. Various Methods for Estimating or Evaluating Quality

Quality Estimation/ Evaluation Method (Methodological Elements)	Quality Indicators	Implementation	Application Scope	Pros	Cons
• Conformance to specification	• Defective rate/ Failure rate/ First pass yield rate: Measuring the percentage rate/ ppm rate of the product out of specification	• Calculating the actual ratio of the number of the product conforming to specifications to the number of defectives	• Manufacturing process	• Easiest to be understood and applied • Zero defective rate/ zero defects are attractive quality performance	• Defective rate is the lagging indicator used after occurring product deficiencies • Even zero defective rate/ zero defects don't mean good quality • Regarding poor quality, the root causes need to be determined, analyzed, and eliminated
• Process control • Process capability analysis	• Process capability indices: Measuring the capability of a (stable) process satisfying the specification	• Using the mean value and standard deviation of product characteristic values to estimate the capability of a process achieving the specification	• Manufacturing process	• Estimating the degree of product characteristic values achieving the specification (simultaneously considering mean value and standard deviation)	• The data must be from a stable process and conform with normal or specific distributions • It's not easy to directly perceive the improvement of quality performance by process capability values • Regarding poor quality, the root causes need to be determined, analyzed, and eliminated

(*Continued*)

TABLE 5.2. (*Continued*)

Quality Estimation/ Evaluation Method (Methodological Elements)	Quality Indicators	Implementation	Application Scope	Pros	Cons
• Various reliability tests (including tests of functions, life, burn-in, etc.)	• Various reliability data: Measuring product's probability of performing its intended function under the certain time interval and certain operating conditions	• Using the simulated random samples from assumed distribution and accelerating tests to estimate the quality and reliability of products	• Product design • Manufacturing	• Various visible tests with extreme conditions are the best to make people have confidence on quality assurance	• The simulated samples were generated from the assumed probability distribution, and more samples and many hours are needed to observe random variation • Regarding the product characteristics which fail in reliability tests, the cause-and-effect relationships are usually utilized to adjust the performance to achieve testing specifications. However, this way tends to improve quality and reliability with "Debug-Fix" approach and fails to provide assurance for the conditions other than test conditions • Needing the investments on advanced testing equipment

• Experimental strategy for robust optimization • Signal-to-noise ratio (SN ratio) • Orthogonal array (OA)	• SN ratio: Measuring the robustness of product function or characteristics against various factors causing variability (environmental and usage conditions, deterioration, and manufacturing variability)	• Designing experiments and compulsorily introduce variability in the experiments to evaluate the functional robustness of product under various unknown conditions. At technology development or product/ process design stages, OA can be used to inspect the reproducibility of the robustness	• Technology development • Product design • Process design • Measurement systems optimization • Transaction item evaluation	• SN ratio is the leading indicator used to evaluate robustness under various unknown conditions/ before the occurrence of quality problems. It only needs short time and few samples (can also use test pieces or computer simulations) • It's easy to directly perceive the improvement of quality performance with the value of SN ratio: every 6 units (decibel, db) increased in SN ratio, the variability reduced by 50% • No need to find out and eliminate the root causes causing poor quality, but to achieve robustness with design approach	• In practice, SN ratio is harder to be understood and accepted to use • Possibly harder to identify the ideal input/ output relationship (ideal function) needed in evaluating dynamic characteristics • Possibly needing to develop new measurement technology (in other words, possibly lack the needed measurement capability)

predict quality before quality problems occur. Thus, before knowing if there are more effective leading indicators, they can just use these substitute indicators to conduct quality assessment. The above approach usually needs to assume that the data are specific probability distributions that are helpful to establish statistical models; but in practice, such consideration may not be practical to cause the predicted value of quality performance to be unbelievable. The SN ratio described in this chapter is the leading evaluation indicator of quality. R&D managers are responsible to urge designers to adopt the SN ratio totally to implement the functionality evaluation of technology or product, so as to improve the disadvantage of evaluating quality too late (after the occurrence of problems or at the downstream stages) and to focus on how to maximize what's right and what we want. Defective rate, process capability, or reliability testing data are used for double confirmation. Table 5.2 compares various methods for estimating or evaluating quality.

5.4.2 Using the SN Ratio

In the following situations, we can use the SN ratio to implement the comparative assessment of functionality:

1. *The Comparison between New Design and Current Design:* Evaluating if the functionality of new design is better than current design or reference model to confirm the design maturity and the effects of learning curve.
2. *Product Competitive Analysis or Benchmarking:* Evaluating the robustness of own company's product and the competitor's product under customer usage conditions to reasonably determine its market positioning and price or to conduct benchmarking.
3. *The Comparison of Quality of Materials or Components among Different Suppliers:* Evaluating the robustness of the materials or components from different suppliers under the conditions of product development and manufacturing of own company, so as to reasonably determine the materials or components from which the supplier is adopted.

Functionality evaluation is to assess the robustness of signal/response relationship of technology or product under various sources of variability (i.e., noises) and usage conditions (i.e., signals); thus we need to collect the response values in various levels of signal and noise and then calculate the SN ratios of those data. Table 5.3 is the form for data collection and arrangement.

In Table 5.3, signal factor is four levels (M_1, M_2, M_3, and M_4) while noise factor is two levels (N_1 and N_2). Under the same signal level and noise level, we collect two entries of response value (R_1 and R_2), thus we have 16 entries of data ($y_1, y_2, \ldots, y_{16}$). As described in Section 2.1, noise factor is the factor causing response values to have variability. Noise factor includes environmental conditions, aging, deterioration, manufacturing variability, and so on. Although usually there are many noise factors, when we conduct functionality

TABLE 5.3. Data Set for Functionality Evaluation

	M1				M2				M3				M4			
	N_1		N_2		N_1		N_2		N_1		N_2		N_1		N_2	
	R_1	R_2	R_1	R_2	R_1	R_2	R_1	R_2	R_1	R_2	R_1	R_2	R_1	R_2	R_1	R_2
Evaluated Design	y1	y2	y3	y4	y5	y6	y7	y8	y9	y10	y11	y12	y13	y14	y15	y16

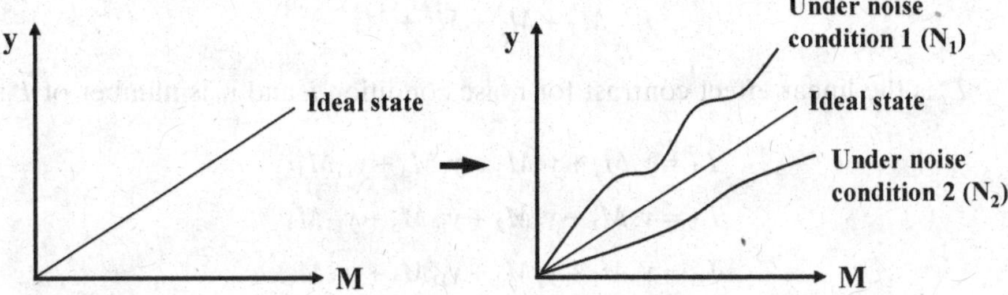

Figure 5.10. Ideal functionality and the functionality under noise conditions.

evaluation, we wouldn't collect all the noise factors and their response values under various possible levels, but we "compound" the factor-level combination of important noises to be two levels.

1. N_1: The noise conditions making response values tend to be low. For example, for a certain technology or product, its important noise factors include temperature, humidity, pressure, aging, and so on. Under the conditions such as high temperature, high humidity, high pressure, and aging, response performance is very poor. Thus, we "compound" the conditions of high temperature, high humidity, high pressure and aging to be the level 1 of noise factor (N_1).
2. N_2: The noise conditions making response values tend to be high. For example, under the conditions such as low temperature, low humidity, low pressure, and non-aging, response performance is very good. Thus, we "compound" the conditions of "low temperature, low humidity, low pressure, and non-aging" to be the level 2 of noise factor (N_2).

The technique to compound noises is very important for evaluating functionality. Simply speaking, its concept is that if functional variability can achieve robustness under these two extreme levels, then, under the factor-level combination of other various possible noises, response values wouldn't have variability.

Figure 5.10 shows the ideal input/output relationship (i.e., the linear relationship of ideal energy transformation) and the input/output relationship with variability under noise conditions (i.e., the disturbed energy transformation caused by noises).

Take the data set in Table 5.3, for example; when we analyze data to evaluate functionality, the more precise calculation process for SN ratio is as follows:

The total sum of squares of all the data points is given by

$$S_T = y_1^2 + y_2^2 + \cdots + y_{15}^2 + y_{16}^2$$

The sum of squares of the signal factor levels:

$$r = M_1^2 + M_2^2 + M_3^2 + M_4^2$$

L_i is the linear effect contrast for noise condition i, and r_0 is number of L_i:

$$L_1 = y_1M_1 + y_5M_2 + y_9M_3 + y_{13}M_4$$

$$L_2 = y_2M_1 + y_6M_2 + y_{10}M_3 + y_{14}M_4$$

$$L_3 = y_3M_1 + y_7M_2 + y_{11}M_3 + y_{15}M_4$$

$$L_4 = y_4M_1 + y_8M_2 + y_{12}M_3 + y_{16}M_4$$

S_T is total sum of squares used as a measure of overall variability in the data, and it can be divided into three components:

- S_β, the sum of squares due to signals (i.e., the signal portion of the variability), given by

$$S_\beta = \frac{(L_1 + L_2 + L_3 + L_4)^2}{r \times r_0}$$

- $S_{\beta\times N}$, the sum of squares due to the interaction between β and noises, given by

$$S_{\beta\times N} = \frac{L_1^2 + L_2^2 + L_3^2 + L_4^2}{r} - S_\beta$$

- S_e, the sum of squares due to error (i.e., the error portion of the variability), given by

$$S_e = S_T - (S_\beta + S_{\beta\times N})$$

In Table 5.3, n is 16 (i.e., $y_1, y_2, \ldots, y_{16}$). Error variance is estimated by

$$V_e = \frac{S_e}{n - r_0}$$

Variance of the harmful part (i.e., $S_{\beta\times N}$ and S_e) is estimated by

$$V_N = \frac{S_{\beta\times N} + S_e}{n-1}$$

Thus,

$$\beta = \sqrt{\frac{1}{r\times r_0}(S_\beta - V_e)} \cong \frac{1}{r\times r_0}(L_1 + L_2 + L_3 + L_4)$$

In the above equation, $[1/(r\times r_0)](S_\beta - V_e)$ is an unbiased estimate of β^2, and $[1/(r\times r_0)](L_1 + L_2 + L_3 + L_4)$ is an unbiased estimate of β.

The SN ratio is estimated by the following equation:

$$\text{SN} = 10\log\left[\frac{\frac{1}{r\times r_0}(S_\beta - V_e)}{V_N}\right]$$

The larger the SN ratio, the smaller the functional variability.

5.5 EXAMPLES

We use two examples to illustrate how to conduct the comparative assessment of functionality.

5.5.1 Two Measurement Systems

The most commonly seen method used to analyze measurement systems is measurement systems analysis (MSA), whose purpose is to analyze measurement variation consisting of various types of measurement error. With the analyzed result of MSA, operational personnel needs to evaluate if they have to reduce variation to be in the acceptable range and take proper actions.

In MSA, the measurement errors composing total measurement variation can be classified into five types:

1. *Bias.* Bias is the difference between the average value of measured samples and the reference average value. The smaller the bias, the higher the accuracy.
2. *Repeatability.* Repeatability is the variation in measurements obtained with one measurement equipment when used several times by one appraiser while measuring the identical characteristic on the same sample

(measurement samples, measurement equipment, and appraisers are all the same). The smaller the observed range, the higher the precision.

3. *Reproducibility.* Reproducibility is the variation in the average of the measurements made by different appraisers using the same measurement equipment when measuring the identical characteristic on the same sample (measurement samples and measurement equipment are the same, but appraisers are not). Since reproducibility is defined as the consistency among the measured values (each appraiser regards the measured value obtained by himself/herself as the true value) obtained by the appraisers, it is related to precision.
4. *Stability.* Stability is the total variation in the measurements obtained with a measurement system on the same samples when measuring a single characteristic over an extended time period (measurement samples, measurement equipment, and appraisers are the same, but time is not). Because stability means that whether measurement variation after an extended time period is the same as measurement variation obtained at current time, it is related to accuracy.
5. *Linearity.* Linearity is the difference in the bias values through the expected operating range of the gauge (measurement equipment and appraisers are the same, but measurement samples are the ones with different true values within the operating range of gauge). Since linearity is an indicator used to evaluate whether the bias values keep constant within the operating range of gauge, it is related to accuracy.

The commonly seen gauge R&R study only analyzes repeatability and reproducibility, but does not consider bias, stability, and linearity.

The following example is the evaluation of measurement equipment using the SN ratio, not MSA. The ideal function of measurement equipment is measured values equal to true values; that is, output response is the measured value of the measured object, and the true value of the measured object is input signal. That ideal function may not be achieved in practice, because, in real life, there are various sources of variability causing measurement variation. Such sources include appraisers, time, environmental conditions, and aging of equipment. Unlike MSA, here we regard these sources of variability as one single factor: noise factor. By means of the integrated measurement indicator, the SN ratio, what we compare is the functionality, "making measured value equal to true value," of measurement equipment—that is, "the 1:1 proportional relationship between measured value and true value."

In this example, the measurement characteristic we are concerned with is the so-called gain value produced by electric current. We try to compare which one of the measurement systems, A or B, is better to help measurement personnel solve the wrong judgment problems caused by the difference of measurement results between measurement systems. We conducted the experiment with the standard samples whose true values are respectively 1400, 1500, 1600, and 1700 as well as to collect two entries of response value (R_1 and R_2) under the same true value and noise level. Although there are many noise factors influencing measurement systems, we only compound the important noises to be the levels of N_1 and N_2. Table 5.4 shows the experimental data and their descriptive statistics (mean and standard deviation).

By observing descriptive statistics, we know that when the two systems measure different samples, measurement system A performs better in standard deviation than measurement system B, while the performance of the two systems in mean value has no significant difference. Under this situation we tend to choose measurement system A; however, while in the condition that mean value and standard deviation have no significant difference, it would be hard to judge with descriptive statistics only. According to Section 5.4.2, we use the SN ratio to conduct comparative assessment of measurement systems.

For measurement system A, the total sum of squares of all the data points is given by

$$S_T = 1397^2 + 1396^2 + \cdots + 1702^2 + 1702^2 = 38,558,171$$

The sum of squares of the signal factor levels are expressed as

$$r = 1400^2 + 1500^2 + 1600^2 + 1700^2 = 9,660,000$$

$$L_1 = 1397(1400) + 1495(1500) + 1592(1600) + 1690(1700) = 9,618,500$$

$$L_2 = 1396(1400) + 1497(1500) + 1595(1600) + 1692(1700) = 9,628,300$$

$$L_3 = 1404(1400) + 1502(1500) + 1602(1600) + 1702(1700) = 9,675,200$$

$$L_4 = 1403(1400) + 1503(1500) + 1603(1600) + 1702(1700) = 9,676,900$$

r_0 is the number of linear contrast for noise conditions, hence it is 4 in this case.

S_T is the total sum of squares used as a measure of overall variability in the data, and it can be divided into three components as follows:

$$S_\beta = \frac{(9,618,500 + 9,628,300 + 9,675,200 + 9,676,900)^2}{9,660,000 \times 4} = 38,557,843.72$$

$$S_{\beta \times N} = \frac{9,618,500^2 + 9,628,300^2 + 9,675,200^2 + 9,676,900^2}{9,660,000} - 38,557,843.72$$

$$= 292.08$$

$$S_e = 38,558,171 - 38,557,843.72 - 292.08 = 35.20$$

Since n is 16, variance is estimated by

$$V_e = \frac{35.20}{16 - 4} = 2.93$$

TABLE 5.4. Experimental Data of Two Measurement Systems

	M1						M2					
	1400						1500					
	N_1		N_2				N_1		N_2			
	R_1	R_2	R_1	R_2	Mean	Std. Dev.	R_1	R_2	R_1	R_2	Mean	Std. Dev.
Measurement System A	1397	1396	1404	1403	1400.00	4.08	1495	1497	1502	1503	1499.25	3.86
Measurement System B	1389	1387	1403	1406	1396.25	9.64	1493	1496	1505	1507	1500.25	6.80

	M3						M4					
	1600						1700					
	N_1		N_2		Mean	Std. Dev.	N_1		N_2		Mean	Std. Dev.
	R_1	R_2	R_1	R_2			R_1	R_2	R_1	R_2		
Measurement System A	1592	1595	1602	1603	1598.00	5.35	1690	1692	1702	1702	1696.50	6.40
Measurement System B	1593	1592	1601	1602	1597.00	5.23	1695	1694	1703	1705	1699.25	5.56

and

$$V_N = \frac{292.08 + 35.20}{16 - 1} = 21.82$$

Thus,

$$\beta = \sqrt{\frac{38{,}557{,}843.72 - 2.93}{9{,}660{,}000 \times 4}} = 0.99894$$

$$\text{SN} = 10\log\left[\frac{\frac{1}{9{,}660{,}000 \times 4}(38{,}557{,}843.72 - 2.93)}{21.82}\right] = -13.3976$$

Applying the same procedure to the assessment of measurement system B, we can know that β is 0.99887 and the SN ratio is −16.2927.

The unit of SN ratio is decibel (dB), and a 6-dB gain in SN ratio is equivalent to 50% reduction in functional variability. Therefore, in terms of measurement variability, measurement system A is better than system B by about 28.43%.

5.5.2 Two Designs

For a company designing and producing transformer products, the most customer complaints it faces are the problems of output voltage shift and overheat arising when transformers operate. In order to solve these problems, the company has conducted many analyses and improvement in production plants for how to improve voltage shift and how to reduce heat. The frequently seen approach is to set improving voltage shift and reducing heat as experiment objectives and, according to the measured values of voltage shift and heat, analyze and modify the production and assembly conditions for the stabilization of voltage and minimization of heat. However, there is no significant improvement with the problems; even when voltage shift and overheat are improved, the performance of the other quality characteristics becomes worse as a result of some changes of production and assembly condition. In fact, the occurrences of voltage shift and overheat do not all come from the manufacturing or assembly deficiencies in the plant; after conducting failure analyses, we find that most cases can be attributed to the design problem of transformer. In other words, despite the fact that the transformer designed by design engineers has passed the verification tests at the upstream of product development, we cannot ensure that design quality and the design problems may occur at the downstream stages.

To improve this situation, design engineers develop a new design by changing design parameters such as wire size in the secondary side, winding position, winding spacing, between-layer stack layout, core structure, core material type, and so on. They hope to effectively solve the quality problems caused by the old design. From the viewpoint of functionality assessment, when we want to evaluate if the new design is better in quality than the old design, design engineers need to define the "ideal function" of transformer and to evaluate, based on the degree the transformer perform its ideal function, its energy transformation efficiency to estimate the energy loss causing voltage shift and overheat. The higher the degree the transformer performs its ideal function, the fewer the opportunities for the occurrence of quality problems including voltage shift and overheat—that is, the better the quality of the transformer. The difference between this functionality evaluation approach and the current evaluation approaches used in the factory and in the R&D department lies in the fact that this approach focuses on the ideal function of the transformer (the design intent of the transformer), not the measurement for the characteristic values of voltage shift and overheat. Regarding enhancing design quality, functionality evaluation is a "forward-looking" thinking. Figure 5.5 shows the P-diagram of transformer, and the input/output proportional relationship between turns ratio and output voltage expresses the ideal function of transformer. We use that ideal function to conduct functionality assessment for the new design and old design: We set the turns ratio of 0.025, 0.05, 0.075, and 0.1 as the input values as well as collect two entries of response output values (R_1 and R_2) under the same load condition and levels of noises. Although there are many noise factors influencing the performance of the transformer, we only compound the important noises to be the levels of N_1 and N_2. Table 5.5 shows the experimental data.

For new design, the total sum of squares of all the data points is given by

$$S_T = (6.2)^2 + (5.1)^2 + \cdots + (18.7)^2 + (19.1)^2 = 3013.78$$

The sum of squares of the signal factor levels:

$$r = (0.025)^2 + (0.05)^2 + (0.075)^2 + (0.1)^2 = 0.02$$

$$L_1 = 6.2(0.025) + 11(0.05) + 15.5(0.075) + 20.9(0.1) = 3.96$$

$$L_2 = 5.1(0.025) + 10.6(0.05) + 16.3(0.075) + 21.3(0.1) = 4.01$$

$$L_3 = 4.6(0.025) + 9(0.05) + 13.9(0.075) + 18.7(0.1) = 3.48$$

$$L_4 = 4.1(0.025) + 9.4(0.05) + 14.3(0.075) + 19.1(0.1) = 3.56$$

r_0 is the number of linear contrast for noise conditions; hence it is 4 in this case.

S_T is the total sum of squares used as a measure of overall variability in the data, and it can be divided into three components as follows:

$$S_\beta = \frac{(3.96 + 4.01 + 3.48 + 3.56)^2}{0.02 \times 4} = 3000$$

TABLE 5.5. Experimental Data of Two Designs

	M1				M2			
	0.025				0.050			
	N_1		N_2		N_1		N_2	
	R_1	R_2	R_1	R_2	R_1	R_2	R_1	R_2
New Design	6.2	5.1	4.6	4.1	11.0	10.6	9.0	9.4
Old Design	6.7	6.2	4.1	3.1	12.4	11.6	8.4	7.6

	M3				M4			
	0.075				0.100			
	N_1		N_2		N_1		N_2	
	R_1	R_2	R_1	R_2	R_1	R_2	R_1	R_2
New Design	15.5	16.3	13.9	14.3	20.9	21.3	18.7	19.1
Old Design	17.2	16.7	12.6	13.5	21.7	22.9	17.9	17.5

$$S_{\beta\times N} = \frac{(3.96)^2 + (4.01)^2 + (3.48)^2 + (3.56)^2}{0.02} - 3000 = 11.89$$

$$S_e = 3013.78 - 3000 - 11.89 = 1.89$$

Since n is 16, variance is estimated by

$$V_e = \frac{1.89}{16-4} = 0.1575$$

and

$$V_N = \frac{11.89 + 1.89}{16-1} = 0.9187$$

Thus,

$$\beta = \sqrt{\frac{3000 - 0.1575}{0.02 \times 4}} = 199.99475$$

$$\text{SN} = 10\log\left[\frac{\frac{1}{0.02\times 4}(3000-0.1575)}{0.9187}\right] = 46.38879$$

Applying the same procedure to the assessment of old design, we can know that β is 200.01353 and SN is 39.7825. Since a 6-dB gain in SN ratio is equivalent to 50% reduction in functional variability, we can know that the functional variability of the new design is smaller than that of the old design by about 53.38%.

What's worth mentioning is: Conducting evaluation of the ideal function doesn't involve the target values or specification values of transformers because making output voltage and turns ratio a robust proportional relationship is a technology. The targets or specifications of a product are defined after product planning is finished, but we can finish the evaluation of technological quality before product planning by utilizing a new evaluation approach; and further, we can tremendously enhance the competitiveness measured by quality, time, and cost.

6 Functionality Design

6.1 R&D AND ROBUST ENGINEERING

Technology development and product development, in short form, are generally called R&D activities. One of the most principal jobs of R&D is to evaluate functionality; the SN ratio we introduced can provide R&D personnel with an effective leading evaluation indicator at the upstream stage to predict as early as possible the robustness of technology or product under unknown usage conditions. Another principal job of R&D is to design functionality—that is, to design-in the functional robustness to technology or product. In fact, the job of functionality design itself must include functionality evaluation tasks to ensure the effectiveness of R&D. At the same time when R&D personnel design-in the functional robustness to technology or product, they must ensure no increment of material cost and development time; thus the effectiveness of the functionality design method greatly influences R&D personnel's operating model. Once R&D personnel adopt other more effective methods, the R&D paradigm would be shifted.

The robust engineering (RE) method described in Part II of this book is an engineering methodology used to optimize the functionality of technology or product. The core of RE is the parameter design, which is the method used to efficiently and effectively determine the optimal combination of design parameters for achieving functional robustness. In technology development, in order to ensure engineered quality, R&D personnel would define some quality characteristics and measure them, and then they would improve the technology according to the performance of quality characteristics. Trying to improve the defect phenomena of quality characteristics is the approach of firefighting, whereas the research on how to prevent the functionality of technology from the influence of various sources of variability to avoid the defect phenomena of various quality characteristics is the approach of proactive fire prevention. The purpose of the RE method is to realize fire prevention; therefore, in this book, it is also regarded as technological strategy. Simply speaking, RE is an engineering strategy used in an R&D laboratory to efficiently improve quality in the market, as shown in Figure 6.1.

Quality Strategy for Research and Development, First Edition. Ming-Li Shiu, Jui-Chin Jiang, and Mao-Hsiung Tu.

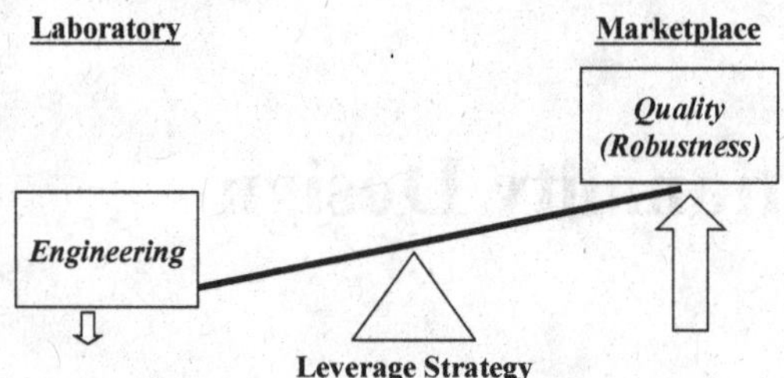

Figure 6.1. RE as an engineering leverage strategy for quality.

The positioning of RE in R&D activities is:

1. *Evaluation Technology:* RE is an evaluation technology of functionality, not a method of solving problems. To solve problems, we still have to depend on specialized technology and related knowledge. Nevertheless, functionality evaluation technology can efficiently confirm the feasibility and effectiveness of technology solutions with regard to cost and quality.
2. *A Methodology for Robust Technology Development and Product Design:* RE is a methodology of functionality design, not function design. Function design is the method to fulfill customer requirements and create the market. Nevertheless, RE is the method to develop the design countermeasures to make technology or product robust against various sources of variability which can simultaneously realize cost leadership and the quality competitiveness of a product in the market and enhance customers' order allocation ratio (deepening the market). Based on this advantage, the company can leverage and re-allocate resources to create more added values or, further, to create a new market.

The broadly defined RE includes system design, parameter design, and tolerance design. There is a relationship of implementation sequence among the three:

1. *System Design:* In order to have certain functions to serve our needs in life or work, we need a system as the mechanism or enabler for providing that function. There may be many types of systems (mechanisms) of realizing functions. Some of them have existed in the world, while some have not been developed yet. R&D personnel have to fully apply their knowledge, experiences, integration capability, and creative capability to develop a new system that doesn't exist in the world but whose ability and approach of realizing the function are better than those of the current system. An innovative system can be protected by patents to form the R&D barriers for the competitors and even may create an industry and its competing field.

2. *Parameter Design:* When a system is developed, R&D personnel have to, under this existing system architecture, determine the optimal levels of system parameters. Since each design parameter in the system has a feasible setting range where it won't deteriorate system functionality under the laws of physics, parameter design is to determine the setting level of each parameter in its range to optimize the entire system and enable it to retain functional robustness under various usage conditions and sources of variability.
3. *Tolerance Design:* We expect that, when the optimal level combination of system parameters are determined, system functionality is improved. However, if we cannot improve functionality even using parameter design, we have to improve quality by upgrading components; this method is called tolerance design. Upgrading components means to increase cost; thus tolerance design is to seek the trade-off between quality and cost. If parameter design has effectively improved system functionality, then tolerance design is not necessary.

The QFD methodology introduced in Part I of this book is related to the methods of function design and system design. In Part II, we focus on parameter design methodology; hence the term RE in this book indicates parameter design.

6.2 PARAMETER DESIGN FOR ROBUSTNESS

Design itself can be regarded as a (proactive) countermeasure; R&D personnel achieve the functionality or robustness of technology or product by this type of countermeasure. Ingeniously enabling energy transformation to achieve (very close to) ideal function is an engineering means and represents the performance of technology development capability: The better the technology development capability, the more able the R&D personnel are to implement engineering optimization and control of energy transformation. RE is the determination method of the parameter conditions for achieving functionality. In this section, we interpret how the concepts, tools, and process steps are helpful to achieve robustness.

6.2.1 Key Concepts

There are several unique and useful concepts in RE; these key concepts become the guiding ideas of developing or applying the RE method:

1. Cost First. Although "quality first" is the most important philosophy of managing quality, cost is at the first place for seeking R&D productivity which is the most advantageous for market competition. More precisely speaking, we have to pursue high quality along with lowest possible cost; this is the guiding idea of quality engineering. In the viewpoint of quality engineering, R&D personnel have various possible approaches to develop technology and design product; however, in order to deal with the competition, they have to seek, among all these possible approaches, the one which is the most economic and lowest cost and assure high quality simultaneously. This is also the biggest difference between engineering and science. The purpose of science is to describe natural phenomena and to establish models, and the cost issue is not an important consideration; while the purpose of engineering is to develop the things that don't exist in nature but involve commercial competition and comparison, hence the needed cost and resources are the determining factors of competitiveness. In the same market segment and under the same perceived product quality, the cost difference between the competitors and us determines profit and success.

The purpose of the RE method (parameter design) is to make R&D personnel realize high product quality by the components of low cost and high variability. When quality requirements cannot be achieved, we would use the tolerance design method to evaluate the effects of improving quality by cost increment (e.g., upgrade components and parts) to decide whether we need to increase cost to make up for the possible amount of quality loss in the future. Despite the fact that cost is the most important guiding idea of quality engineering, RE may be the most powerful weapon among the methods of improving functionality and quality. RE is a strategy of "killing multiple birds with one stone" (simultaneously improving cost and quality).

2. Utilization of Nonlinearity. The relationship between the levels of control factors and output values is already hard to grasp, but the relationship is even more complicated after considering noises and signal factors; thus determining parameter conditions usually consumes a lot of R&D time and cost. RE is applied to technology development; to make it easier to explain how that method creates utility on technology itself, we use a conceptual model of internal structure of a technology (as shown in Figure 6.2) to describe what technology is and how RE improves technology development.

In Figure 6.2, we explain technology as the system formed by many input/output relationships consisting of control-factor levels in design space and their output responses. These relationships are very complicated, so technology development is not an easy job. The form of the input/output relationships in technology can be classified into nonlinear and linear. In this figure, the relationship between control factor A and output response is nonlinear, while the relationship between control factor B and output response is linear. Regarding the target value we want to achieve, the mean value of the distribution shaped by the output response of A1 level would hit the target value but has big variability. In order to reduce the variability, R&D personnel usually change to adopt the components of higher grade. Regarding control factor B, selecting B2 level can make output response equal to target value. To sum up, R&D personnel will adopt A1B2 as the level setting of control factors for

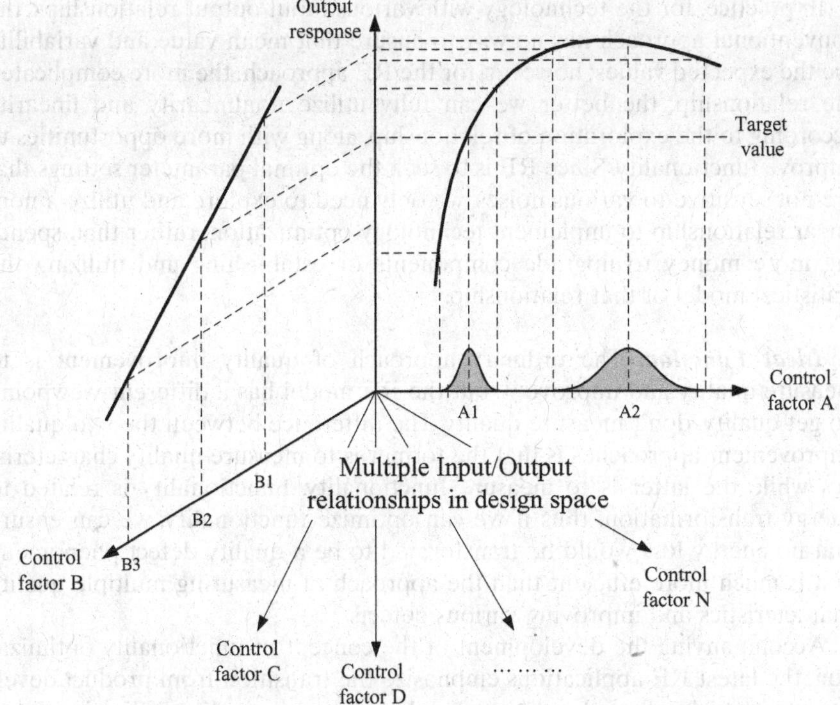

Figure 6.2. Internal structure of a technology.

technology optimization (In practice, they still have to determine the levels of control factors C, D, . . ., and N).

However, in the viewpoint of RE, the utilization of nonlinearity is the key to efficient (i.e., low cost) technological optimization. By applying RE, we can explore that the variability of output response in the A2 level is much lower than the variability of output response in the A1 level. Regarding the principle of "small functional variability is the target of technology optimization," we will choose the A2 level as the level setting of control factor A. But on the other hand, the output response of the A2 level is higher than the target value. At this time, we can use the level setting of control factor B, which has a linear relationship with output response to tune the mean value; that is, we will choose the B3 level so that the response value obtained by choosing the A2 level will shift to the target value. Thus, the optimal factor-level combination obtained by applying the RE method is A2B3, which is fully different from A1B2 obtained via the conventional approach. The advantage of choosing A2B3 lies in the fact that we can choose the components (A2) that are cheaper and with larger variability to achieve the output with higher quality (small variability); then, by choosing B3, which influences only the mean value but not the variability (input/output relationship is linear), we can adjust the output with small variability to be the target value.

In practice, for the technology with various input/output relationships, the conventional approach has no way to ensure that mean value and variability are the expected values; however, for the RE approach, the more complicated the relationship, the better we can fully utilize nonlinearity and linearity according to the exploration of relationship, along with more opportunities to improve functionality. Since RE is to seek the optimal parameter settings that are not sensitive to various noises, we only need to explore and utilize a nonlinear relationship to implement technology optimization, rather than spending more money to upgrade components or establishing and utilizing the statistical model of that relationship.

3. Ideal Function. The ordinary approach of quality improvement is to measure quality and improve it, but the RE model has a different viewpoint: To get quality, don't measure quality. The difference between the two quality improvement approaches is that the former is to measure quality characteristics while the latter is to measure functionality. Functionality is related to energy transformation; thus if we can optimize functionality, we can ensure that no energy loss would be transformed to be a quality defect phenomena that is much more efficient than the approach of measuring multiple quality characteristics and improving various defects.

Accompanying the development of the concept of functionality optimization, the latest RE applications emphasize the transition from product development to technology development—that is, to transit the application focus of RE from the past robust product design (realizing specific target value) to robust technology development (ensuring a technology achieving dynamic target value and that it can be applied to an entire family of products).

4. Two-Step Optimization. It is very hard to deal with the problem of variability; thus 80% of R&D time might be consumed on the reduction of variability. For R&D personnel, since it's not easy to deal with variability, they often use the approach of upgrading the grade of materials or components to conduct variability control and set the specification range as the consideration of whether they accept the functional performance: If the measured values (or the mean value of many samples) are in the specification, they accept that design; if not, they conduct design changes. Nevertheless, such approach dealing with variability, besides increasing the cost of materials or components, fails to ensure quality (the robustness related to variability) because of focusing on mean value.

The so-called two-step optimization is to implement functionality improvement with two steps according to the exploration of nonlinear and linear relationships: First reduce the variability, and then tune the mean value to the target value. Therefore, we choose, according to experimental analysis, the factor level with large SN ratio to reduce variability, and we also choose the factor level with large mean or β to conduct the tuning of the mean value (or sensitivity) based on our needs. We additionally note that because dynamic

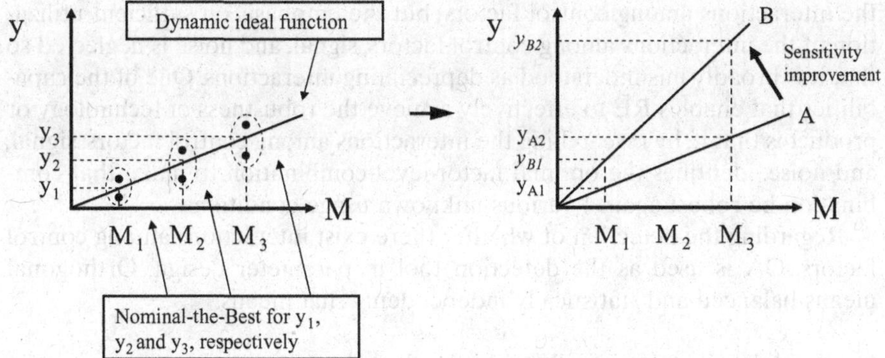

Figure 6.3. Sensitivity improvement of dynamic functionality.

response can be viewed as formed by a series of static nominal-the-best characteristics, the tuning of mean value of all these NTB characteristics can be regarded as the adjustment of sensitivity of dynamic functionality, as shown in Figure 6.3.

In RE, we use the approach of two-step optimization to ensure quality. Although it seems to be a very understandable concept, most R&D personnel, in practice, tune the mean value to be within the target specification first, and they conduct variability reduction according to the result of verification tests. In this chapter, we use examples to explain how to conduct two-step optimization in the RE experimental process.

6.2.2 Key Tools

RE is to enhance R&D efficiency by some effective tools; in this section, these key tools are explained.

1. SN Ratio. Although, in the beginning, the SN ratio was developed by Dr. Taguchi's research on measurement systems, accompanying new development and new application, the SN ratio has become the effective indicator to evaluate technology (functionality). Improving the SN ratio means to improve both linearity and variability of the input/output relationship of technology; hence, the larger the SN ratio, the more robust the performance of technology or product. RE can be regarded as the leading-edge technique for maximizing the SN ratio.

2. Orthogonal Array (OA). One of the most frequently seen misunderstandings about RE is that it doesn't value the interactions among factors that R&D personnel consider very important. Actually, RE doesn't depreciate the interactions; instead it greatly emphasizes interactions. The reason why there is such misunderstanding is that, in the viewpoint of RE, it is not a good way to utilize the interactions among control factors; but the emphasis on sufficient utilization of the interactions among control factors, signal, and noise is neglected so that it is broadly misunderstood as depreciating interactions. One of the capabilities that enables RE to effectively achieve the robustness of technology or product is that it, by researching the interactions among control factors, signal, and noise, identifies the optimal factor-level combination to make that combination be robust against various unknown usage conditions.

Regarding the detection of whether there exist interactions among control factors, OA is used as the detection tool in parameter design. Orthogonal means balanced and statistically independent. That means:

- Each level occurs equally within each column.
- For each level within one column, each level within any other column also occurs equally.

OA is expressed as $L_a\ (b^c)$:

- *L* stands for Latin square.
- *a* stands for the trial number or experimental runs.
- *b* stands for the number of levels.
- *c* stands for the number of control factors.

For example, $L_{18}\ (2^1 \times 3^7)$ means that this OA can allocate eight factors in total, among which one factor has two levels while the other seven are of three levels, and we need to implement 18 factor-level combinations, as shown in Table 6.1.

Take the 12th trial, for example; we conducted that experimental run with the factor-level combination formed by A2, B1, C3, D2, E2, F1, G1, and H3 and recorded the result. When all the 18 trials are finished, we can conduct experimental analysis. Once the reproducibility of the experiment conclusion based on OA is insufficient, we can know that there are strong interactions among control factors, and those interactions would make the combined effect formed by various factors hard to be predicted; and further, create the chaos of R&D in product performance, manufacturing process, cost, and time control. Therefore, in RE, OA is the indispensable tool used to quickly evaluate different design concepts (the various factor-level combinations of control factor).

3. Test Piece and Simulation Tools. Test pieces or computer simulation tools are helpful to reinforce the "precedency" of technology development before product planning (at this time, there is no product information or physical product). Since RE's optimization target is functionality, not quality, R&D personnel can use test pieces or simulation approaches to implement technology and product development without physical products. When completing optimization, R&D personnel can conduct a confirmation experiment with a

TABLE 6.1. L_{18} Orthogonal Array

L_{18}	A 1	B 2	C 3	D 4	E 5	F 6	G 7	H 8	Result
1	1	1	1	1	1	1	1	1	
2	1	1	2	2	2	2	2	2	
3	1	1	3	3	3	3	3	3	
4	1	2	1	1	2	2	3	3	
5	1	2	2	2	3	3	1	1	
6	1	2	3	3	1	1	2	2	
7	1	3	1	2	1	3	2	3	
8	1	3	2	3	2	1	3	1	
9	1	3	3	1	3	2	1	2	
10	2	1	1	3	3	2	2	1	
11	2	1	2	1	1	3	3	2	
12	2	1	3	2	2	1	1	3	
13	2	2	1	2	3	1	3	2	
14	2	2	2	3	1	2	1	3	
15	2	2	3	1	2	3	2	1	
16	2	3	1	3	2	3	1	2	
17	2	3	2	1	3	1	2	3	
18	2	3	3	2	1	2	3	1	

physical model. In practice, despite the fact that various computer simulation tools have been widely applied, the improvement of R&D efficiency is limited to the time shortening by using such simulation tools. Therefore, if R&D personnel can combine RE with computer simulation, R&D efficiency can be improved with a breakthrough.

6.2.3 Process Steps

The application itself of RE is a process of experiment planning and execution. Through this experiment, we effectively study the interactions between design parameters and various sources of variability, and we use this information to achieve functionality optimization. The process of implementing RE can be divided into the following seven steps.

1. Identify the System to Be Optimized. Only with the joint operation of many subsystems or mechanisms (functional blocks) in product architecture, technology can elaborate functionality to realize the design intent of product. When we are developing technology and intending to realize the design intent of product with that technology, the development work we are doing is to define the various functional blocks or modules needed by a product to perform its function and to define the interrelationship among them to confirm that the design intent of product can be achieved. If, in the design process of the product, we can understand what form the technology exists in, it would be helpful to think about how to systematically and effectively deploy the needed technology and identify the bottleneck-engineering that would hinder product development. Mechanisms (functional blocks) are the visible connection between product and technology; that is, technology is applied to the product in the way of storing it in the systems or mechanisms to make the product perform its design intent. More simply speaking, each mechanism in a product is the enabler of technology. They are used to realize various technologies within the boundary of product architecture. Therefore, R&D personnel have to, before conducting technology optimization, first identify the systems or mechanisms where the technologies are—that is, define the scope of optimization.

Ideally, the functionality of each technology in a system needs to be optimized at the origin stage of development; however, in practice, R&D personnel usually define the systems or mechanisms in which problems happened and seek to use RE or other methods to improve them when technological bottlenecks or quality problems occurs. If that paradigm can be changed, then R&D efficiency would be enhanced tremendously.

2. Define Ideal Function and Its Input Signal and Output Response. Ideal function is the design intent of a system or mechanism. It consists of desired output response and the input signal used to adjust it. Energy transformation is often used to express an ideal function, because a technological system, in nature, is a tool facilitating the occurrence of energy transformation process. Although sometimes it is not easy to define the ideal function, in practice, maybe the development of "measurement capability" requires more effort. The input signal and output response of the ideal function are formulated according to the nature of technology, so their measurement items are usually different from the items of quality characteristic form, hence the measured values may be limited to the usage or capability of current measurement systems (appraisers and equipment) and can't be provided because the data cannot be collected and analyzed. Therefore, for a successful RE experiment and method application, ideal function formulation and measurement capability development are the most important parts in experimental planning.

The formulation of the ideal function sometimes requires many hours' worth technical discussions to decide what to measure. The understanding of the nature of technology and the creativity of identifying what to measure is very important for R&D personnel. A proper ideal function formulation is helpful for R&D personnel to solve quality problems efficiently and effectively. Table 6.2 shows the examples of the ideal functions used in practical cases. In addition, Section 6.5 of this book explains, using case studies, how to formulate an ideal function and apply RE process.

TABLE 6.2. Examples of Ideal Function

System	*M*: Input Signal	*y*: Output Response
Measurement system	True value	Measured value
Injection molding	Mold dimension	Product dimension
NC machining	Programmed dimension	Product dimension
Imaging	Master line width	Product line width
Fan/Cooler	Motor voltage	Air flow
Current sensor	Current	Gain
Transformer	Turns ratio	Voltage
Brake	Line pressure	Braking torque
Paper feeder	Roller rotation	Paper travel distance

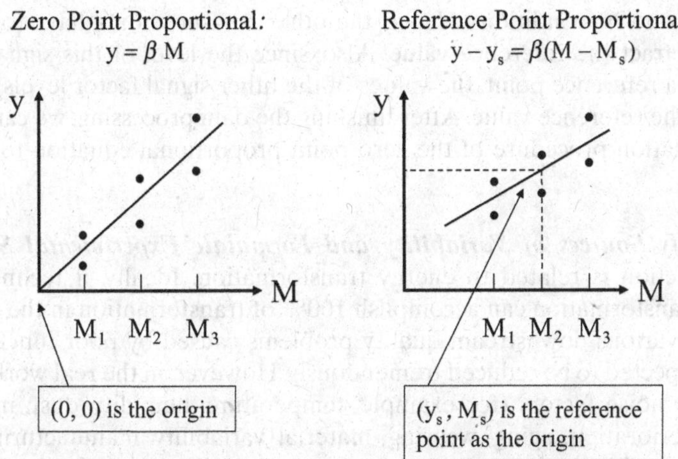

Figure 6.4. Two types of SN ratios for dynamic response.

To conduct the experiment for whether functionality is approaching to the ideal state, we need to identify several different levels of input signal to confirm, under different levels, if the variability, linearity, and sensitivity of output response approach the ideality. In view of this, we should select the level values of signal factor in a range that is as broad as possible so that we can, in a large enough design space, seek the characteristics in input/output relationship which can be used to reduce variability and to improve linearity and sensitivity and then tremendously improve technological quality with such characteristics.

The equations of SN ratio for dynamic response can be classified into two types: zero point proportional equation and reference point proportional equation, as shown in Figure 6.4. Viewing them from the nature of energy transformation and technology, we generally use zero point proportional equation. This is because when the input energy is zero, the output energy is also zero, and the ideal input/output relationship should be a straight line pass through the origin (0, 0). Nevertheless, in some situations, what we care about may be an energy transformation where the input and output energy are very big so that the energy transformation around the origin is relatively meaningless for us. Therefore, we turn to use reference point proportional equation. As implied by the name, when using reference point proportional equation, we use a certain reference point (y_s, M_s) as the role of the origin in zero point proportional equation. We usually use a certain level of signal factor as the reference point.

Whether using a zero point proportional equation or a reference point proportional equation, their calculation procedures are the same. The only difference is that, when using the latter one, the reference value is the mean value of the measured values of the signal factor used as the reference point under N_1 and N_2 conditions. Hence, the other measured data have to respectively subtract the reference value. Also, since the level of this signal factor becomes a reference point, the values of the other signal factor levels have to subtract the reference value. After finishing the data processing, we can follow the calculation procedure of the zero point proportional equation to get the SN ratio.

3. Identify Sources of Variability and Formulate Experimental Strategy. Ideal function is related to energy transformation. Ideally, if technological energy transformation can accomplish 100% of transformation in the process, then the various downstream quality problems caused by poor functionality can be expected to be reduced tremendously. However, in the real world, there are many noise factors (for example, temperature, humidity, dust, magnetic field, deterioration, aging, wearing, material variability, manufacturing variability, etc.) which we can't control so that the function fails to realize its ideality. The more serious the interference of these factors for functionality, the greater the number of quality problems.

In order to make functionality insensitive to these noises, we need to develop experimental noise strategy. Its purpose is to compulsorily introduce these noises into the experiment to study how to make functionality insensitive to these noises. Therefore, R&D personnel have to consider the important customer usage conditions (environmental conditions, deterioration, and aging) and production conditions (the between product variability) to formulate effective noise strategy, so as to enable the robustness of function under sources of variability to be sufficiently analyzed to acquire valuable design information. Noise strategy is important for RE experiment, but we only need to collect the response value under two types of noise level: low-response noise condition (N_1) and high-response noise condition (N_2).

4. Define Design Parameters and Their Levels. In design space, the design parameters whose levels can be freely determined by R&D personnel are called *control factors*. R&D personnel, by determining the optimal conditions of control factors, make functionality insensitive to noise factors and can adapt output response by usage intent (signal factors). Judging from this we can know that, while conducting an RE experiment, R&D personnel should, as possible, select the level values of control factor which are largely different from each other in the design space to fully explore the nonlinear relationship in design space and, further, to achieve functional robustness.

5. Allocate and Conduct Experiment. When various types of factor and their levels are determined, we need to allocate the factors into the orthogonal array (OA) and then conduct an experiment according to OA allocation. The OAs suggested by Dr. Taguchi to use include L_{12}, L_{18}, L_{36}, and L_{54} (see Appendix A). L_{12} is suitable for all the experiments with the factor level as two levels, while L_{18}, L_{36}, and L_{54} are suitable for all the experiments with the factor level which is the combination of two levels and three levels. L_{36} and L_{54} experiments are generally conducted by computer simulations.

These OAs all have the same features: The interactions among factors are averagely distributed to each column so that the interactions wouldn't be confounded with main effects to mislead the conclusion. The main purpose of using these OAs is that we first hypothesize that there are no interactions among main factors, and then we judge if interactions exist by the final result of experiment. If interactions exist, it means there would be poor reproducibility in experiment conclusion. Thus, unlike the other OAs (shown in Appendix B), the columns of these OAs can only allocate main factors. Besides, for the factors that are not easy to change levels (need to consume more time or cost), we can allocate them in the columns on the left of OA to reduce their frequencies of level change in the experiment. After finishing OA allocation, R&D personnel, according to that allocation, measure and collect the result data of various factor-level combinations.

6. Analyze Data and Conduct Two-Step Optimization and Confirmation Experiment. When finishing the experiment, we calculate the SN ratio and β according to the procedure described in Section 5.4.2 and calculate the SN ratio values and β values (i.e., factorial effects) of each factor level, respectively. According to the procedure of two-step optimization:

- First, for each factor, we choose the level with the largest SN ratio as its optimal level setting (i.e., reduce variability first).
- Second, we choose the factors that only influence sensitivity (or mean value in static case) but not (or less) SN ratio as adjustment factors (i.e., adjust the sensitivity to target value).

Response graphs are used to express the factorial effects of each factor level. As shown in Figure 6.5, we explain how we select the factors and their levels:

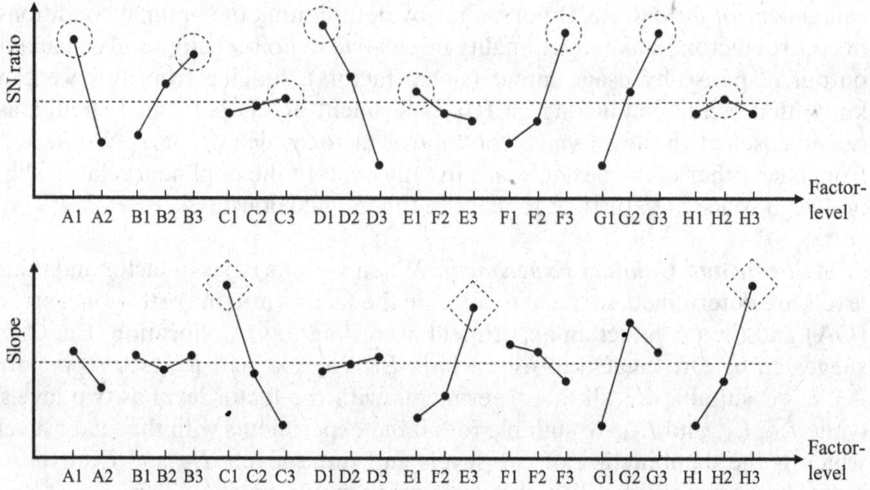

Figure 6.5. Response graphs for SN ratio and slope.

- First, we choose the level setting with a larger SN ratio:

  A1B3C3D1E1F3G3H2

- If there is a target value for sensitivity, we need to select adjustment factors to tune sensitivity to target value. Adjustment factors are those factors that almost never influence SN ratio but influence slope: C, E, and H. Assume that now we need to tune sensitivity higher; we can selectively utilize C, E, or H whose levels are C1, E3, and H3 according to the situations. When it needs to tune sensitivity to be the highest value, we can simultaneously utilize C, E, and H to conduct sensitivity improvement.

Response graphs are drawn based on the calculated values of SN ratio and slope of each factor level. We will have a detailed description about the steps for calculation and graph drawing in the case studies in Section 6.5.

Once the optimal design condition is determined, R&D personnel have to conduct the confirmation experiment according to that optimal factor-level combination under N_1 and N_2 conditions to evaluate the reproducibility of the experiment result. If the former experiment is conducted by computer simulation, then, at this step, we suggest to conduct it with physical model. Before the confirmation, we should obtain the predicted value of SN ratios of the optimal design and initial design based on the equation formulated in Section 5.4.2 and calculate the actual SN ratios according to the result of confirmation experiment. The general principle of confirming experiment successful is that the difference between predicted SN ratio gain (the predicted SN ratio of optimal design minus the predicted SN ratio of initial design) and actual gain (the actual SN ratio of optimal design minus the actual SN ratio of initial design) is less than 30%.

The common causes of experiment failures are the poor selection of what to measure (i.e., ideal function) causing strong interactions among control factors or the noise strategy needs to be refined. We should sufficiently consider the formulation of ideal function and noise strategy to ensure the success of RE experimental strategy. It's helpful to understanding how to define more proper ideal function and noise strategy by referring to and learning the published RE case studies in various technology fields.

7. Standardization and Future Plans. A successful experimental conclusion of RE has reproducibility and universality. The so-called reproducibility means that the conclusion of small-scale experiment conducted in the lab can be directly applied to large-scale production. Universality means that the optimized technology can be applied to an entire family of products. Therefore, documenting and standardizing the experimental conclusion of RE is very beneficial for R&D knowledge management and experience sharing. The key point of documenting and standardizing an RE experiment lies in inducing control factors into four types.

- Influencing variability only.
- Influencing sensitivity (or mean value in static case) only.
- Influencing variability and sensitivity (or mean value in static case) simultaneously.
- Having no influence on both variability and sensitivity (or mean value in static case).

The information is helpful for enhancing R&D efficiency to reduce the time needed for evaluating and selecting the optimal levels of control factors and, further, to shorten total cycle time. Especially while the variability can be controlled, R&D personnel can, utilizing adjustment factors (the factors only influencing sensitivity or mean value, not SN ratio), easily tune the sensitivity or mean value to different target values. In addition, since the levels of signal factor may be considered the input values of present and future generation technologies or products, over a certain time, R&D personnel may review again if signal factor levels need to be modified to comply with the requirements of technology development and can optimize new system parameters. When there is new development of measurement technology, R&D personnel possibly can identify more effective ideal function to conduct functionality optimization.

All the above-mentioned R&D advantages are based on the information of technology and product development acquired from RE experiment. Unlike the traditional knowledge management database that piles up a lot of technical information only suitable for individual cases and situations, that information is the accumulated R&D technical information and knowledge which is easy to be utilized, shared, and horizontally deployed.

6.3 COMMON PROBLEMS OF RE APPLICATION IN PRACTICE

In practice, RE or the so-called Taguchi Methods is familiar to most quality practitioners. However, although RE has been widely known and promoted, its popularity doesn't mean that it has been correctly applied. In fact, many quality improvement personnel applying Taguchi Methods had failed experiences or the experimental results are the same as those obtained through empirical rules. Hence, the effectiveness of Taguchi Methods in improving the efficiency of product or process design is doubted or not recognized.

Based on our observations, the frequently seen problems of RE application are described as follows.

1. Use Quality Characteristics as What to Measure. Many RE application cases involve solving and improving existing problems; hence quality improvement personnel have to identify the "problems" needed to be solved at present before applying the methods. The so-called problems are usually described as deficiencies of quality characteristics such as noises and vibrations. The improvement of the measured values of quality characteristics such as noises and vibrations becomes the targets of RE application. We may select one important quality characteristic to make improvement, and we may try to simultaneously enhance the performance of multiple quality characteristics. The issue that arose from above-mentioned facts is, How can RE make improvements for multiple targets?

Many books or research treatises are of the opinion that RE is only suitable for improving a single quality characteristic; for the product or process with multiple quality characteristics, it cannot be applied or is limited. In view of such, the common approach in practice is to simultaneously collect the measured values of various quality characteristics and choose the factor-level combination that can conform to the requirements of key or most quality characteristics as the optimal parameter condition. In the research field, different scholars proposed many complex quantitative methods to focus on discussing how to solve this kind of problem. No matter which kind of method, it seems that quality improvement needs a trade-off. In other words, it's not easy for us to simultaneously improve all quality characteristics: In the case of multiple quality characteristics, the improvement of some quality characteristics must cause the deterioration of some quality characteristics. However, the so-called multiple quality characteristics actually is the case that quality improvement personnel must face, this is because, for a product or process, it's almost impossible to rely on the measurement of a single quality characteristic to judge quality good or bad. This shows that dealing with multiple quality characteristics is a basic requirement for judging if RE is effective.

How does RE deal to multiple quality characteristics? The correct thinking direction is to focus on "how to define and measure ideal function," not to answer "how to analyze the multiple characteristics." More specifically speaking, technology is related to energy transformation; hence the better the functionality of realizing ideal function, the less the energy used to create various quality problems; that is, we can solve all quality problems by the approach of "killing multiple birds (multiple quality characteristics) with one stone" while optimizing functionality. In the viewpoint of RE, there is not the issue of how to analyze multiple quality characteristics, since the phenomenon of improving some quality characteristics causing the deterioration of some quality characteristics is the result of "exchanging one type of lost energy for another type in energy transformation." For example, after reducing noises, according to the law of the conservation of energy, the energy originally used to create noises is to be exchanged to create other quality problems such as vibrations. If the functionality (energy transformation) can be optimized, the multiple quality characteristics can be improved simultaneously. The concept of RE described above is one of the biggest differences among this book and other books. In fact, for Japan's quality engineering field and the companies guided by Dr. Taguchi and his team, such RE concept was widely applied since the 1990s. However, until now, the applications in other places were still based on the old concept of Taguchi Methods, using static characteristics (Nominal-the-Best, Larger-the-Better, and Smaller-the-Better), which was popular in the 1980s. The "dynamic characteristics" that represent the true core of RE are not emphasized and widely applied. The above-mentioned old concept becomes a general recognition and trend not easy to be changed by the continual international transfer and circulation of such researches (widely seen in related books, literatures, and the guidance of professors) and practices (widely seen in application cases and the guidance of consultants) for many years until now. Perhaps it's more appropriate to view this difference as the evolution of Taguchi Methods, but there is no denying that the most powerful part of RE, the concept and method of optimizing dynamic characteristics, must be emphasized nowadays.

According to the viewpoint of Dr. Taguchi, the so-called quality problems are the problems perceived at the customer site and described in the customer's language, and the so-called "engineering" means that the downstream quality problems must be solved in a laboratory. Therefore, RE is an optimization technology used at the laboratory research stage for effectively predicting and solving various unknown problems at the customer site. Once the core of RE transits from emphasizing (multiple) quality characteristics to ideal function, the applications of the method itself simultaneously transit from "solving occurred quality problems" or "firefighting" to "predicting various unknown quality problems" or "fire prevention." Viewing from another perspective, which quality characteristics and what specifications of a product should be set are to be defined and measured after a product is "formed" or "materialized" by following the design process. At the technology development stage when a product is not materialized or even not in the planning stage, we cannot define or develop quality characteristics. However, technology development is the main application field intended for RE; hence we can know that RE is not intended to set quality characteristics as the objects to be optimized and study how to solve the occurred quality problems. The strategy of RE is to realize the state as close as possible to ideal function, and at the same time, to solve various quality characteristic deficiencies that may occur in the future; in other words, make improvement before the occurrence of problems.

2. Not Considering the Noise Strategy. Design (of a product) itself is a kind of countermeasure, and this countermeasure is used to solve the possible quality problems occurred under various unknown customer usage conditions. RE is an effective method for developing proactive prevention countermeasures. The core of realizing the robustness of technology or product under various sources of variability is to sufficiently utilize the interactions between control factors and noises to improve functionality. Therefore, an RE experiment without considering noise strategy cannot be used to effectively develop robust technologies. However, the use of OA is often emphasized in RE application, but the importance of allocating noise factors in OA is overlooked. Since the sources of variability which influence product's functional performance are many, we cannot identify or evaluate all possible noise factors. The most recommended noise strategy is to "compound" the key noise factors into two levels: N_1 and N_2 (the noise conditions make the response value tend to be low or high respectively). We utilize OAs to study how to set control factors to make technology or product design insensitive to the influences of noises, so as to achieve functional robustness.

3. Few Control Factors. The common approach of problem solving is to analyze or find out the causes of variability and to prevent the recurrence of problems by eliminating the causes. Nevertheless, it is usually not easy to analyze and find out the root causes of variability. The prerequisite of correctly finding out the cause is that the cause appears again in the period of analysis; hence when we conduct root cause verification, the resulting quality problems can be reproduced and help to judge whether this cause is the root. If the situation described above doesn't happen, the identification of the possible causes relies on empirical rules and intuition, and quality improvement personnel draw up countermeasures for all the possible causes. The worst situation is that the countermeasures are ineffective or make the system chaotic due to erroneously judging causes.

The difficulty of managing quality lies in dealing with variability; that is, it's hard to predict and prevent the occurrence of variability. In the viewpoint of

RE, the causes of variability described above are noise factors. What we need is not to try to control or eliminate these sources of variability (they may are uncontrollable or hard to be controlled) but to utilize the control factors whose levels we can freely set to make technology or product itself be insensitive to the impact of sources of variability; after that, we can have freedom from the occurrence of various quality problems. In other words, it's not necessary for us to try to find out those causes of resulting variability, what's important is how to prevent the recurrence of the quality problems. RE is a powerful strategy for solving the problems that have not yet happened or are unknown. Since the setting of control factors is the problem prevention countermeasure itself, too few control factors would make the improvement space limited. On the contrary, the larger the system and more control factors, the more interactions can be sufficiently utilized and the larger the improvement space. We suggest that the number of control factors is at least eight to achieve the least number of factors needed by the OA (L_{18} is the recommended OA that needs the least factors) recommended by Dr. Taguchi.

4. Take Advantage of the Interactions among Control Factors. The most common misunderstanding for RE is that this method does not value, and even neglects, interactions. However, it's not true; RE values and sufficiently utilizes interactions. The reason why such a misunderstanding exists is that the types of interaction are not distinguished. Interaction can be classified as two types:

- Interactions among control factors.
- Interactions between control factors and signal factors or noise factors.

In the viewpoint of RE, the former should be avoided and the latter is the item that must be utilized to realize robustness. Many R&D personnel are taught to value and utilize the interactions among control factors. For instance, assume that we have two factors, A and B, and that A1 and B1 levels have the effect of increasing response value respectively, but when we expect to have substantial effects on response value by combining A1 and B1 in one design, the response value is lower instead due to the strong interaction between factors A and B. At this time, the design basis used by R&D personnel is the interaction effects of the two, rather than the individual effect. Assume that in this example we need to change the setting to be A1B2 so that the response value can be increased, hence we can know that the interactions can make the main effects of control factors lose "additivity"; that is, the effects of A1 and B1 on response value are not accumulated, but may cause unpredictable performance.

If we do not consider that the approach of valuing and utilizing the interactions among control factors should avoid being used, R&D personnel probably cannot assure that the research conclusions in laboratory can be reproduced at downstream large-scale manufacturing and customer usage conditions. Viewing from the example mentioned above, we select A1B2 as the optimal parameter combination according to the interaction effects of the two factors. Assume that the downstream manufacturing conditions cannot be set or selected as B2 level (suppose that they must be set or selected as B1) for some reason, then A1B1 combination would make the response value worse as described above. When factor B needs to be set as other levels (such as B3 or B4, etc.), then the effects of the combination of factors A and B determined at the manufacturing stage on response value cannot be predicted and may further cause a product's quality problems. Therefore, the approach of utilizing the interactions among control factors is considered disadvantageous in RE. Nevertheless, the interactions between control factors and signals or noises play a very important role in RE. Such sufficient utilization of interactions helps to realize functional robustness of technology or product.

5. Inappropriate Selection of Orthogonal Arrays. The purpose of using OAs generally is considered to enhance the experimental efficiency. Nevertheless, this is a benefit but not the purpose of using OAs. The actual purpose of using OAs is to check if there are strong interactions among control factors—that is, to check if the optimal factor-level combination has downstream reproducibility. If the experimental results obtained according to OAs cannot be reproduced in a confirmation experiment, most reasons may come from the interactions among control factors.

Some OAs are recommended by Dr. Taguchi since these OAs can effectively check the reproducibility. Their common feature is that the interaction effects are skillfully distributed to each column equally while designing the arrays, so as to make the effects of the main factors and the distributed interactions be confounded; that is, the two kinds of effects are mixed. The experimental logic of using OAs is that: We assume that there is no interaction among control factors and boldly only allocate main factors into the OA, and then we verify if the improvement of SN ratio equals what we predicted by a confirmation experiment. If the confirmed gain equals the predicted SN ratio gain, this means that the main effects of control factors are additive; if not, it means that there are interactions among control factors. To sum up, since we want to verify that there are no interactions among control factors, we only allocate the main factors into the OA. If we use the OAs recommended by Dr. Taguchi, the check of reproducibility is more effective.

6. Not Conducting the Confirmation Experiment. The purpose of confirmation experiment is to confirm if the experimental conclusions have reproducibility—that is, to confirm if the interactions among control factors are significant. If we directly use the optimal factor-level combination from the analysis according to orthogonal arrays and not check it with confirmation experiment, it cannot be ensured that the functional variability can be improved. One important thing, if we conduct optimization experiments by computer simulations, we should confirm the results with hardware.

7. Focusing on Manufacturing Operations. The frequently seen RE applications are process improvement implemented in factories; that is, RE is not considered as the method used at the R&D stage. If we view the enemy that RE needs to resist—namely, noise factors—we can know that improving manufacturing operations only can resist "between product noise" (material variability and manufacturing variability), but only the personnel of the technology development and product design can develop countermeasures to resist "outer noise" (environmental and usage conditions) and "inner noise" (aging, deterioration, or wearing). If the positioning of RE applying to technology and product development is not sufficiently recognized, it will be hard to transit its applications in practice from product manufacturing to R&D.

Table 6.3 summarizes the above-mentioned problems and the resulting consequences.

Besides, the authors sum up the common recognition and concern about RE, and they propose brief explanations as follows:

1. *RE is the same as the frequently used "design of experiments" (DOE).* RE is totally different from DOE in philosophy and experimental strategy. The main difference between the two is the difference between emphasizing "the modeling of the relationship among factors" and "the engineering strategy for developing the optimal design under the considerations of quality and cost." Since the philosophies are different, the methodologies themselves

TABLE 6.3. Common Problems of RE Application

Problems	Leads to . . .
Use quality characteristics as what to measure	• Debug engineering and quality improvement trade-off decisions • Incapable of transiting Taguchi Method application from firefighting to fire prevention and from product development to technology development
Not considering the noise strategy	• Incapable of engineering/optimizing the robustness
Few control factors	• Very limited space for improvement
Take advantage of the interactions among control factors	• Incapable of assuring reproducibility
Inappropriate selection of orthogonal arrays	• Insufficient check of reproducibility
Not conducting the confirmation experiment	• Insufficient confirmation of reproducibility
Focusing on manufacturing operations	• Insufficient countermeasures for all types of noise • Incapable of transiting Taguchi Method application from product manufacturing to R&D

developed based on them are also different. Simply speaking, DOE emphasizes mean effects and scientific modeling, while RE focuses on variability and engineering practices.

2. *RE is just the well-known Taguchi Method.* RE is the method developed by Dr. Taguchi. The concepts described in this book is the evolutionary Taguchi Methods. What is reflected by RE described in this book is Dr. Genichi Taguchi's latest thinking, so it's different from the traditional Taguchi Methods. RE emphasizes the optimization for dynamic response (i.e., evaluation and design of functionality), rather than the optimization for single quality characteristic (i.e., what the traditional Taguchi Methods do).

3. *Taguchi Methods are methods of optimal parameter determination and quality improvement applied to manufacturing processes.* RE can be applied not only to process design, but also to product development and technology development. In fact, the most obvious benefit can be obtained by applying it to technology development: The developed robust technology can be deployed to the whole product family, which uses the same generic technology.

4. *I own profound technological knowledge and many years of technology development experiences; why don't I know this method?* Although RE is regarded as the methodology for technology development, it was actually developed in the quality engineering field. That's why technology experts might not be familiar with this method.

5. *In my field, technology development is always conducted following up some techniques and models; could RE be better than this development model which has been adopted for a long time and improved continuously?* Basically, RE is positioned as an evaluation technology in R&D practice, rather than the specialized technology used to solve technical problems. To solve technical problems still relies on the specialized technological knowledge of R&D personnel. Evaluation technology and specialized technology can be regarded as the two indispensable technologies in R&D, but the relationship between them is not that they substitute each other.

If the restrictions on various resources such as cost and time during R&D don't need to be considered, R&D personnel can effectively conduct technology development with their technological knowledge and experience. The benefit of RE to R&D lies mainly on that it can enhance the efficiency in functionality evaluation (i.e., evaluation of the functionality of a technology or product itself under various customer usage conditions and sources of variability), cost and time.

6. *The so-called ideal function might have different perspectives and definitions varying with different R&D personnel; that way, it will make technology development have different experimental planning and results.* Yes, there might be different definitions on ideal function by different perspectives of R&D personnel. However, they usually reach consensus after discussion because ideal function is regarded as a way to describe the nature of a technology, and

they shouldn't have many different opinions about that nature after discussion. Sometimes, R&D personnel conduct experiments and result comparisons using ideal functions with two or three different definitions.

The ideal functions defined in different periods may be different due to the advancement of measurement technology. The more the advance of measurement technology, the higher the feasibility of conducting experiments according to the ideal function we defined. Otherwise, we might need to seek the substitute for the ideal function initially defined.

7. *We might fail to measure the most ideal input/output relationship because of insufficiency of measurement technology.* In this situation, R&D personnel have to define the substitutes for input signal or output response. In fact, many breakthroughs of product or process technologies are derived by the advancement of measurement technology.

8. *After implementing RE, can reliability tests really be canceled entirely?* In the viewpoint of RE, reliability testing is not efficient. Ideally, applying RE is conducive to develop technology and product with the "no trials, no tests, no inspections" approach. However, in practice, many reliability tests are for responding to the requirements raised by customers, so they are inevitable. We expect to complete technology optimization through RE and then use reliability tests to conduct verification.

9. *I tried this method, but the result was a failure.* The most common causes of failure in applying RE are that:

- Using quality characteristics as targets of optimization (which easily cause strong interactions among control factors).
- Noise strategy is not sufficient.

If RE is correctly used, the failure of experiments means the early confirmation of feasibility of technological capability, that is, if R&D personnel continue to develop the product according to this design concept, they will face failure. Therefore, the failure of experiments is regarded as the early confirmation of poor design concepts. We further explain this by the following words that may not be easy to be understood: An experiment is conducted for the sake of failure. The objective of conducting experiments is not to confirm the known items, but to study the unknown items. The technical reviews conducted after the unexpected failure of an optimization experiment are truly technology development. Technology development is not an easy job, so the earlier the technological feasibility or design concept is evaluated, the more time is gained for conducting reviews and developments. Briefly, the key to determine the victory of technology development is functionality evaluation.

10. *I tried this method, but there was no big difference between optimized parameter combination and the engineering judgment of senior designers.* Although senior designers' knowledge and experience can provide a good

guidance, their role should be to create higher value in managing R&D teams and shifting R&D paradigm, not to play one-man hero or firefighting hero in problem solving. Senior designers are responsible to seek more effective methodologies to accelerate the learning of junior designers.

11. *The major reason to apply RE is to improve quality; so under the circumstance that R&D personnel have capability to develop high-quality technologies with stable performance, they don't need to use the RE method.* The core concept of RE is "cost first," not emphasizing quality improvement. Applying RE is to improve quality under the situations of reducing or not increasing cost, as well as to enhance the competitiveness in the market through this.

Table 6.4 shows the summary of the mentioned above.

6.4 ROBUST TECHNOLOGY DEVELOPMENT

To deal with the competition, R&D personnel need to develop strategies to implement a competitive technology development. Simply speaking, the competition in the market means the competition of quality and price (more precisely, cost). The higher the quality and the lower the price or cost, the higher the value the product can deliver and the more competitive it is in the market. Back to the radical aspect, if we describe the competition in quality and cost in the market with the language in Section 1.1, we would say that this is the level of technological competition. Success or failure of this competition has been determined at the stage of developing technological strategy. As described above in this book, the so-called technological strategy is the universal framework for rationalizing and efficientizing the company's technology development in various product fields; the implementation of that strategy is called robust technology development in this book. Although robust technology development is not a concept that appeared for the first time in this book, it can be regarded as the activities of applying RE to technology development fields. We want to, in this section, emphasize the important elements of this strategy and interpret more clearly the difference between robust technology development and conventional model to make R&D managers have clearer cognition and promote the shift of R&D paradigm as soon as possible.

The four key strategic elements of robust technology development are described as follows:

1. *Cost first.* In the field of quality, we usually say "quality first." However, in the viewpoint of robust technology development, cost is the first concern. Although robust technology development is also strategy and methodology for improving quality, its purpose is to pursue the high quality under cost as the first concern, rather than to pursue quality first but simultaneously cause possible cost increment. If we achieve high quality but increase cost

TABLE 6.4. Useful Recognition of RE

Conventional Recognition/ Common Doubt	Useful Recognition of RE
RE is the same as the frequently used DOE.	• RE is totally different from DOE in philosophy and experimental procedure. DOE emphasizes mean effects and scientific modeling, while RE focuses on variability and engineering practices.
RE is just the well-known Taguchi Methods.	• Yes, it is the Taguchi Methods, but is the evolutionary Taguchi Methods. • RE emphasizes the optimization for dynamic response more than the traditional Taguchi Methods do.
Taguchi Methods are methods of optimal parameter determination and quality improvement applied to manufacturing processes.	• RE can be applied not only to process design, but also to product development and technology development. In fact, the most obvious benefit can be obtained by applying it to technology development.
I own profound technological knowledge and many years of technology development experiences, why don't I know this method?	• Although RE is regarded as the methodology of technology development, at the very beginning, it was developed in quality engineering field, so technology experts might not be familiar with this method.
In my field, technology development is always conducted following up some techniques and models; could RE be better than this development model which has been adopted for long time and improved continuously?	• Basically, RE is positioned as an evaluation technology in R&D practice, rather than the specialized technology used to solve technical problems. To solve technical problems still relies on the technological knowledge of R&D personnel. • If the restrictions on various resources such as cost and time during R&D don't need to be considered, R&D personnel can effectively conduct technology development with their technological knowledge and experience. The benefit of RE for R&D lies mainly in that it can enhance the efficiency in cost and time.
The so-called ideal function might have different perspectives and definitions varying with different R&D personnel; that way, it will make technology development have different experimental planning and results.	• Yes, ideal function might have different definitions varying with R&D personnel's different perspectives. However, since ideal function is the nature of technology, they usually can reach consensus about the nature after discussion.

(*Continued*)

TABLE 6.4. (*Continued*)

Conventional Recognition/ Common Doubt	Useful Recognition of RE
We might fail to measure the most ideal input/output relationship because of insufficiency of measurement technology.	• In this situation, R&D personnel have to define substitute input signal or output response. In fact, many breakthroughs of product or process technologies are from the development of measurement technology.
After implementing RE, can reliability tests really be canceled entirely?	• Ideally, applying RE can let us develop technology and product with the "no trials, no tests, no inspections" approach; but in practice, many reliability tests are for responding to the requirements raised by customers, so they are inevitable. We expect to complete technology optimization through RE and then use reliability tests to conduct validation/verification.
I tried this method, but the result was a failure.	• The most frequently seen failure cause in RE application is to set quality characteristics as optimization targets. Such an approach always brings strong interactions among control factors to make the experiment have no reproducibility. Poor noise strategy is another common cause. • After applying the RE method correctly, experiment failure means the early confirmation of insufficiency of technological capability. Thus, experiment failure is regarded as early confirmation of poor design concepts.
I tried this method, but there was no big difference between optimized parameter combination and the engineering judgment of senior designers.	• Although senior designers' knowledge and experience can provide good guidance, their role should be to create higher value in managing R&D teams and shifting R&D paradigm, not to play "one-man hero" or "firefighting hero" in problem solving. Senior designers are responsible for seeking effective methodologies to accelerate the learning of junior designers about R&D experience.
The major reason to apply RE is to improve quality; thus, under the circumstance that R&D personnel have capability to develop high-quality technologies with stable performance, they don't need to use RE method.	• The core concept of RE is "cost first," not to emphasize quality improvement. Applying RE is to improve quality on the condition of lowering or not increasing cost and enhance market competitiveness through this.

simultaneously, then at the same time we realizing "quality first," we still face the realistic challenge of competition. Once we can achieve high quality by means of the materials or components of low cost (wide tolerance, high variability) or even just using computer simulation or test pieces, it means that the productivity (to achieve higher quality with lower cost and less time) of R&D work itself is enhanced, likewise the competitiveness. In robust technology development, the key that we can achieve high quality with low cost lies in the interactions between design parameters and various usage conditions; this is effectively utilized to make the product performance insensitive to the influence of various sources of variability (including high variability of materials or components).

2. *To get quality, don't measure quality.* Ideally, we can implement robust technology development with the approaches of no trials, no tests, and no inspections to achieve high quality. Learning from this concept, even if we don't know which quality problems of the developed technology may occur (i.e., not through trials, tests, and inspections), we still can implement technology improvement directly. Conventionally, we get quality (i.e., identify quality characteristics that need to be improved and try to improve them to enhance quality) by measuring quality (i.e., simultaneously measure multiple quality characteristics). Nevertheless, the concept of robust technology development is that "to get quality, don't measure quality" (Taguchi et al., 2005); the quality here refers to quality characteristics. As mentioned above, when we can, through functionality evaluation, predict the actual quality in mass production and market usage, we can improve multiple quality characteristics simultaneously before the occurrence of quality problems; that is the strategy of "killing multiple birds with one stone," which is also the reason why robust technology development approach is more efficient than conventional R&D approach. In other words, this method can allow R&D personnel to evaluate design ideas and concepts reasonably and efficiently before the occurrence of quality problems; if we can detect the failure of design concept as early as possible by this method, we can determine the insufficiency of technological capability as early as possible and take proper action to avoid various possible occurred quality problems.

3. *Two-step optimization.* Two-step optimization means to improve variability first and then hit the target value. In this book, R&D work is viewed as the robustness development (RD) of technology and product. Therefore, we need to improve the issue of variability which is hard to be dealt with first while the sensitivity or mean value are more easily to achieve target value by tuning. We use "the interactions between design parameters and various usage conditions (best and worst cases)" to effectively improve variability. After minimizing variability, we can, based on target specification, determine if the increment of variability is tolerated with regard to further acquiring the opportunity of lowering cost. Besides, this approach is helpful for R&D personnel to escape from the debug cycle of Design–Build–Test–Fix.

4. *Considering the nature of technology development: Precedency, universality, and reproducibility.* In the viewpoint of robust technology development, technology development has to effectively realize the following three aspects:

- *Precedency (Technology Readiness):* The functional robustness of technology itself has to be achieved before product planning; when R&D personnel determine product specification after customer specification acquirement or product planning, they need only to tune that technology to satisfy various target values.
- *Universality (Application Flexibility of Technology):* Since one generic technology can be applied to an entire product family, R&D personnel need only to conduct once the optimization of generic technology which can be applied to an entire family of products (the products with different specifications or target values). Consequently, they can tremendously reduce duplicated development, time, and effort.
- *Reproducibility:* The research conclusion acquired by R&D personnel from test pieces, small-scale experiments, and limited conditions must be able to reproduce its effects on actual products, in large-scale manufacturing and various customer usage conditions, otherwise, when we use the technology in the places other than laboratory, we would need to re-optimize technology because of its failure.

Table 6.5 summarizes the four strategic elements.

In order to fully interpret the concepts of robust technology development, we have to analyze the difference between robust technology development and conventional model in technology development practice. First, in practice, the most frequently seen design procedure is that R&D personnel seek to shoot the design target and then control variability. However, in the viewpoint of RE, we have to improve the variability first and then seek to hit the target value. Although we have emphasized the importance of two-step optimization, we still need to explain in detail why that conventional procedure has to be changed. While comparing the two, we can discover that the former needs to adapt different product specifications to make response values hit different target values and simultaneously control the variability of response values near these targets. However, the latter prevents variability from being influenced by the change of target values; thus, for the entire product family with different product target values, R&D personnel only need to conduct the tuning of the response value and do not need to re-conduct robust optimization for each product. The reason why this can be realized is because, through RE, R&D personnel can effectively grasp and utilize the individual effect of each parameter on sensitivity (or mean value) tuning and variability reduction of the result to quickly lower variability or tune response value to the target value. Conventionally, R&D personnel just know if design parameters have influence on the sensitivity (or mean value) of result; however, its influence on variability can be known by means of reliability and other technical tests.

TABLE 6.5. Strategic Elements of Robust Technology Development

Strategic Elements	Description	
Cost first	• Use low-cost/wide-tolerance/high-variability materials or components, computer simulation and test pieces to achieve quality, that is, enhance the productivity of R&D work itself (using less time and less cost to realize higher quality). • Effectively use the interactions among design parameters and various usage conditions to make product performance be insensitive against various sources of variability, that is, ensure the robustness (high quality with small variability) of product performance under various unknown usage conditions.	
To get quality, don't measure quality	Evaluate functionality (functional variability) to improve and predict the actual quality in mass production and market usage, that is, make improvements before quality problems occur	• Kill multiple birds (simultaneously improve multiple quality characteristics) with one stone (measure and improve functionality). • Reasonably and efficiently evaluate design ideas/concepts (we should discover the failure of experiment as early as possible to know early the deficiency of technological capability and cope with it as early as possible) • Achieve no trials, no tests, and no inspections.
Two-step optimization	Improve variability first and then hit target value.	• Effectively utilize the interactions among design parameters and various usage conditions to improve functionality. • Escape from Debug cycle (Design–Build–Test–Fix). • Improve variability (robustness) first to obtain the opportunity of cost reduction (that is, under the loosener target specification, we can loosen variability controls to reduce cost). • Regard R&D work as robustness development (RD) of products.

(*Continued*)

TABLE 6.5. (*Continued*)

Strategic Elements		Description
Precedency, universality, and reproducibility	Precedency (Technology readiness)	Technology readiness means that, before product planning, the robustness of technology itself has already been achieved. When acquiring customer specification or determining product specification after product planning, R&D personnel only need to tune the technology to satisfy target values, rather than needing to conduct robust optimization for every product.
	Universality (Application flexibility of technology)	Since the generic technology of a whole product family is the same, the optimization for generic technology only needs to be conducted one time, and it can be applied to the whole product family (products with different specifications or target values), so as to greatly reduce the development time and repeated efforts.
	Reproducibility	The effects of the research conclusions obtained from test pieces, small-scale experiments, and limited conditions can be reproduced in actual products, in large-scale manufacturing, and under various customer usage conditions.

If test results don't meet the requirements, then design changes are conducted; but since R&D personnel fail to clarify which parameter can be used to reduce variability and not to influence sensitivity (or mean value), they get involved in debug cycle. In addition, R&D personnel usually implement improvement with the cost-increasing approaches, such as replacing higher-grade materials or components with higher price and lower variability, to control the variability which is too large to achieve target specification.

One of the important concepts of robust technology development is to improve functionality first and then adjust sensitivity (or mean value) to target value (i.e., two-step optimization); hence R&D personnel's design work stops only when the minimization of variability is completed. To explain this, we need to distinguish the difference between design work and design change. Strictly speaking, design change is the action conducted only when design work is not well done or causes problems. Ideally, a perfect design doesn't need to conduct design change. The achievement of a perfect design means that design work must be able to ensure the variability minimization of technology or product and to hit the target simultaneously; that is, when design work is completed, the variability is surely improved and target value is achieved. However, in practice, the commonly seen situation is that design work stops after the achievement of the target value. After this, the activity to control the variability around the target value is conducted by the approach of "design change." Such approach possibly makes the performance of technology or product deviate from the target value. Therefore, once again (or repeatedly), R&D personnel have to conduct design change to adjust performance to the target value and control the variability around this target value. If the above-mentioned situation occurs repeatedly, R&D personnel would get involved in the debug cycle. In this book, the above-mentioned process is not considered as the should-be and nature of design work. In fact, design change is caused by the procedure of achieving target value first and then trying to control variability. If we seek to conduct design work according to the procedure of two-step optimization (i.e., whatever the target value is, seeking variability minimization first and then adjusting sensitivity or mean value to the target value), we can determine whether the variability is allowed to be increased (i.e., reducing cost) by target specification after minimizing variability. Judging from this, the advantages of two-step optimization not only reduce design changes, but also discover the opportunities of reducing cost. In addition, once we minimize variability, we can assess the feasibility of achieving strict target specification; however, the conventional approach is only able to assess the feasibility through the debug cycle. For the technology development that is assessed to be highly feasible, no matter how large the technological breakthrough R&D personnel obtained in laboratory research, its reproducibility at downstream is the most important. Since one technology's application conditions one product's manufacturing conditions, and one product's usage conditions in the market vary a lot from laboratory, many sources of variability might cause functional variability in technology. Also, if there are significant interactions among design parameters, the effects will make functional performance abnormal. To avoid such problem, we can utilize orthogonal arrays to efficiently confirm the significance of interactions among design parameters, so as to ensure that the technology developed under laboratory conditions can be reproduced at downstream.

After developing technologies, R&D personnel have to manage to enhance technological quality—that is, enhance its robustness. The common method is that R&D personnel find out the actual quality problems by reliability and other technical tests and then utilize the cause-and-effect relationship between design parameters and test conditions to propose the specific technological

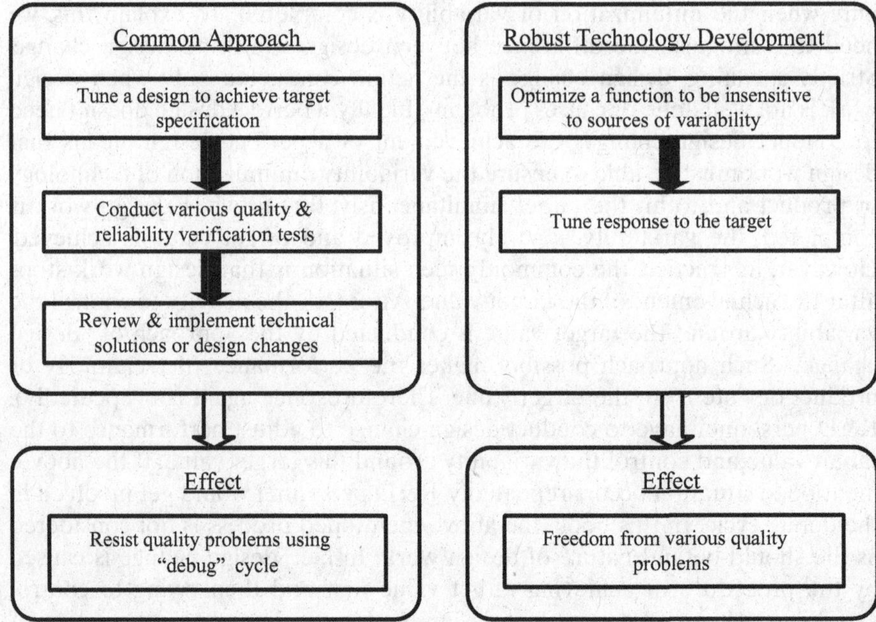

Figure 6.6. Common approach versus robust technology development.

countermeasures to pass the test conditions. Although that design change is effective in passing the tests, something undeniable is that the technological countermeasures can't guarantee whether the technology can pass other unknown conditions; these unknown conditions include the conditions of various sources of variability and various customer usage conditions. In fact, the quality must be built-in by the RE method when R&D personnel first study the technological function; it's very inefficient to conduct modifications according to test results after technology development. Figure 6.6 shows the difference between the common model and robust technology development. Unlike the conventional approach, RE is helpful for R&D personnel to effectively determine, before the occurrence of quality problems, the design concepts that can be robust against various unknown usage conditions (not limited to test conditions); hence they don't need to utilize the cause-and-effect relationship dealing with test conditions to seek specific design changes.

R&D personnel's review on the causes of technological failure varies with the adoption of different design philosophies and methods. Conventionally, R&D personnel's review on the causes of technological failure is usually expressed using the following description: The change of element A creates the deterioration of element B to cause the functional failure of product.

Based on such perspective, R&D personnel will focus on the minimization of quality problems; that is, by means of measuring multiple quality characteristics, identify the characteristic items that need to be improved and find out the root causes for solving problems. However, such improvement approaches usually need a trade-off, because the improvement of one quality characteristic often causes the deterioration of another quality characteristic. Unlike the conventional approach, in robust technology development, the review on the causes of technological failure is usually expressed as follows: The functional failures are caused by the insufficient robustness of product against elements A and B. Based on such perspective, R&D personnel will focus on technological intent (the function that we hope to realize through technology), not problems. R&D personnel avoid the occurrence of various quality problems by maximizing of the technological functionality.

Viewing from the recognition of quality, the quality concept hiding behind the above-mentioned debug cycle and conventional approach is: Quality is achieved by inspections and tests (even though they are precise and advanced). This is because design changes are conducted according to the quality problems occurring in various tests; thus if there are no quality problems, then there are no design changes. Moreover, the quality problem that is the hardest to deal with is the problem related to variability, and the most common design change approach is to change used materials or components (usually, upgrading the grade of materials or components) to reduce variability. The frequent usage of this approach makes quality improvement usually regarded as demanding a cost increment; further more, this logic also implies that R&D personnel fail to deal with the issue of material or component deficiencies by design approach. In the viewpoint of robust technology development, in order to win the competition in the market, R&D personnel have to improve quality under the condition that no cost increment is a pragmatic requirement. Thus, the good design methods must be able to deal with the issue of material or component deficiencies; even if the materials or components provided by the supply sources have the tendency toward high variability and easy deterioration, R&D personnel still have to develop the products with the quality which is higher than that of the products adopting the materials from higher-grade suppliers. The quality concept hiding behind robust technology development is that "quality is realized by design"; we can apply effective design methods to achieve high quality (low functional variability) and high reliability with materials and components that are cheap and of high variability.

Table 6.6 shows the fundamental difference in technology development practice between conventional model and robust technology development. R&D managers can regard that difference as the further understanding of why an R&D paradigm needs change and which model it will become, and they also can perceive the necessity of an R&D paradigm shift based on this.

Robust technology development can be considered as the core of quality engineering in the new century.

TABLE 6.6. Paradigm Shift of R&D

Item	Conventional Model	Robust Technology Development
Technology development	• Seeking to hit design target first and then control variability; thus we need to adapt to different product specifications to conduct tuning of response values and variability control. • Evaluating whether quality and reliability of product would have variations while influenced by certain factor we don't want to control. If there is variation after assessment, then we had better control it (by spending cost), otherwise we neglect it. • Only knowing whether design parameters have influence on the mean value of result, but its influence on variability is known through reliability tests. If test result doesn't meet the requirements, then design changes are conducted. However, since which parameters can be used to reduce variability but not influence mean value cannot be clarified, we get involved in the debug cycle of development. • Once target value is achieved, design work is stopped. • Regarding the control of variability around target specification, it is usually improved by cost-increasing approaches of changing high-grade materials/ components with high price and small variability.	• Improving variability first and then seeking to hit target value; thus, regarding the entire family of products with different product target values or specifications, we only need to conduct the adjustment of response values, not to conduct robust optimization for every product. • Not focusing on evaluating whether quality and reliability of products would have variations while influenced by certain factors we don't want to control, but focusing on how to make product quality and reliability not influenced by that factor through designing—that is, to be robust against that noise. • Grasp the individual effect of parameters on the adjustment of sensitivity or mean value, as well as on the reduction of variability; hence we can quickly and effectively reduce variability or tune response value to the target. • Design work is stopped after completing the minimization of variability; thus we can loosen variability controls according to wider target specification or can assess the feasibility of achieving more strict target specification.

TABLE 6.6. (*Continued*)

Item	Conventional Model	Robust Technology Development
	• Since there might be significant interactions among design parameters, the technology developed under laboratory conditions cannot be assured to have reproducibility at downstream. • May not timely evaluate the possible quality problems caused by material/component changes by suppliers.	• Applying effective design methods to achieve high quality/high reliability with the materials/components which are cheap and with high variability. • The significance of interactions among design parameters can efficiently be confirmed to ensure reproducibility. • If the materials/components changed by suppliers are the control factor levels in an effective experiment, then their effects on sensitivity or mean value and variability can be evaluated by experimental response graphs.
Technological quality improvement	• Only after reliability tests, field tests, or the occurrences of actual quality problems, we can know if there exist technological problems and then use the cause-and-effect relationship between design parameters and test conditions to propose the specific technological countermeasures or design changes dealing with test conditions (but the adaption to other unknown conditions cannot be ensured).	• Before occurring quality problems, the design concept that can be robust against various unknown usage conditions (not limited to test conditions) is effectively evaluated and determined. Therefore, the cause-and-effect relationships dealing with test conditions do not need to be utilized to seek specific design changes. • Achieve no trials, no tests, and no inspections. • The review on the causes of technological failure is usually expressed as: Because the product is not sufficiently robust against elements A and B, etc.

(*Continued*)

TABLE 6.6. (*Continued*)

Item	Conventional Model	Robust Technology Development
	• When the design itself can't be robust against various sources of variability, the caused variability can be regarded as "designed-in." Consequently, we need to conduct various trials, tests, and inspections at downstream stages to identify quality problems and improve them. • The review on the causes of technological failure is usually expressed as: The change of element A creates the deterioration of element B; and further, to cause the functional failure of product. • Focus on the minimization of quality problems; that is, by means of measuring multiple quality characteristics, identify the characteristic items that need to be enhanced and make improvements. The commonly seen situation is that the improvement of one quality characteristic often causes the deterioration of another quality characteristic, hence the trade-off is usually needed.	• Focusing on technological intent, which is the function we wish to realize through technology. We avoid the occurrence of various quality problems by conducting the optimization/ maximization of such functionality.
Quality recognition	• Quality is got by (precise and advanced) inspections and tests. • Quality improvement surely increases cost. This logic implies the issue that design methods fail to deal with the deficiency of materials/ components.	• Quality is obtained by design.

TABLE 6.6. (*Continued*)

Item	Conventional Model	Robust Technology Development
		• In order to win the competition in market, it is a very practical requirement for R&D to enhance quality without increasing cost. This means that good design methods can effectively deal with the issue of material/component deficiency. Even though materials/components from supply sources are with high variability and tend to be easy to deteriorate, R&D personnel still can develop products with higher quality than those using the materials from higher grade suppliers.
Design capability indicator	• Using the downstream lagging indicators as the substitute indicators for measuring quality/engineers' design capability.	• Using upstream leading indicators to effectively assess engineers' design capability and predict quality.

6.5 CASE STUDIES

This section uses several case studies to demonstrate the applicability of RE. These cases cover the development of electrical technology, measurement technology, and machining technology. We illustrate the case studies with the process steps described in Section 6.2.3.

6.5.1 Optimization of a Current–Voltage Conversion Circuit

Current–voltage conversion is a very common function in power circuits; however, improving the efficiency and variability of the conversion is not an easy design task. The issue does not lie in design complexity, but lies in how to realize optimal functionality of current–voltage conversion at the lowest possible cost.

In this case, we apply the RE method to optimize the technology of current–voltage conversion. The so-called optimization indicates that we can utilize the components with the same grade as that of the engineer-proposed design (i.e., initial design developed by conventional approach) to realize better functionality (i.e., functional performance closer to the ideal state) than that of the engineer-proposed design. In other words, this technology can be improved by how the circuit design is. Furthermore, if the functionality realized by the optimal design can achieve a sufficient level, the control mechanism used to compensate the effects of energy loss in engineer-proposed circuit is not needed any more, and consequently the system cost is reduced. This case study demonstrates the power of the RE method in reducing cost. There are two opportunities for cost reduction:

- The first opportunity is that using the same grade components to realize better functionality than that of engineer-proposed circuit design. Such type of cost down should be precisely regarded as cost effective.
- The second opportunity is to remove the control mechanism (which is used to compensate the effects of poor function) from a system when the functionality is perfected by optimal circuit design.

Next, we interpret the case study step by step:

1. Identify the System to Be Optimized. In circuit design, the current-to-voltage conversion is a common means used to achieve a certain functional objective. In this case, R&D engineers identify a circuit in the electrical architecture of a certain product to be the subsystem to be optimized. The reason why engineers chose this circuit is that it is responsible for the most important current–voltage conversion of overall product. Whenever that conversion fails to achieve functional objective at every operation, there will be a functional reliability problem happening to the overall product. Thus, such circuit can be regarded as the design highlight of the overall product architecture.

2. Define Ideal Function and Its Input Signal and Output Response. As described above, the function of this circuit is current-to-voltage conversion. Thus, if we explain it by the terminology of RE, current is input signal and voltage means output response. The conversion relationship of the two can be described as: Output voltage = $\beta \times$ Input current (β denotes slope). Figure 6.7 shows the ideal relationship between input current and output voltage.

3. Identify Sources of Variability and Formulate Experimental Strategy. In the real world, the ideal relationship shown in Figure 6.7 hardly happens. This is because many noise factors might bring functional variability to the circuit. A circuit is composed of many components and printed circuit board (PCB); hence the part-to-part variability of each component and PCB will influence that current-to-voltage conversion. The occurrence of component variability

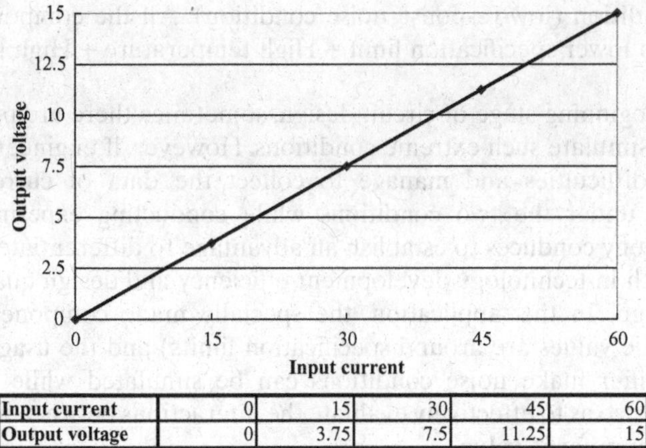

Input current	0	15	30	45	60
Output voltage	0	3.75	7.5	11.25	15

Figure 6.7. Ideal function.

comes not only from supplier's production variability, but also from the influences of environmental conditions such as temperature and humidity on components.

With regard to a component supplier's production variability and the variability of product usage conditions, it's difficult for R&D engineers to eliminate or control their occurrence. Generally speaking, engineers can only count on quality inspection personnel to ensure that the characteristic values of materials provided by suppliers are all within the defined or standard specifications. Also, they might limit customers to specific product usage conditions. Another common design approach is to add feedback mechanisms or compensation mechanisms (which may be hardware or firmware) into product architecture. Such mechanisms are further categorized as feed-forward control and feedback control. The function of them is to activate feedback or compensation to adjust the output performance to a normal state while detecting something abnormal in output performance. Adding feedback or compensation mechanisms means increasing system cost.

In this case, R&D engineers expect to design, with RE method, a circuit that is "insensitive" to component variability and usage condition variability. To reach this objective, R&D engineers need to, based on engineering knowledge and experience, identify main noise conditions and put them into experimental evaluation to find out a robust design through exploring the interaction relationships among different component values, input current, and various noise conditions. The noise factors that mostly influence performance variability in this case are defined as component variability, temperature, and humidity conditions. The two extreme noise conditions are:

- N_1 condition (high-response noise condition): All the component values tend to upper specification limit + Low temperature + Low humidity.

- N_2 condition (low-response noise condition): All the component values tend to lower specification limit + High temperature + High humidity.

At the beginning stage of circuit design, sometimes there are practical difficulties to simulate such extreme conditions. However, if engineers can overcome the difficulties and manage to collect the data of current–voltage conversion under the two conditions while conducting experiments, such design strategy conduces to establish an advantage to differentiate their competitors both in technology development efficiency and design quality of that circuit design. In this application, the specially made components (whose characteristic values are around specification limits) and the usage of heater and humidifier make noise conditions can be simulated while conducting experiments, so as to effectively evaluate the interactions between noise effects and each component value.

4. Define Design Parameters and Their Levels. In this circuit, there are eight components whose component values can be freely selected by R&D engineers. These eight components are design parameters. For each component, if there are three component values that can be freely selected, there are $3^8 = 6561$ parameter combinations in whole design space. The differences in design capability among different R&D engineers are reflected in that which an engineer can, from the design space of so many possible combinations, efficiently identify the optimal design minimizing functional variability.

To efficiently identify the optimal design, R&D engineers have two different strategies:

- Determine the component values of critical components first based on engineering judgment, so as to narrow the design space. Since the influence of the rest of the noncritical components on circuit is limited, it's not difficult for R&D engineers to determine these component values. They wouldn't spend much time even using the trial-and-error approach.
- The second strategy is to simultaneously determine the values of all components on circuit by RE's approaches of experimental planning and data analysis.

The biggest difference between the design logic of the second strategy and conventional approach is that the conventional approach utilizes the cause-and-effect relationship between design parameters and output performance to design the circuit, while the RE method utilizes the interaction relationship among design parameters, input current, and noise conditions to design the circuit. In this case, we apply the second strategy to design that circuit, and we check if the design quality is better as compared with the engineer-proposed design (initial design) developed by adopting the first strategy.

According to the design logic of the second strategy, we list all the design parameters, A–H, on this circuit which can be controlled and determined by

TABLE 6.7. Factors and Levels

	A	B	C	D	E	F	G	H
Level 1	3.3	4.7	51	470	3,000	4,700	68,000	3,600
Level 2	6.2	10	100	680	5,600	10,000	120,000	8,200
Level 3		20	150	1,000	8,200	15,000	220,000	13,000

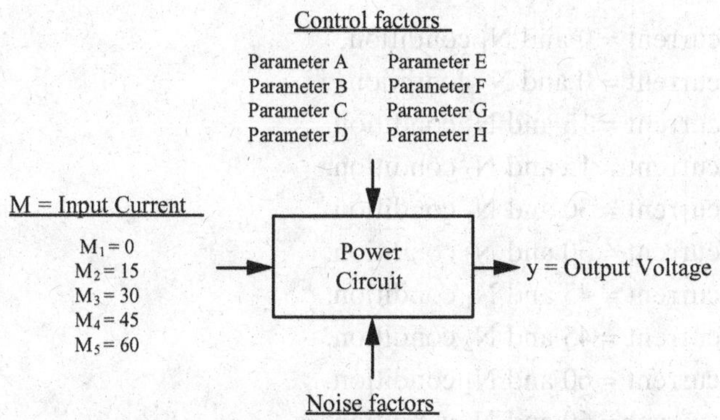

Figure 6.8. P-diagram.

R&D engineers. After determining design parameters, R&D engineers have to determine 2–3 possible values of each component. All the combinations of these component values or parameter levels compose the overall design space, and we expect to use RE method to efficiently find out the optimal combination that can optimize design quality. In this case, while determining the component values expected to be evaluated by an experiment, R&D engineers are based on the following points to choose parameter levels:

- Feasible range within design space
- Common part BOM (bill of materials)
- No more extra cost than initial design

Therefore, R&D engineers define design parameters A–H and their levels as shown in Table 6.7 (parameter A is a two-level component).

5. ***Allocate and Conduct Experiment.*** Based on the above, we can describe the experimental strategy the by P-diagram, as shown in Figure 6.8.

L_{18} orthogonal array (see Appendix A) can be used for this experiment, and the assignment of factors is shown as the inner array in Table 6.8. R&D engineers need to collect the data of output voltage under noise conditions (N_1 and N_2) while changing different input current values (M_1, M_2, M_3, M_4, M_5). The overall experimental layout (inner array and outer array) is shown in Table 6.8.

Take the first run of 18 runs as an example, where R&D engineers insert various components on PCB according to factor-level combinations first and then collect the data of output voltage under 10 different conditions.

- Input current = 0 and N_1 condition.
- Input current = 0 and N_2 condition.
- Input current = 15 and N_1 condition.
- Input current = 15 and N_2 condition.
- Input current = 30 and N_1 condition.
- Input current = 30 and N_2 condition.
- Input current = 45 and N_1 condition.
- Input current = 45 and N_2 condition.
- Input current = 60 and N_1 condition.
- Input current = 60 and N_2 condition.

Table 6.9 shows all the collected data of 18 runs.

6. Analyze Data and Conduct Two-Step Optimization and Confirmation Experiment. Take the first and second runs of 18 runs as examples, we use the equations explained in Section 5.4.2 to calculate SN ratio and β values. The calculation of the first run is as follows:

$$S_T = (0.744)^2 + (0.636)^2 + (3.861)^2 + \cdots + (18.992)^2 + (19.792)^2 = 1274.890$$

$$r = 0^2 + 15^2 + 30^2 + 45^2 + 60^2 = 6750$$

$$L_1 = (0.744 \times 0) + (3.861 \times 15) + (8.584 \times 30) + (14.224 \times 45) + (18.992 \times 60) = 2095.052$$

$$L_2 = (0.636 \times 0) + (3.237 \times 15) + (8.022 \times 30) + (12.480 \times 45) + (19.792 \times 60) = 2038.347$$

r_0 is the number of linear contrast for noise conditions, hence it is 2 in this case.

$$S_\beta = \frac{(2095.052 + 2038.347)^2}{6750 \times 2} = 1265.555$$

$$S_{\beta \times N} = \frac{(2095.052)^2 + (2038.347)^2}{6750} - 1265.555 = 0.238$$

TABLE 6.8. Experimental Layout

									M1		M2		M3		M4		M5	
									0		15		30		45		60	
	A	B	C	D	E	F	G	H	N1	N2	N1	N2	N1	N2	N1	N2	N1	N2
1.	3.3	4.7	51	470	3,000	4,700	68,000	3,600										
2.	3.3	4.7	100	680	5,600	10,000	120,000	8,200										
3.	3.3	4.7	150	1,000	8,200	15,000	220,000	13,000										
4.	3.3	10	51	470	5,600	10,000	220,000	13,000										
5.	3.3	10	100	680	8,200	15,000	68,000	3,600										
6.	3.3	10	150	1,000	3,000	4,700	120,000	8,200										
7.	3.3	20	51	680	3,000	15,000	120,000	13,000										
8.	3.3	20	100	1,000	5,600	4,700	220,000	3,600										
9.	3.3	20	150	470	8,200	10,000	68,000	8,200										
10.	6.2	4.7	51	1,000	8,200	10,000	120,000	3,600										
11.	6.2	4.7	100	470	3,000	15,000	220,000	8,200										
12.	6.2	4.7	150	680	5,600	4,700	68,000	13,000										
13.	6.2	10	51	680	8,200	4,700	220,000	8,200										
14.	6.2	10	100	1,000	3,000	10,000	68,000	13,000										
15.	6.2	10	150	470	5,600	15,000	120,000	3,600										
16.	6.2	20	51	1,000	5,600	15,000	68,000	8,200										
17.	6.2	20	100	470	8,200	4,700	120,000	13,000										
18.	6.2	20	150	680	3,000	10,000	220,000	3,600										

TABLE 6.9. Data Collection

									M1		M2		M3		M4		M5	
									0		15		30		45		60	
	A	B	C	D	E	F	G	H	N1	N2	N1	N2	N1	N2	N1	N2	N1	N2
1.	3.3	4.7	51	470	3,000	4,700	68,000	3,600	0.744	0.636	3.861	3.237	8.584	8.022	14.224	12.480	18.992	19.792
2.	3.3	4.7	100	680	5,600	10,000	120,000	8,200	0.840	0.579	2.665	2.450	6.055	5.636	9.939	9.293	15.264	13.399
3.	3.3	4.7	150	1,000	8,200	15,000	220,000	13,000	0.744	0.636	3.861	3.237	8.584	8.022	14.224	12.480	18.992	19.792
4.	3.3	10	51	470	5,600	10,000	220,000	13,000	1.403	1.068	2.477	2.231	5.011	4.583	7.840	7.220	10.985	10.156
5.	3.3	10	100	680	8,200	15,000	68,000	3,600	0.352	0.206	3.039	2.773	7.298	6.755	12.222	11.371	17.765	16.569
6.	3.3	10	150	1,000	3,000	4,700	120,000	8,200	1.813	1.652	3.258	3.058	6.748	6.253	10.215	9.829	14.547	13.810
7.	3.3	20	51	680	3,000	15,000	120,000	13,000	1.096	0.953	1.628	1.491	3.146	2.925	4.821	4.513	6.673	6.273
8.	3.3	20	100	1,000	5,600	4,700	220,000	3,600	2.143	1.965	2.299	2.127	3.945	3.663	5.955	5.285	7.564	7.051
9.	3.3	20	150	470	8,200	10,000	68,000	8,200	0.983	0.660	9.813	9.050	23.691	22.058	39.752	37.134	57.832	54.113
10.	6.2	4.7	51	1,000	8,200	10,000	120,000	3,600	0.484	0.388	1.018	0.910	2.138	1.949	3.399	3.124	4.806	4.436
11.	6.2	4.7	100	470	3,000	15,000	220,000	8,200	0.603	0.489	2.149	1.967	4.862	4.505	7.968	7.416	11.451	10.683
12.	6.2	4.7	150	680	5,600	4,700	68,000	13,000	1.021	0.834	5.302	4.961	12.388	11.693	20.546	19.455	29.712	28.181
13.	6.2	10	51	680	8,200	4,700	220,000	8,200	2.238	2.016	2.454	2.247	4.253	3.919	6.156	5.696	8.384	7.634
14.	6.2	10	100	1,000	3,000	10,000	68,000	13,000	0.658	0.532	2.520	2.359	5.880	5.448	9.440	8.995	13.586	12.979
15.	6.2	10	150	470	5,600	15,000	120,000	3,600	0.454	0.322	3.810	3.477	9.138	8.417	15.298	14.137	22.229	20.577
16.	6.2	20	51	1,000	5,600	15,000	68,000	8,200	0.622	0.481	1.561	1.413	3.377	3.127	5.437	5.081	7.741	7.268
17.	6.2	20	100	470	8,200	4,700	120,000	13,000	3.241	2.847	7.249	6.674	15.389	14.283	24.586	21.375	34.853	32.515
18.	6.2	20	150	680	3,000	10,000	220,000	3,600	1.054	0.949	1.886	1.746	3.826	3.564	4.815	5.599	8.407	7.864

$$S_e = 1274.890 - (1265.555 + 0.238) = 9.097$$

$$V_e = \frac{9.097}{10-2} = 1.137$$

$$V_N = \frac{0.238 + 9.097}{10-1} = 1.037$$

Thus,

$$\beta = \frac{2095.052 + 2038.347}{6750 \times 2} = 0.306$$

and

$$\text{SN} = 10 \log \left[\frac{\frac{1}{6750 \times 2}(1265.555 - 1.137)}{1.037} \right] = -10.443$$

The calculation of the second run is as follows:

$$S_T = (0.840)^2 + (0.579)^2 + (2.665)^2 + \cdots + (15.264)^2 + (13.399)^2 = 680.244$$

$$r = 0^2 + 15^2 + 30^2 + 45^2 + 60^2 = 6750$$

$$L_1 = (0.840 \times 0) + (2.665 \times 15) + (6.055 \times 30) + (9.939 \times 45) + (15.264 \times 60)$$
$$= 1584.742$$

$$L_2 = (0.579 \times 0) + (2.450 \times 15) + (5.636 \times 30) + (9.293 \times 45) + (13.399 \times 60)$$
$$= 1427.945$$

r_0 is number of linear contrast for noise conditions. It is 2 in this case.

$$S_\beta = \frac{(1584.742 + 1427.945)^2}{6750 \times 2} = 672.317$$

$$S_{\beta \times N} = \frac{(1584.742)^2 + (1427.945)^2}{6750} - 672.317 = 1.821$$

$$S_e = 680.244 - (672.317 + 1.821) = 6.105$$

$$V_e = \frac{6.105}{10-2} = 0.763$$

$$V_N = \frac{1.821 + 6.105}{10-1} = 0.881$$

Thus,

$$\beta = \frac{1584.742 + 1427.945}{6750 \times 2} = 0.223$$

and

$$SN = 10\log\left[\frac{\frac{1}{6750 \times 2}(672.317 - 0.763)}{0.881}\right] = -12.481$$

Using this calculation method, we can yield the individual β and SN ratio of 18 runs, as shown in Table 6.10.

We do not choose the parameter combination with the best data result in Table 6.10 to be the optimal condition. The analyzed β's and SN ratios shown in Table 6.10 are the raw data used to further analyze the factorial effects.

How do we analyze these data of the SN ratio and β? We take advantage of the feature of "data are obtained by a balanced experimental layout with orthogonality" to assess the effects of each parameter. Take Factor A as an example:

- The effect of level 1 of Factor A on the SN ratio is assessed as the average value of the SN ratios corresponding to that level.
- The effect of level 2 of Factor A on the SN ratio is assessed as the average value of the SN ratios corresponding to that level.

We use Figure 6.9 to more clearly illustrate this concept.

We apply the same approach to calculate the effects of Factor B on the SN ratio, as shown in Figure 6.10:

- The effect of level 1 of Factor B on the SN ratio is assessed as the average value of the SN ratios corresponding to that level.
- The effect of level 2 of Factor B on the SN ratio is assessed as the average value of the SN ratios corresponding to that level.
- The effect of level 3 of Factor B on the SN ratio is assessed as the average value of the SN ratios corresponding to that level.

Applying the same calculation method, we can yield the effects of each factor on the SN ratio; and we can know which factors' influences on SN ratio are more significant and which factors' influences on SN ratio are insignificant. Take Factor B for instance; the effects of its three levels on SN ratio are respectively −10.972, −12.395, and −13.258. The range of them is −10.972 (maximum) minus −13.258 (minimum) and equals 2.286. After we yield each

TABLE 6.10. Data Analysis

									M1		M2		M3		M4		M5			
									0		15		30		45		60			
	A	B	C	D	E	F	G	H	N1	N2	N1	N2	N1	N2	N1	N2	N1	N2	β	S/N
1.	3.3	4.7	51	470	3,000	4,700	68,000	3,600	0.744	0.636	3.861	3.237	8.584	8.022	14.224	12.480	18.992	19.792	0.306	–10.443
2.	3.3	4.7	100	680	5,600	10,000	120,000	8,200	0.840	0.579	2.665	2.450	6.055	5.636	9.939	9.293	15.264	13.399	0.223	–12.481
3.	3.3	4.7	150	1,000	8,200	15,000	220,000	13,000	0.744	0.636	3.861	3.237	8.584	8.022	14.224	12.480	18.992	19.792	0.306	–10.443
4.	3.3	10	51	470	5,600	10,000	220,000	13,000	1.403	1.068	2.477	2.231	5.011	4.583	7.840	7.220	10.985	10.156	0.171	–12.178
5.	3.3	10	100	680	8,200	15,000	68,000	3,600	0.352	0.206	3.039	2.773	7.298	6.755	12.222	11.371	17.765	16.569	0.269	–11.138
6.	3.3	10	150	1,000	3,000	4,700	120,000	8,200	1.813	1.652	3.258	3.058	6.748	6.253	10.215	9.829	14.547	13.810	0.229	–12.013
7.	3.3	20	51	680	3,000	15,000	120,000	13,000	1.096	0.953	1.628	1.491	3.146	2.925	4.821	4.513	6.673	6.273	0.106	–13.721
8.	3.3	20	100	1,000	5,600	4,700	220,000	3,600	2.143	1.965	2.299	2.127	3.945	3.663	5.955	5.285	7.564	7.051	0.124	–18.214
9.	3.3	20	150	470	8,200	10,000	68,000	8,200	0.983	0.660	9.813	9.050	23.691	22.058	39.752	37.134	57.832	54.113	0.876	–11.114
10.	6.2	4.7	51	1,000	8,200	10,000	120,000	3,600	0.484	0.388	1.018	0.910	2.138	1.949	3.399	3.124	4.806	4.436	0.074	–11.491
11.	6.2	4.7	100	470	3,000	15,000	220,000	8,200	0.603	0.489	2.149	1.967	4.862	4.505	7.968	7.416	11.451	10.683	0.175	–10.544
12.	6.2	4.7	150	680	5,600	4,700	68,000	13,000	1.021	0.834	5.302	4.961	12.388	11.693	20.546	19.455	29.712	28.181	0.456	–10.427
13.	6.2	10	51	680	8,200	4,700	220,000	8,200	2.238	2.016	2.454	2.247	4.253	3.919	6.156	5.696	8.384	7.634	0.134	–17.844
14.	6.2	10	100	1,000	3,000	10,000	68,000	13,000	0.658	0.532	2.520	2.359	5.880	5.448	9.440	8.995	13.586	12.979	0.210	–9.991
15.	6.2	10	150	470	5,600	15,000	120,000	3,600	0.454	0.322	3.810	3.477	9.138	8.417	15.298	14.137	22.229	20.577	0.335	–11.208
16.	6.2	20	51	1,000	5,600	15,000	68,000	8,200	0.622	0.481	1.561	1.413	3.377	3.127	5.437	5.081	7.741	7.268	0.120	–10.661
17.	6.2	20	100	470	8,200	4,700	120,000	13,000	3.241	2.847	7.249	6.674	15.389	14.283	24.586	21.375	34.853	32.515	0.534	–11.919
18.	6.2	20	150	680	3,000	10,000	220,000	3,600	1.054	0.949	1.886	1.746	3.826	3.564	4.815	5.599	8.407	7.864	0.127	–13.918

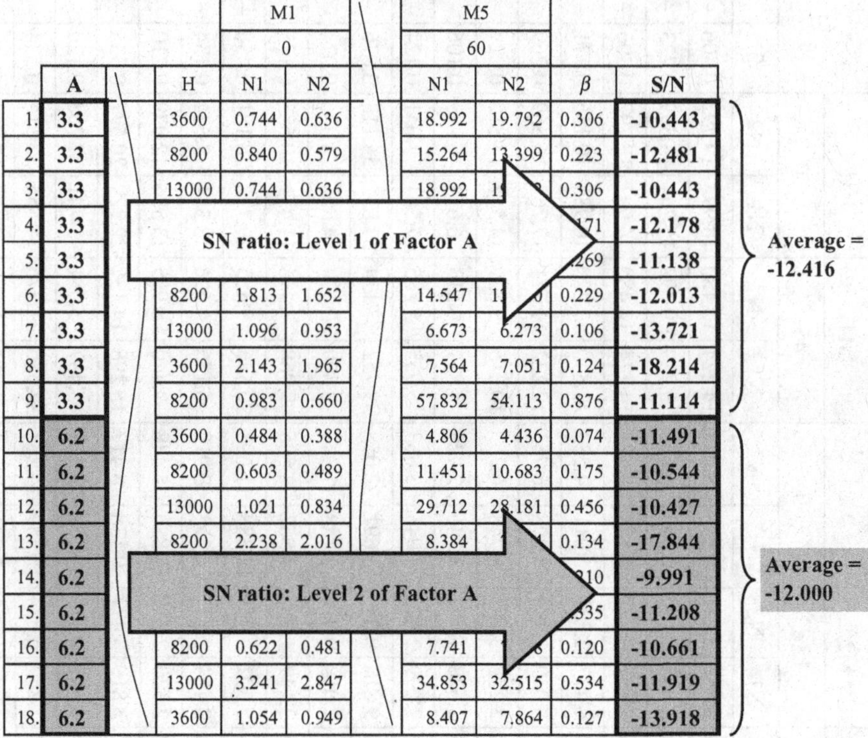

Figure 6.9. Effects of Factor A on the SN ratio.

factor's range value with the same calculation approach, we can know what the rank of 2.286 is. The higher the rank of some factor, the larger the range value of that factor—that is, the more significant its influence on the SN ratio. On the contrary, if the rank of some factor is the lowest, the influence of that factor on the SN ratio is regarded as the smallest among all the factors.

We use Table 6.11, which is called the *response table*, to illustrate the information described above. The table shows that the influence of Factor G on the SN ratio is the largest one and that the influence of Factor A on the SN ratio is the smallest.

The approach of calculating factorial effects on β is the same as that of calculating factorial effects on the SN ratio. Take Factor A as an example:

- The effect of level 1 of Factor A on β is assessed as the average value of the β values corresponding to that level.
- The effect of level 2 of Factor A on β is assessed as the average value of the β values corresponding to that level.

Figure 6.11 illustrates this concept.

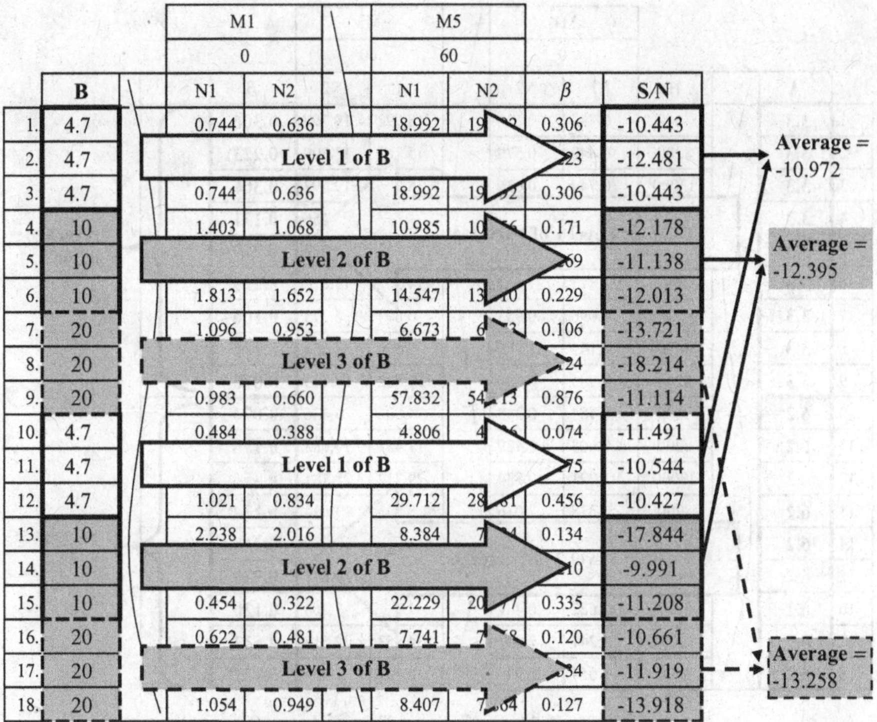

Figure 6.10. Effects of Factor B on the SN ratio.

TABLE 6.11. Response Table for the SN Ratio

	A	B	C	D	E	F	G	H
Level 1	−12.416	−10.972	−12.723	−11.234	−11.772	−13.477	−10.629	−12.735
Level 2	−12.000	−12.395	−12.381	−13.255	−12.528	−11.862	−12.139	−12.443
Level 3		−13.258	−11.521	−12.136	−12.325	−11.286	−13.857	−11.447
Range	0.416	2.286	1.202	2.020	0.757	2.191	3.228	1.289
Rank	8	2	6	4	7	3	1	5

Using the same calculation method, we can obtain the effects of each factor on β; and we can know which factors' influences on β are more significant and which factors' influences on β are insignificant. Take Factor A for instance; the effects of its two levels on β are respectively 0.290 and 0.241. The range of the two is 0.290 minus 0.241 and equals 0.049. According to the range value of

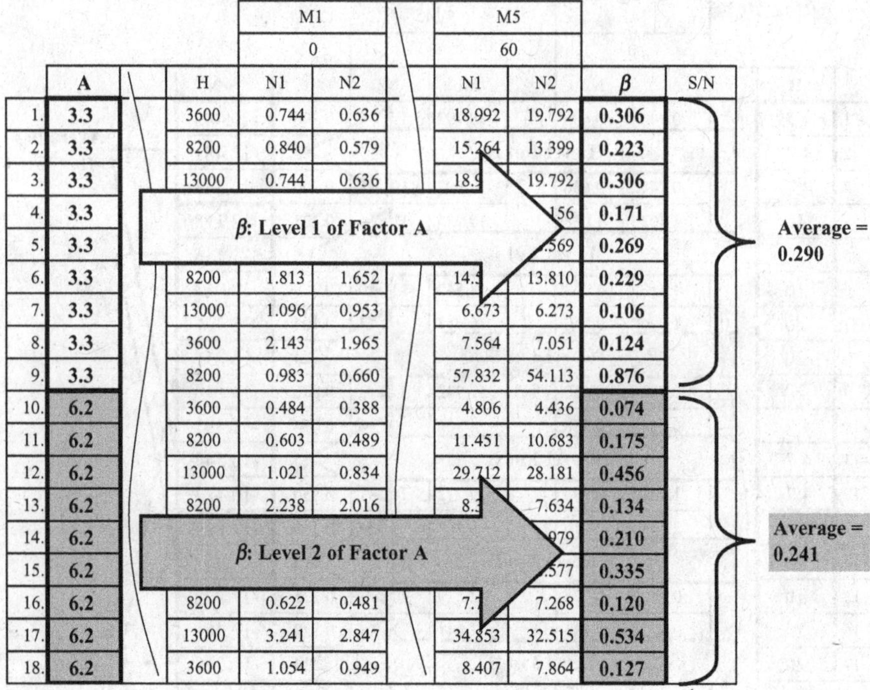

Figure 6.11. Effects of Factor A on β.

TABLE 6.12. Response Table for β

	A	B	C	D	E	F	G	H
Level 1	0.290	0.257	0.152	0.400	0.192	0.297	0.373	0.206
Level 2	0.241	0.225	0.256	0.219	0.238	0.280	0.250	0.293
Level 3		0.315	0.388	0.177	0.366	0.218	0.173	0.297
Range	0.049	0.090	0.237	0.223	0.173	0.079	0.200	0.091
Rank	8	6	1	2	4	7	3	5

each factor, Table 6.12 (response table) shows the ranking of the effects of each factor on β: The effect of Factor C on β is the largest one whereas the effect of Factor A on β is the smallest.

According to the response tables for the SN ratio and β, we can draw the so-called "response graphs" for the SN ratio and β, as shown in Figures 6.12 and 6.13.

When we can obtain the effects of each factor by means of the above-mentioned analysis, we can, according to the factorial effects on variability and

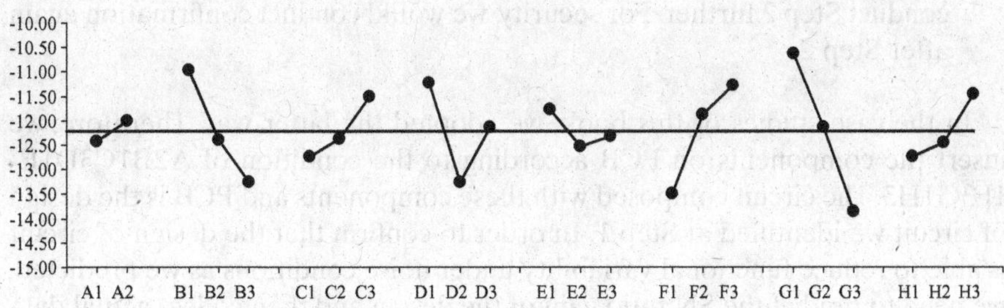

Figure 6.12. Response graph for the SN ratio.

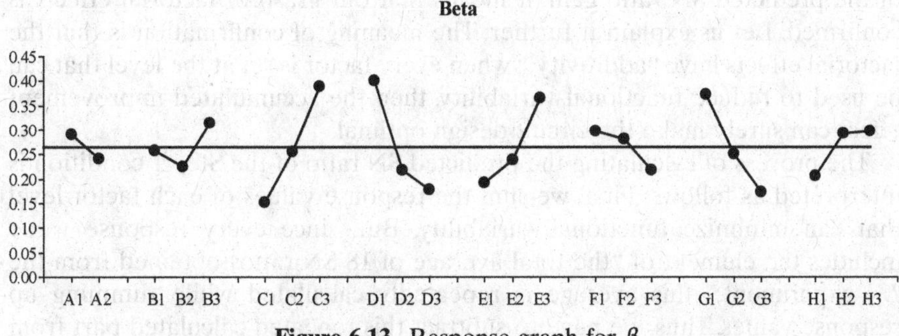

Figure 6.13. Response graph for β.

slope, identify the optimal condition. This process of identifying the optimum can be divided into two steps; hence, it is called *two-step optimization*:

- Step 1: Choose the level with the largest SN ratio to reduce the functional variability under noise conditions.
- Step 2: Choose the factors that influence only β but not (or less) the SN ratio as adjustment factors to fine-tune the slope to be placed on the target.

In this case, while conducting Step 1, we choose A2B1C3D1E1F3G1H3 (A = 6.2, B = 4.7, C = 150, D = 470, E = 3000, F = 15,000, G = 68,000, and H = 13,000) as the factor-level combination that can minimize functional variability because these levels have the largest SN ratios. As for the activities following Step 1, we have two alternatives:

- Conduct Step 2 and then implement the confirmation experiment to confirm if the optimal condition we identified has reproducibility under noise conditions.
- Implement confirmation experiment first and then confirm if the factor-level combination we identified at Step 1 is able to reduce functional

variability under noise conditions as we predicted. Once confirmed, we conduct Step 2 further. For security, we would conduct confirmation again after Step 2.

In the case studies of this book, we adopted the latter way. Therefore, we insert the components on PCB according to the condition of A2B1C3D1E-1F3G1H3. The circuit composed with these components and PCB is the design of circuit we identified at Step 1. In order to confirm that the design of circuit is able to reduce functional variability under noise conditions as we predicted, we need to predict the SN ratio gain of the design and then collect actual data of output voltage under various conditions of input current and noises. If the actual SN ratio gain calculated from this actual data is within the ±30% range of the predicted SN ratio gain, it means that our grasp of factorial effects is confirmed. Let us explain it further: The meaning of confirmation is that the factorial effects have "additivity"; when every factor is set at the level that can be used to reduce functional variability, then the accumulated improvement effect can surely make the circuit design optimal.

The process of calculating the predicted SN ratio of the Step 1 condition is interpreted as follows: First, we sum the response values of each factor level that can minimize functional variability. But, since every response value includes the element of "the total average of 18 SN ratios obtained from the L_{18} experiment," this average is repeatedly calculated while summing up response values. Thus, we have to subtract this repeated calculated part from the sum of response values, and we have the predicted value of the SN ratio. In this case, because there are eight factors, the above-mentioned average (which is −12.208) is repeatedly calculated seven times. To sum up, the predicted SN ratio is calculated as follows:

$$\begin{bmatrix}(-12.000)+(-10.972)+(-11.521)+(-11.234)+(-11.772)+\\(-11.286)+(-10.629)+(-11.447)\end{bmatrix}$$
$$-[(8-1)\times(-12.208)]=-5.402$$

Using the same calculation approach, the predicted SN ratio of initial design is −13.546. Hence, the predicted SN ratio gain is 8.144.

Table 6.13 shows the factor-level condition and collected data in confirmation experiment.

TABLE 6.13. Collected Data in the Confirmation Experiment

								M1		M2		M3		M4		M5	
								0		15		30		45		60	
A	B	C	D	E	F	G	H	N1	N2	N1	N2	N1	N2	N1	N2	N1	N2
6.2	4.7	150	470	3,000	15,000	68,000	13,000	0.026	0.012	1.369	1.256	2.578	2.327	3.972	3.578	5.034	4.623

According to the actual data, we calculate the SN ratio of the Step 1 condition as follows:

$$S_T = (0.026)^2 + (0.012)^2 + \cdots + (5.034)^2 + (4.623)^2 = 90.806$$

$$r = 0^2 + 15^2 + 30^2 + 45^2 + 60^2 = 6750$$

$$L_1 = (0.026\times 0) + (1.369\times 15) + (2.578\times 30) + (3.972\times 45) + (5.034\times 60) = 578.655$$

$$L_2 = (0.012\times 0) + (1.256\times 15) + (2.327\times 30) + (3.578\times 45) + (4.623\times 60) = 527.040$$

r_0 is 2 in this case.

$$S_\beta = \frac{(578.655 + 527.040)^2}{6750\times 2} = 90.560$$

$$S_{\beta\times N} = \frac{(578.655)^2 + (527.040)^2}{6750} - 90.560 = 0.197$$

$$S_e = 90.806 - (90.560 + 0.197) = 0.048$$

$$V_e = \frac{0.048}{10-2} = 0.006$$

$$V_N = \frac{0.197 + 0.048}{10-1} = 0.027$$

Thus,

$$\text{SN} = 10\log\left[\frac{\frac{1}{6750\times 2}(90.560 - 0.006)}{0.027}\right] = -6.094$$

Since we have the actual data (as shown in Table 6.14) of initial design (engineer-proposed design: A2B2C2D2E2F2G2H2), we can compute the

TABLE 6.14. Data Collection for Initial Versus Optimal Design

	M1		M2		M3		M4		M5	
	0		15		30		45		60	
	N1	N2	N1	N2	N1	N2	N1	N2	N1	N2
Initial design	0.788	0.672	2.867	2.461	6.044	5.739	9.904	9.017	15.682	12.449
Optimal design	0.014	0.001	4.114	3.676	8.002	7.435	12.435	10.948	16.147	14.639

actual SN ratio and obtain –13.825. Therefore, the actual SN ratio gain is 7.731 (–6.094 minus –13.825). The additivity of factorial effects is confirmed because the difference between 7.731 and 8.144 (which is the predicted gain mentioned above) is within ±30% of 8.144. However, the difference in sensitivity (slope) between that performance and the ideal relationship of input current and output voltage shown in Figure 6.7 is quite big. We need to adjust that slope to make it achieve the expected sensitivity and not to cause the change in the SN ratio; this is the focus of Step 2.

We can use the data in Table 6.13 to calculate β as follows:

$$\beta = \frac{578.655 + 527.040}{6750 \times 2} = 0.082$$

Since the slope of the ideal relationship between input current and output voltage is 0.25, we can know that actual β is much smaller than expected β; hence we need to adjust β to be 0.25 without influencing functional variability. Judging from Figures 6.12 and 6.13, Factor E can be used as an adjustment factor because its influence on the SN ratio is insignificant, but it is significant on β. At Step 1, we set it as E1 (E = 3000); while at Step 2, in order to make β increase from 0.082 to 0.25 (increase 0.168 in sensitivity), we change Factor E from E1 to E3 (E = 8200). This is because, judging from Table 6.12, the range of the effects of Factor E is 0.173, which is very close to 0.618 that we need. Also, the influence of the change from E1 to E3 on the SN ratio is estimated to be 0.55 obtained from –11.772 minus –12.325 (which can be estimated from Table 6.11) and it's very insignificant.

According to the process of two-step optimization described above, the optimal factor-level combination is finalized as A2B1C3D1E3F3G1H3 (A = 6.2, B = 4.7, C = 150, D = 470, E = 8200, F = 15,000, G = 68,000 and H = 13,000). One point worth mentioning is that the parameter combination obtained through two-step optimization might not be found in the 18 runs in Table 6.8; instead it might be a parameter combination that has never been tried in the L_{18} experiment. In order to confirm that the optimal condition can simultaneously improve functional variability and slope, we need to collect data again and compare them with the initial design (engineer-proposed design: A2B2C2D2E2F2G2H2).

Table 6.14 shows the data of output voltage under different input currents and noise conditions.

Based on the equations of calculating the SN ratio and β, we can get these values of initial design and optimal design respectively:

- *Initial Design:* SN ratio = –13.825 and β = 0.220.
- *Optimal Design:* SN ratio = –6.264 and β = 0.258.

The unit used by the SN ratio is the decibel (dB). The approach of interpreting the SN ratio is that whenever the SN ratio is improved by 6 dB, the

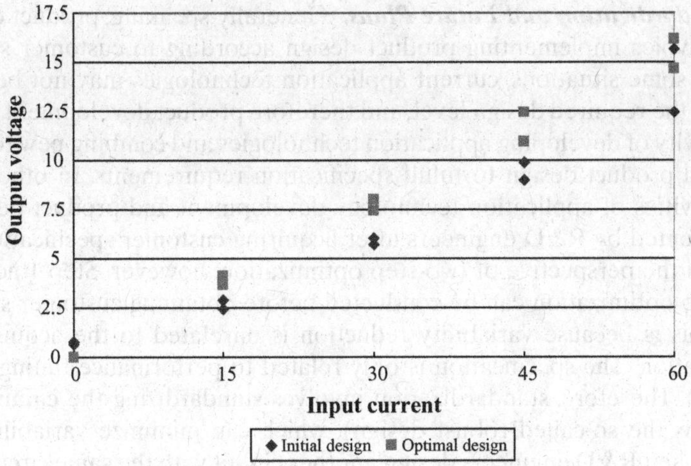

Figure 6.14. Initial design versus optimal design.

functional variability is reduced by 50%. In this case, the SN ratio is improved by 7.561 dB; hence we can estimate that, under different input currents and noise conditions, the variability of output voltage is reduced by 58.25%. In the aspect of sensitivity, the slope of the relationship between input current and output voltage is closer to ideal slope. We can see such difference in Figure 6.14.

Although, in terms of design quality, an optimal design is better than an engineer-proposed design (initial design) by more than 50%, the engineer-proposed design can still satisfy the customer specification; that is, that design may realize zero defect, which means that the functional variability of initial design would not exceed the customer specification with relatively looser range. This highlights a phenomenon: When R&D engineers regard customer specification as a basis to judge if a design is good or not, then as soon as they put the design performance into the specification, their design work is finished. R&D engineers will conduct design changes if problems are found when verifying more samples. According to such design logic, the optimal condition in this case is very hard to find. Because, at Step 1 of two-step optimization, the sensitivity of the factor-level combination we identified deviates the expected slope too much, R&D engineers will try to adjust the slope to be close to the target by design changes and then conduct the fine-tuning around the target. However, in this way, functional variability varies with the design changes found by the trial-and-error approach and becomes difficult to be predicted; even the problem of a large variability can't be found until implementing downstream verification tests. At this time, much time and quality cost have been spent. In view of this, two-step optimization is very powerful to the realization of concurrently improving both variability and sensitivity.

7. Standardization and Future Plans. Generally speaking, product development involes implementing product design according to customer specification. In some situations, current application technologies may not be able to achieve the required design level, and therefore product development includes the activity of developing application technologies and combing new technologies and product design to fulfill specification requirements. In other words, the activities of application technology development and product design are implemented by R&D engineers after acquiring customer specification.

From the perspective of two-step optimization, however, Step 1 activity of two-step optimization can be conducted before obtaining customer specification. This is because variability reduction is unrelated to the acquisition of specification. The specification is only related to performance tuning (Step 2 activity). Therefore, standardization involves standardizing the circuit design (which is the so-called robust design), which can minimize variability. As a result, when R&D engineers design another circuit with the same circuit architecture or product family, they don't need to do such an experiment again, but just implement necessary tuning or adjustment according to different specifications. This is the power of two-step optimization. The robust design in this case is the factor-level combination, A2B1C3D1E1F3G1H3 (A = 6.2, B = 4.7, C = 150, D = 470, E = 3000, F = 15,000, G = 68,000, and H = 13,000).

Regarding the future plan, if, in the future, the trend of input current under customer's usage conditions will be larger and larger or smaller and smaller, the levels of the signal factor can be modified to be those input values of next-generation technology. Then we can conduct robust optimization of the new system.

6.5.2 Robust Engineering of a Voltage Adjustment Component

In order to ensure the performance stability of an electrical product, component design may be a key issue. In the condition of this case, R&D engineers design the critical components by themselves, rather than purchase them from component suppliers for fear that the quality of the critical components cannot be assured.

There are 10 parameters in the design of the component discussed in this case study. Most parameters are related to component architecture because this architecture is very complicated. While determining parameter levels, there is a parameter with as many as 12 levels. It means we need to modify the orthogonal array used by this experiment so that we can, under the condition of not increasing the size of the orthogonal array, explore the design parameter with so many levels. Thereinafter, we explain the case study and the technique of orthogonal array modification step-by-step:

1. Identify the System to Be Optimized. In this case, *system* means a mechanism composed of a circuit, several components, and coils. This mechanism can make an electrical device operate its function of voltage adjustment.

2. Define Ideal Function and Its Input Signal and Output Response. To further interpret the above, the function of this hardware mechanism is to adjust the voltage to the value we expected through the turns ratio (which is the ratio of the number of turns in one side of the mechanism to the number of turns in the other side) among coils. R&D engineers determine the number of coils on different positions of the hardware based on the specification requirements first; when the number of coils on different positions is determined, the turns ratio among coils can be obtained by calculation.

In this case, the hardware mechanism has five types of turns ratio; and therefore when users supply input energy, the mechanism generates five types of output voltage simultaneously. In other words, the above-mentioned technical means of voltage adjustment is the turns ratio. In an ideal situation, a good mechanism design can make input energy be precisely transformed to the expected voltage value by the corresponding turns ratio, as shown in Figure 6.15.

When R&D engineers implement a mechanism design based on their own engineering experience, the design capability of different R&D engineers will be reflected in the differences of functional variability between turns ratio and output voltage. For example, in Figure 6.16, the design capability of Engineer A is better than that of Engineer B because the design developed by Engineer A can make the relationship between turns ratio and output voltage closer to the ideal function; that is, the design quality is superior.

3. Identify Sources of Variability and Formulate Experimental Strategy. Under the environmental conditions of a laboratory, it might be very easy for R&D engineers to develop a design with small functional variability. However, the challenge they are facing is how to develop a design that can perform its

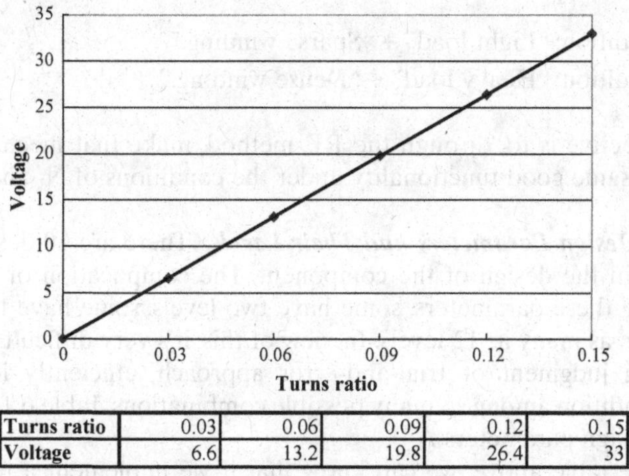

Turns ratio	0.03	0.06	0.09	0.12	0.15
Voltage	6.6	13.2	19.8	26.4	33

Figure 6.15. Ideal function.

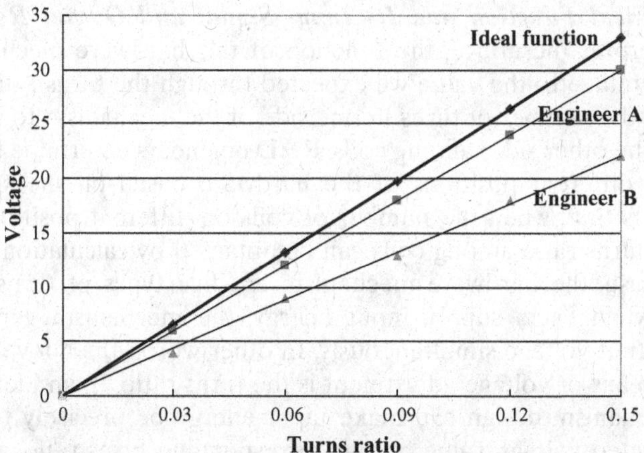

Figure 6.16. Capability of designing for functionality.

functionality under various unknown conditions—that is, as close to ideal function as possible. R&D engineers always spend much time and cost to solve the problems related to variability.

In view of this, we have to, by means of a systematic experiment, allow the effects of various sources of variability to be evaluated; then we can easily identify a design that is the most insensitive to the sources of variability. In this case, R&D engineers define two extreme conditions as the noise conditions. Under the two conditions, the variability of the output voltage transformed from the same turns ratio is the largest. If the design developed by us can reduce the functional variability under these two conditions, the users will enjoy excellent functionality under other application conditions. The noise conditions are:

- N_1 condition: "Light load" + "Sparse winding."
- N_2 condition: "Heavy load" + "Dense winding."

Our objective is to, through the RE method, make that mechanism have almost the same good functionality under the conditions of N_1 and N_2.

4. Define Design Parameters and Their Levels. There are 10 design parameters, A–J, in the design of the component. The complication of this case is that, among these parameters, some have two levels, some have three levels, and one has as many as 12 levels. In view of this, it's very difficult to, through engineering judgment or trial-and-error approach, efficiently identify the optimal condition among so many possible combinations. Table 6.15 shows the levels of design parameters A–J.

From the table above, we can know that if we implement a full factorial experiment to search the optimal condition, there are $12^1 \times 3^8 \times 2^1 = 157{,}464$

TABLE 6.15. Factors and Levels

	A	B	C	D	E	F	G	H	I	J
Level 1	X1 on left side / Y2 on left side	Large	10	Dense	Thick	Thick	Dense	Left	Left	Left
Level 2	X1 on left side / Y2 on right side	Medium	20	Sparse	Medium	Medium	Medium	Center	Center	Center
Level 3	X1 on right side / Y2 on left side	Small	30		Thin	Thin	Sparse	Right	Right	Right
Level 4	X1 on right side / Y2 on right side									
Level 5	Y2 on left side / X1 on left side									
Level 6	Y2 on left side / X1 on right side									
Level 7	Y2 on right side / X1 on left side									
Level 8	Y2 on right side / X1 on right side									
Level 9	(X1 + Y2) on left side									
Level 10	(X1 + Y2) on right side									
Level 11	(Y2 + X1) on left side									
Level 12	(Y2 + X1) on right side									

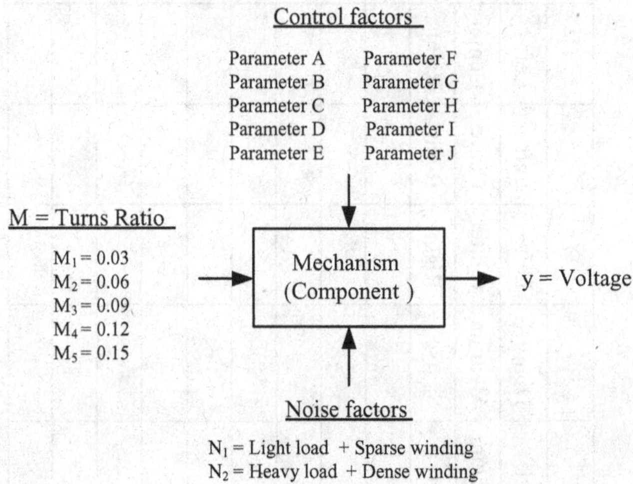

Figure 6.17. P-diagram.

combinations in whole design space. We expect, through the RE method, to substantially improve the efficiency of finding out the optimal condition: The number of total trials is reduced from 157,464 to 36 trials.

5. Allocate and Conduct Experiment. Based on the above, we describe the experimental strategy by using P-diagram, as shown in Figure 6.17.

L_{36} orthogonal array can be used for this experiment. L_{36} shown in Table 6.16 (also shown in Appendix A) is a recommended orthogonal array for robust engineering. However, since, in this case, there is a parameter with 12 levels, we need to utilize the column-merging technique to implement the modification of orthogonal array: to merge columns 1–4 to be one column and regard the same factor-level combinations in columns 1–4 as the same level. For example, columns 1–4 are merged to be a new Factor A, and the combination of A1B1C1D1 is regarded as level 1 of new Factor A; likewise, the combination of A1B2C2D1 is regarded as level 2 of new Factor A. In this way, we can utilize the column-merging technique to create a new column containing 12 levels.

In addition, in this case, there's a parameter with only two levels. However, L_{36} is an orthogonal array belonging to the 3-level series, so we have to utilize the dummy treatment technique to implement factor assignment: replacing level 3 of the column (where the parameter with two levels is placed) with level 1 or level 2 (choosing the level that R&D engineers regard as possibly more important. In this case, we replace level 3 with level 1 and remark it as level 1′, i.e., level 1 = level 1′).

To sum up, the modification techniques (column-merging and dummy treatment) described above can be illustrated in Figure 6.18. By the way, since there are only 10 parameters in this case, after applying column-merging to L_{36}, there are three empty columns that are not assigned with any design parameters.

TABLE 6.16. L_{36} $(2^3 \times 3^{13})$ **Orthogonal Array**

	A	B	C	D	E	F	G	H	I	J	K	L	M	N	O	P
L_{36}	1	2	3	4	5	6	7	8	9	10	11	12	13	14	15	16
1	1	1	1	1	1	1	1	1	1	1	1	1	1	1	1	1
2	1	1	1	1	2	2	2	2	2	2	2	2	2	2	2	2
3	1	1	1	1	3	3	3	3	3	3	3	3	3	3	3	3
4	1	2	2	1	1	1	1	1	2	2	2	2	3	3	3	3
5	1	2	2	1	2	2	2	2	3	3	3	3	1	1	1	1
6	1	2	2	1	3	3	3	3	1	1	1	1	2	2	2	2
7	2	1	2	1	1	1	2	3	1	2	3	3	1	2	2	3
8	2	1	2	1	2	2	3	1	2	3	1	1	2	3	3	1
9	2	1	2	1	3	3	1	2	3	1	2	2	3	1	1	2
10	2	2	1	1	1	1	3	2	1	3	2	3	2	1	3	2
11	2	2	1	1	2	2	1	3	2	1	3	1	3	2	1	3
12	2	2	1	1	3	3	2	1	3	2	1	2	1	3	2	1
13	1	1	1	2	1	2	3	1	3	2	1	3	3	2	1	2
14	1	1	1	2	2	3	1	2	1	3	2	1	1	3	2	3
15	1	1	1	2	3	1	2	3	2	1	3	2	2	1	3	1
16	1	2	2	2	1	2	3	2	1	1	3	2	3	3	2	1
17	1	2	2	2	2	3	1	3	2	2	1	3	1	1	3	2
18	1	2	2	2	3	1	2	1	3	3	2	1	2	2	1	3
19	2	1	2	2	1	2	1	3	3	3	1	2	2	1	2	3
20	2	1	2	2	2	3	2	1	1	1	2	3	3	2	3	1
21	2	1	2	2	3	1	3	2	2	2	3	1	1	3	1	2
22	2	2	1	2	1	2	2	3	3	1	2	1	1	3	3	2
23	2	2	1	2	2	3	3	1	1	2	3	2	2	1	1	3
24	2	2	1	2	3	1	1	2	2	3	1	3	3	2	2	1
25	1	1	1	3	1	3	2	1	2	3	3	1	3	1	2	2
26	1	1	1	3	2	1	3	2	3	1	1	2	1	2	3	3
27	1	1	1	3	3	2	1	3	1	2	2	3	2	3	1	1
28	1	2	2	3	1	3	2	2	2	1	1	3	2	3	1	3
29	1	2	2	3	2	1	3	3	3	2	2	1	3	1	2	1
30	1	2	2	3	3	2	1	1	1	3	3	2	1	2	3	2
31	2	1	2	3	1	3	3	3	2	3	2	2	1	2	1	1
32	2	1	2	3	2	1	1	1	3	1	3	3	2	3	2	2
33	2	1	2	3	3	2	2	2	1	2	1	1	3	1	3	3
34	2	2	1	3	1	3	1	2	3	2	3	1	2	2	3	1
35	2	2	1	3	2	1	2	3	1	3	1	2	3	3	1	2
36	2	2	1	3	3	2	3	1	2	1	2	3	1	1	2	3

L_{36}	A 1	B 2	C 3	D 4	E 5	F 6	G 7	H 8	I 9	J 10	K 11	L 12	M 13
1	1	1	1	1	1	1	1	1	1	1	1	1	1
2	1	1	1	1	2	2	2	2	2	2	2	2	2
3	1	1	1	1	3	3	3	3	3	3	3	3	3
4	1	2	2	1	1	1	1	1	2	2	2	2	3
5	1	2	2	1	2	2	2	2	3	3	3	3	1
6	1	2	2	1	3	3	3	3	1	1	1	1	2
7	2	1	2	1	1	1	2	3	1	2	3	3	1
8	2	1	2	1	2	2	3	1	2	3	1	1	2
9	2	1	2	1	3	3	1	2	3	1	2	2	3
10	2	2	1	1	1	1	3	2	1	3	2	3	2
11	2	2	1	1	2	2	1	3	2	1	3	1	3
12	2	2	1	1	3	3	2	1	3	2	1	2	1
13	1	1	1	2	1	2	3	1	3	2	1	3	3
14	1	1	1	2	2	3	1	2	1	3	2	1	1
15	1	1	1	2	3	1	2	3	2	1	3	2	2
…													
34	2	2	1	3	1	3	1	2	3	2	3	1	2
35	2	2	1	3	2	1	2	3	1	3	1	2	3
36	2	2	1	3	3	2	3	1	2	1	2	3	1

Column Merging

Dummy Treatment

L_{36}	~~A~~ A 1	~~E~~ B 5	~~F~~ C 6	~~G~~ D 7	~~H~~ E 8	~~I~~ F 9	~~J~~ G 10	~~K~~ H 11	~~L~~ I 12	~~M~~ J 13
1	1	1	1	1	1	1	1	1	1	1
2	1	2	2	2	2	2	2	2	2	2
3	1	3	3	1'	3	3	3	3	3	3
4	2	1	1	1	1	2	2	2	2	3
5	2	2	2	2	2	3	3	3	3	1
6	2	3	3	1'	3	1	1	1	1	2
7	3	1	1	2	3	1	2	3	3	1
8	3	2	2	1'	1	2	3	1	1	2
9	3	3	3	1	2	3	1	2	2	3
10	4	1	1	1'	2	1	3	2	3	2
11	4	2	2	1	3	2	1	3	1	3
12	4	3	3	2	1	3	2	1	2	1
13	5	1	2	1'	1	3	2	1	3	3
14	5	2	3	1	2	1	3	2	1	1
15	5	3	1	2	3	2	1	3	2	2
…										
34	12	1	3	1	2	3	2	3	1	2
35	12	2	1	2	3	1	3	1	2	3
36	12	3	2	1'	1	2	1	2	3	1

Figure 6.18. Modified L_{36} orthogonal array.

Thus, we can delete the last three columns of original L_{36}. The factor assignment is shown as the inner array in Table 6.17. R&D engineers need to collect the data of voltage under N_1 and N_2 while changing different turns ratios: M_1, M_2, M_3, M_4, and M_5. The overall experimental layout (inner array and outer array) is shown as Table 6.17.

Table 6.18 shows all the collected data of 36 runs.

6. Analyze Data and Conduct Two-Step Optimization and Confirmation Experiment. Take the first and second runs of 36 runs as examples. We use the equations for the SN ratio and β to calculate their values. The calculation of the first run is as follows:

$$S_T = (5.178)^2 + (3.915)^2 + \cdots + (26.975)^2 + (24.333)^2 = 3782.98820$$

$$r = (0.03)^2 + (0.06)^2 + (0.09)^2 + (0.12)^2 + (0.15)^2 = 0.04950$$

$$L_1 = (5.178 \times 0.03) + (15.841 \times 0.06) + \cdots + (25.445 \times 0.12) + (26.975 \times 0.15)$$
$$= 10.20945$$

$$L_2 = (3.915 \times 0.03) + (12.821 \times 0.06) + \cdots + (19.923 \times 0.12) + (24.333 \times 0.15)$$
$$= 8.86940$$

r_0 is the number of linear contrasts for noise conditions. It is 2 in this case.

$$S_\beta = \frac{(10.20945 + 8.86940)^2}{0.04950 \times 2} = 3676.79394$$

$$S_{\beta \times N} = \frac{(10.20945)^2 + (8.86940)^2}{0.04950} - 3676.79394 = 18.13871$$

$$S_e = 3782.98820 - (3676.79394 + 18.13871) = 88.05555$$

$$V_e = \frac{88.05555}{10 - 2} = 11.00694$$

$$V_N = \frac{18.13871 + 88.05555}{10 - 1} = 11.79936$$

Thus,

$$\beta = \frac{10.20945 + 8.86940}{0.04950 \times 2} = 192.716$$

and

$$\mathrm{SN} = 10 \log \left[\frac{\dfrac{1}{0.04950 \times 2}(3676.79394 - 11.00694)}{11.79936} \right] = 34.967$$

The calculation of the second run is as follows:

$$S_T = (5.153)^2 + (4.960)^2 + \cdots + (27.424)^2 + (26.709)^2 = 4293.72566$$

$$r = (0.03)^2 + (0.06)^2 + (0.09)^2 + (0.12)^2 + (0.15)^2 = 0.04950$$

TABLE 6.17. Experimental Layout

	A	B	C	D	E	F	G
1.	X1 on left side / Y2 on left side	Large	10	Dense	Thick	Thick	Dense
2.	X1 on left side / Y2 on left side	Medium	20	Sparse	Medium	Medium	Medium
3.	X1 on left side / Y2 on left side	Small	30	Dense	Thin	Thin	Sparse
4.	X1 on left side / Y2 on right side	Large	10	Dense	Thick	Medium	Medium
5.	X1 on left side / Y2 on right side	Medium	20	Sparse	Medium	Thin	Sparse
6.	X1 on left side / Y2 on right side	Small	30	Dense	Thin	Thick	Dense
7.	X1 on right side / Y2 on left side	Large	10	Sparse	Thin	Thick	Medium
8.	X1 on right side / Y2 on left side	Medium	20	Dense	Thick	Medium	Sparse
9.	X1 on right side / Y2 on left side	Small	30	Dense	Medium	Thin	Dense
10.	X1 on right side / Y2 on right side	Large	10	Dense	Medium	Thick	Sparse
11.	X1 on right side / Y2 on right side	Medium	20	Dense	Thin	Medium	Dense
12.	X1 on right side / Y2 on right side	Small	30	Sparse	Thick	Thin	Medium
13.	Y2 on left side / X1 on left side	Large	20	Dense	Thick	Thin	Medium
14.	Y2 on left side / X1 on left side	Medium	30	Dense	Medium	Thick	Sparse
15.	Y2 on left side / X1 on left side	Small	10	Sparse	Thin	Medium	Dense
16.	Y2 on left side / X1 on right side	Large	20	Dense	Medium	Thick	Dense
17.	Y2 on left side / X1 on right side	Medium	30	Dense	Thin	Medium	Medium
18.	Y2 on left side / X1 on right side	Small	10	Sparse	Thick	Thin	Sparse
19.	Y2 on right side / X1 on left side	Large	20	Dense	Thin	Thin	Sparse
20.	Y2 on right side / X1 on left side	Medium	30	Sparse	Thick	Thick	Dense
21.	Y2 on right side / X1 on left side	Small	10	Dense	Medium	Medium	Medium
22.	Y2 on right side / X1 on right side	Large	20	Sparse	Thin	Thin	Dense
23.	Y2 on right side / X1 on right side	Medium	30	Dense	Thick	Thick	Medium
24.	Y2 on right side / X1 on right side	Small	10	Dense	Medium	Medium	Sparse
25.	(X1 + Y2) on left side	Large	30	Sparse	Thick	Medium	Sparse
26.	(X1 + Y2) on left side	Medium	10	Dense	Medium	Thin	Dense
27.	(X1 + Y2) on left side	Small	20	Dense	Thin	Thick	Medium
28.	(X1 + Y2) on right side	Large	30	Sparse	Medium	Medium	Dense
29.	(X1 + Y2) on right side	Medium	10	Dense	Thin	Thin	Medium
30.	(X1 + Y2) on right side	Small	20	Dense	Thick	Thick	Sparse
31.	(Y2 + X1) on left side	Large	30	Dense	Thin	Medium	Sparse
32.	(Y2 + X1) on left side	Medium	10	Dense	Thick	Thin	Dense
33.	(Y2 + X1) on left side	Small	20	Sparse	Medium	Thick	Medium
34.	(Y2 + X1) on right side	Large	30	Dense	Medium	Thin	Medium
35.	(Y2 + X1) on right side	Medium	10	Sparse	Thin	Thick	Sparse
36.	(Y2 + X1) on right side	Small	20	Dense	Thick	Medium	Dense

			M1		M2		M3		M4		M5	
			0.03		0.06		0.09		0.12		0.15	
H	I	J	N1	N2	N1	N2	N1	N2	N1	N2	N1	N2
Left	Left	Left										
Center	Center	Center										
Right	Right	Right										
Center	Center	Right										
Right	Right	Left										
Left	Left	Center										
Right	Right	Left										
Left	Left	Center										
Center	Center	Right										
Center	Right	Center										
Right	Left	Right										
Left	Center	Left										
Left	Right	Right										
Center	Left	Left										
Right	Center	Center										
Right	Center	Right										
Left	Right	Left										
Center	Left	Center										
Left	Center	Center										
Center	Right	Right										
Right	Left	Left										
Center	Left	Left										
Right	Center	Center										
Left	Right	Right										
Right	Left	Right										
Left	Center	Left										
Center	Right	Center										
Left	Right	Center										
Center	Left	Right										
Right	Center	Left										
Center	Center	Left										
Right	Right	Center										
Left	Left	Right										
Right	Left	Center										
Left	Center	Right										
Center	Right	Left										

TABLE 6.18. Data Collection

	A	B	C	D	E	F	G
1.	X1 on left side / Y2 on left side	Large	10	Dense	Thick	Thick	Dense
2.	X1 on left side / Y2 on left side	Medium	20	Sparse	Medium	Medium	Medium
3.	X1 on left side / Y2 on left side	Small	30	Dense	Thin	Thin	Sparse
4.	X1 on left side / Y2 on right side	Large	10	Dense	Thick	Medium	Medium
5.	X1 on left side / Y2 on right side	Medium	20	Sparse	Medium	Thin	Sparse
6.	X1 on left side / Y2 on right side	Small	30	Dense	Thin	Thick	Dense
7.	X1 on right side / Y2 on left side	Large	10	Sparse	Thin	Thick	Medium
8.	X1 on right side / Y2 on left side	Medium	20	Dense	Thick	Medium	Sparse
9.	X1 on right side / Y2 on left side	Small	30	Dense	Medium	Thin	Dense
10.	X1 on right side / Y2 on right side	Large	10	Dense	Medium	Thick	Sparse
11.	X1 on right side / Y2 on right side	Medium	20	Dense	Thin	Medium	Dense
12.	X1 on right side / Y2 on right side	Small	30	Sparse	Thick	Thin	Medium
13.	Y2 on left side / X1 on left side	Large	20	Dense	Thick	Thin	Medium
14.	Y2 on left side / X1 on left side	Medium	30	Dense	Medium	Thick	Sparse
15.	Y2 on left side / X1 on left side	Small	10	Sparse	Thin	Medium	Dense
16.	Y2 on left side / X1 on right side	Large	20	Dense	Medium	Thick	Dense
17.	Y2 on left side / X1 on right side	Medium	30	Dense	Thin	Medium	Medium
18.	Y2 on left side / X1 on right side	Small	10	Sparse	Thick	Thin	Sparse
19.	Y2 on right side / X1 on left side	Large	20	Dense	Thin	Thin	Sparse
20.	Y2 on right side / X1 on left side	Medium	30	Sparse	Thick	Thick	Dense
21.	Y2 on right side / X1 on left side	Small	10	Dense	Medium	Medium	Medium
22.	Y2 on right side / X1 on right side	Large	20	Sparse	Thin	Thin	Dense
23.	Y2 on right side / X1 on right side	Medium	30	Dense	Thick	Thick	Medium
24.	Y2 on right side / X1 on right side	Small	10	Dense	Medium	Medium	Sparse
25.	(X1 + Y2) on left side	Large	30	Sparse	Thick	Medium	Sparse
26.	(X1 + Y2) on left side	Medium	10	Dense	Medium	Thin	Dense
27.	(X1 + Y2) on left side	Small	20	Dense	Thin	Thick	Medium
28.	(X1 + Y2) on right side	Large	30	Sparse	Medium	Medium	Dense
29.	(X1 + Y2) on right side	Medium	10	Dense	Thin	Thin	Medium
30.	(X1 + Y2) on right side	Small	20	Dense	Thick	Thick	Sparse
31.	(Y2 + X1) on left side	Large	30	Dense	Thin	Medium	Sparse
32.	(Y2 + X1) on left side	Medium	10	Dense	Thick	Thin	Dense
33.	(Y2 + X1) on left side	Small	20	Sparse	Medium	Thick	Medium
34.	(Y2 + X1) on right side	Large	30	Dense	Medium	Thin	Medium
35.	(Y2 + X1) on right side	Medium	10	Sparse	Thin	Thick	Sparse
36.	(Y2 + X1) on right side	Small	20	Dense	Thick	Medium	Dense

			M1		M2		M3		M4		M5	
			0.03		0.06		0.09		0.12		0.15	
H	I	J	N1	N2	N1	N2	N1	N2	N1	N2	N1	N2
Left	Left	Left	5.178	3.915	15.841	12.821	22.267	21.579	25.445	19.923	26.975	24.333
Center	Center	Center	5.153	4.960	17.158	16.178	20.364	19.610	27.389	25.917	27.424	26.709
Right	Right	Right	5.154	5.044	16.201	15.825	18.460	17.805	26.199	24.807	26.679	26.097
Center	Center	Right	5.126	4.923	17.131	16.141	21.934	20.330	27.402	25.765	27.450	26.815
Right	Right	Left	5.178	4.968	15.953	15.849	18.935	17.826	25.683	24.768	27.616	26.275
Left	Left	Center	5.170	5.017	15.857	15.583	19.112	17.934	25.853	24.518	27.794	26.443
Right	Right	Left	5.167	3.910	16.089	12.592	22.264	22.170	26.025	19.886	27.571	23.633
Left	Left	Center	5.146	4.973	15.814	15.470	18.139	17.378	25.361	24.315	26.016	24.945
Center	Center	Right	5.178	5.005	17.946	16.464	18.115	17.547	29.949	26.728	26.100	25.628
Center	Right	Center	5.165	5.013	16.405	16.272	21.980	19.547	26.465	26.002	29.662	27.207
Right	Left	Right	5.205	4.865	18.203	17.569	20.435	18.774	29.660	28.330	31.764	28.933
Left	Center	Left	5.205	5.049	16.323	16.067	20.548	18.785	26.287	25.612	30.646	27.185
Left	Right	Right	5.199	5.036	15.833	15.425	18.573	17.451	25.533	24.282	27.897	25.472
Center	Left	Left	5.202	5.044	16.179	15.657	19.006	17.467	26.197	25.018	28.205	25.208
Right	Center	Center	5.148	4.241	16.285	14.184	22.130	20.818	26.282	22.632	31.322	28.651
Right	Center	Right	5.070	4.828	16.716	15.941	18.895	17.512	27.006	25.515	28.412	25.430
Left	Right	Left	5.167	4.405	16.056	14.589	22.214	20.274	25.645	22.080	31.425	28.334
Center	Left	Center	5.167	4.965	17.060	15.970	18.344	17.470	28.036	25.765	26.329	25.437
Left	Center	Center	5.124	4.907	17.725	17.225	20.366	18.843	29.666	27.693	31.669	28.704
Center	Right	Right	5.202	4.926	17.540	16.163	21.194	19.410	28.011	26.062	33.863	29.966
Right	Left	Left	5.202	5.042	16.883	16.351	18.278	17.443	25.806	24.713	26.221	25.443
Center	Left	Left	5.132	4.978	15.925	15.865	21.634	18.933	25.500	24.573	35.010	29.351
Right	Center	Center	5.135	4.933	17.111	16.315	17.965	17.401	25.437	24.300	26.040	25.448
Left	Right	Right	5.287	4.029	15.868	13.377	20.026	18.744	29.748	27.338	29.730	27.236
Right	Left	Right	5.265	5.124	17.011	15.802	20.137	18.755	26.362	24.716	27.063	25.519
Left	Center	Left	5.243	5.056	16.787	15.736	18.202	17.421	25.605	23.432	28.089	24.995
Center	Right	Center	5.323	5.075	16.850	15.974	17.978	17.440	26.271	24.142	29.722	26.866
Left	Right	Center	5.323	5.195	17.247	15.974	19.959	18.813	26.357	21.839	25.900	23.786
Center	Left	Right	5.325	5.203	17.272	15.982	20.770	19.380	28.118	24.863	33.963	29.406
Right	Center	Left	5.268	4.370	15.810	14.509	17.913	17.413	25.926	23.660	25.494	25.262
Center	Center	Left	5.273	5.197	17.255	15.827	17.753	17.517	26.099	19.190	27.563	25.476
Right	Right	Center	5.245	5.073	16.842	15.742	21.540	19.517	25.432	23.464	29.730	27.236
Left	Left	Right	5.298	5.119	16.994	15.901	20.026	18.744	30.039	25.792	30.722	28.758
Right	Left	Center	5.290	5.170	17.163	15.876	20.137	18.755	26.541	25.091	32.851	30.023
Left	Center	Right	5.243	5.056	16.787	15.736	18.202	17.421	27.084	24.622	30.274	30.111
Center	Right	Left	5.329	4.998	16.343	15.939	18.094	17.375	30.001	29.750	29.485	28.743

$$L_1 = (5.153\times 0.03)+(17.158\times 0.06)+\cdots+(27.389\times 0.12)+(27.424\times 0.15)$$
$$= 10.41698$$

$$L_2 = (4.960\times 0.03)+(16.178\times 0.06)+\cdots+(25.917\times 0.12)+(26.709\times 0.15)$$
$$= 10.00067$$

r_0 is 2 in this case.

$$S_\beta = \frac{(10.41698+10.00067)^2}{0.04950\times 2} = 4210.91245$$

$$S_{\beta\times N} = \frac{(10.41698)^2+(10.00067)^2}{0.04950} - 4210.91245 = 1.75063$$

$$S_e = 4293.72566-(4210.91245+1.75063) = 81.06258$$

$$V_e = \frac{81.06258}{10-2} = 10.13282$$

$$V_N = \frac{1.75063+81.06258}{10-1} = 9.20147$$

Thus,

$$\beta = \frac{10.41698+10.00067}{0.04950\times 2} = 206.239$$

and

$$\text{SN} = 10\log\left[\frac{\frac{1}{0.04950\times 2}(4210.91245-10.13282)}{9.20147}\right] = 36.638$$

Using the calculation approach, we can obtain the individual β and SN ratio of 36 runs, as shown in Table 6.19.

The analyzed β's and SN ratios shown in Table 6.19 need to be further analyzed to assess the effects of each parameter. Figures 6.19 and 6.20 illustrate how we analyze the effects of Factors A and B on the SN ratio.

Using the same calculation method, we can yield the effects of each factor on the SN ratio; and we can know which factors' influences on the SN ratio are more significant and which factors' influences on the SN ratio are insignificant. Table 6.20 shows the information described above. This table shows that the influence of Factor A on the SN ratio is the largest one and that the influence of Factor D on the SN ratio is the smallest.

The approach of calculating factorial effects on β is the same as that of calculating factorial effects on the SN ratio. Figure 6.21 illustrate how we analyze the effects of Factor A on β.

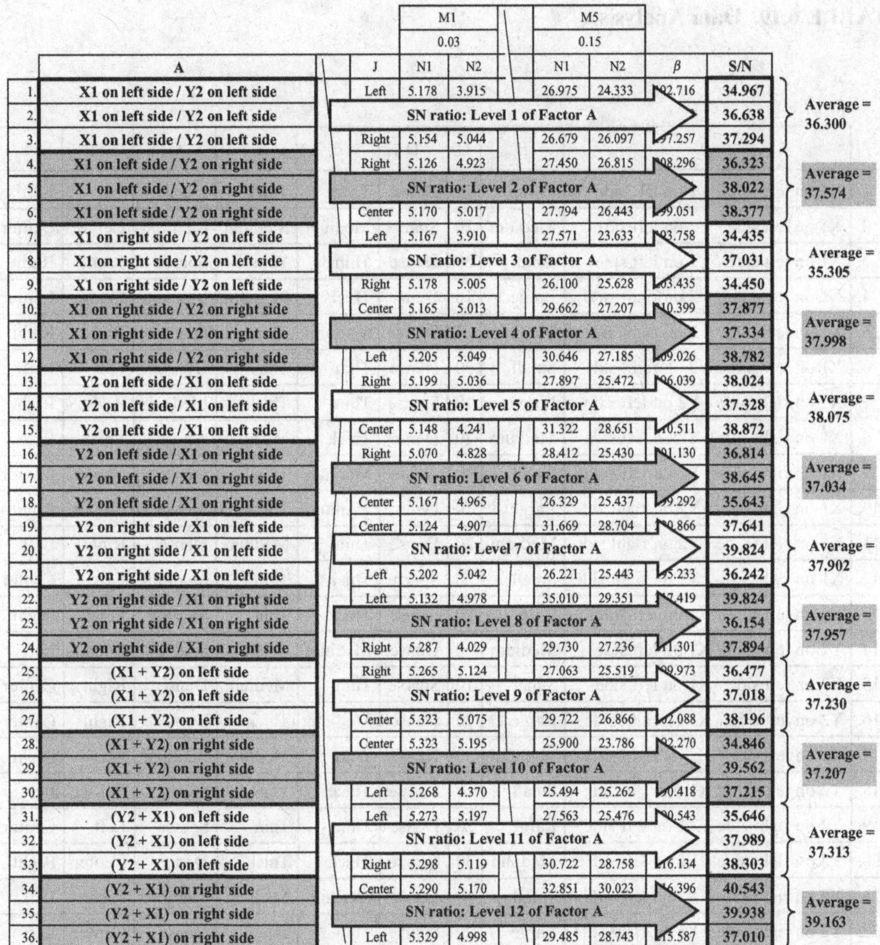

	A	J	M1 (0.03) N1	M1 (0.03) N2	M5 (0.15) N1	M5 (0.15) N2	β	S/N	
1.	X1 on left side / Y2 on left side	Left	5.178	3.915	26.975	24.333	92.716	34.967	
2.	X1 on left side / Y2 on left side	SN ratio: Level 1 of Factor A						36.638	Average = 36.300
3.	X1 on left side / Y2 on left side	Right	5.154	5.044	26.679	26.097	97.257	37.294	
4.	X1 on left side / Y2 on right side	Right	5.126	4.923	27.450	26.815	98.296	36.323	
5.	X1 on left side / Y2 on right side	SN ratio: Level 2 of Factor A						38.022	Average = 37.574
6.	X1 on left side / Y2 on right side	Center	5.170	5.017	27.794	26.443	99.051	38.377	
7.	X1 on right side / Y2 on left side	Left	5.167	3.910	27.571	23.633	93.758	34.435	
8.	X1 on right side / Y2 on left side	SN ratio: Level 3 of Factor A						37.031	Average = 35.305
9.	X1 on right side / Y2 on left side	Right	5.178	5.005	26.100	25.628	03.435	34.450	
10.	X1 on right side / Y2 on right side	Center	5.165	5.013	29.662	27.207	10.399	37.877	
11.	X1 on right side / Y2 on right side	SN ratio: Level 4 of Factor A						37.334	Average = 37.998
12.	X1 on right side / Y2 on right side	Left	5.205	5.049	30.646	27.185	09.026	38.782	
13.	Y2 on left side / X1 on left side	Right	5.199	5.036	27.897	25.472	96.039	38.024	
14.	Y2 on left side / X1 on left side	SN ratio: Level 5 of Factor A						37.328	Average = 38.075
15.	Y2 on left side / X1 on left side	Center	5.148	4.241	31.322	28.651	10.511	38.872	
16.	Y2 on left side / X1 on right side	Right	5.070	4.828	28.412	25.430	01.130	36.814	
17.	Y2 on left side / X1 on right side	SN ratio: Level 6 of Factor A						38.645	Average = 37.034
18.	Y2 on left side / X1 on right side	Center	5.167	4.965	26.329	25.437	99.292	35.643	
19.	Y2 on right side / X1 on left side	Center	5.124	4.907	31.669	28.704	0.866	37.641	
20.	Y2 on right side / X1 on left side	SN ratio: Level 7 of Factor A						39.824	Average = 37.902
21.	Y2 on right side / X1 on left side	Left	5.202	5.042	26.221	25.443	95.233	36.242	
22.	Y2 on right side / X1 on right side	Left	5.132	4.978	35.010	29.351	7.419	39.824	
23.	Y2 on right side / X1 on right side	SN ratio: Level 8 of Factor A						36.154	Average = 37.957
24.	Y2 on right side / X1 on right side	Right	5.287	4.029	29.730	27.236	1.301	37.894	
25.	(X1 + Y2) on left side	Right	5.265	5.124	27.063	25.519	99.973	36.477	
26.	(X1 + Y2) on left side	SN ratio: Level 9 of Factor A						37.018	Average = 37.230
27.	(X1 + Y2) on left side	Center	5.323	5.075	29.722	26.866	02.088	38.196	
28.	(X1 + Y2) on right side	Center	5.323	5.195	25.900	23.786	92.270	34.846	
29.	(X1 + Y2) on right side	SN ratio: Level 10 of Factor A						39.562	Average = 37.207
30.	(X1 + Y2) on right side	Left	5.268	4.370	25.494	25.262	90.418	37.215	
31.	(Y2 + X1) on left side	Left	5.273	5.197	27.563	25.476	90.543	35.646	
32.	(Y2 + X1) on left side	SN ratio: Level 11 of Factor A						37.989	Average = 37.313
33.	(Y2 + X1) on left side	Right	5.298	5.119	30.722	28.758	16.134	38.303	
34.	(Y2 + X1) on right side	Center	5.290	5.170	32.851	30.023	6.396	40.543	
35.	(Y2 + X1) on right side	SN ratio: Level 12 of Factor A						39.938	Average = 39.163
36.	(Y2 + X1) on right side	Left	5.329	4.998	29.485	28.743	15.587	37.010	

Figure 6.19. Effects of Factor A on the SN ratio.

Using the same calculation approach, the effects of each factor on β are obtained. Table 6.21 shows the rank of the effects of each factor on β: the effect of Factor A on β is the largest among all the factors, and the effect of Factor I on β is the smallest.

Based on the response tables for the SN ratio and β, we can draw the response graphs for SN ratio and β, as shown in Figures 6.22 and 6.23.

Once we have the information about the factorial effect analysis, we can conduct two-step optimization:

- Step 1: Choose the level with the largest SN ratio to reduce the functional variability under noise conditions.

TABLE 6.19. Data Analysis

	A	B	C	D	E	F	G	H	I
1.	X1 on left side / Y2 on left side	Large	10	Dense	Thick	Thick	Dense	Left	Left
2.	X1 on left side / Y2 on left side	Medium	20	Sparse	Medium	Medium	Medium	Center	Center
3.	X1 on left side / Y2 on left side	Small	30	Dense	Thin	Thin	Sparse	Right	Right
4.	X1 on left side / Y2 on right side	Large	10	Dense	Thick	Medium	Medium	Center	Center
5.	X1 on left side / Y2 on right side	Medium	20	Sparse	Medium	Thin	Sparse	Right	Right
6.	X1 on left side / Y2 on right side	Small	30	Dense	Thin	Thick	Dense	Left	Left
7.	X1 on right side / Y2 on left side	Large	10	Sparse	Thin	Thick	Medium	Right	Right
8.	X1 on right side / Y2 on left side	Medium	20	Dense	Thick	Medium	Sparse	Left	Left
9.	X1 on right side / Y2 on left side	Small	30	Dense	Medium	Thin	Dense	Center	Center
10.	X1 on right side / Y2 on right side	Large	10	Dense	Medium	Thick	Sparse	Center	Right
11.	X1 on right side / Y2 on right side	Medium	20	Dense	Thin	Medium	Dense	Right	Left
12.	X1 on right side / Y2 on right side	Small	30	Sparse	Thick	Thin	Medium	Left	Center
13.	Y2 on left side / X1 on left side	Large	20	Dense	Thick	Thin	Medium	Left	Right
14.	Y2 on left side / X1 on left side	Medium	30	Dense	Medium	Thick	Sparse	Center	Left
15.	Y2 on left side / X1 on left side	Small	10	Sparse	Thin	Medium	Dense	Right	Center
16.	Y2 on left side / X1 on right side	Large	20	Dense	Medium	Thick	Dense	Right	Center
17.	Y2 on left side / X1 on right side	Medium	30	Dense	Thin	Medium	Medium	Left	Right
18.	Y2 on left side / X1 on right side	Small	10	Sparse	Thick	Thin	Sparse	Center	Left
19.	Y2 on right side / X1 on left side	Large	20	Dense	Thin	Thin	Sparse	Left	Center
20.	Y2 on right side / X1 on left side	Medium	30	Sparse	Thick	Thick	Dense	Center	Right
21.	Y2 on right side / X1 on left side	Small	10	Dense	Medium	Medium	Medium	Right	Left
22.	Y2 on right side / X1 on right side	Large	20	Sparse	Thin	Thin	Dense	Center	Left
23.	Y2 on right side / X1 on right side	Medium	30	Dense	Thick	Thick	Medium	Right	Center
24.	Y2 on right side / X1 on right side	Small	10	Dense	Medium	Medium	Sparse	Left	Right
25.	(X1 + Y2) on left side	Large	30	Sparse	Thick	Medium	Sparse	Right	Left
26.	(X1 + Y2) on left side	Medium	10	Dense	Medium	Thin	Dense	Left	Center
27.	(X1 + Y2) on left side	Small	20	Dense	Thin	Thick	Medium	Center	Right
28.	(X1 + Y2) on right side	Large	30	Sparse	Medium	Medium	Dense	Left	Right
29.	(X1 + Y2) on right side	Medium	10	Dense	Thin	Thin	Medium	Center	Left
30.	(X1 + Y2) on right side	Small	20	Dense	Thick	Thick	Sparse	Right	Center
31.	(Y2 + X1) on left side	Large	30	Dense	Thin	Medium	Sparse	Center	Center
32.	(Y2 + X1) on left side	Medium	10	Dense	Thick	Thin	Dense	Right	Right
33.	(Y2 + X1) on left side	Small	20	Sparse	Medium	Thick	Medium	Left	Left
34.	(Y2 + X1) on right side	Large	30	Dense	Medium	Thin	Medium	Right	Left
35.	(Y2 + X1) on right side	Medium	10	Sparse	Thin	Thick	Sparse	Left	Center
36.	(Y2 + X1) on right side	Small	20	Dense	Thick	Medium	Dense	Center	Right

	M1		M2		M3		M4		M5			
	0.03		0.06		0.09		0.12		0.15			
J	N1	N2	N1	N2	N1	N2	N1	N2	N1	N2	β	S/N
Left	5.178	3.915	15.841	12.821	22.267	21.579	25.445	19.923	26.975	24.333	192.716	34.967
Center	5.153	4.960	17.158	16.178	20.364	19.610	27.389	25.917	27.424	26.709	206.239	36.638
Right	5.154	5.044	16.201	15.825	18.460	17.805	26.199	24.807	26.679	26.097	197.257	37.294
Right	5.126	4.923	17.131	16.141	21.934	20.330	27.402	25.765	27.450	26.815	208.296	36.323
Left	5.178	4.968	15.953	15.849	18.935	17.826	25.683	24.768	27.616	26.275	198.571	38.022
Center	5.170	5.017	15.857	15.583	19.112	17.934	25.853	24.518	27.794	26.443	199.051	38.377
Left	5.167	3.910	16.089	12.592	22.264	22.170	26.025	19.886	27.571	23.633	193.758	34.435
Center	5.146	4.973	15.814	15.470	18.139	17.378	25.361	24.315	26.016	24.945	191.740	37.031
Right	5.178	5.005	17.946	16.464	18.115	17.547	29.949	26.728	26.100	25.628	203.435	34.450
Center	5.165	5.013	16.405	16.272	21.980	19.547	26.465	26.002	29.662	27.207	210.399	37.877
Right	5.205	4.865	18.203	17.569	20.435	18.774	29.660	28.330	31.764	28.933	222.632	37.334
Left	5.205	5.049	16.323	16.067	20.548	18.785	26.287	25.612	30.646	27.185	209.026	38.782
Right	5.199	5.036	15.833	15.425	18.573	17.451	25.533	24.282	27.897	25.472	196.039	38.024
Left	5.202	5.044	16.179	15.657	19.006	17.467	26.197	25.018	28.205	25.208	198.563	37.328
Center	5.148	4.241	16.285	14.184	22.130	20.818	26.282	22.632	31.322	28.651	210.511	38.872
Right	5.070	4.828	16.716	15.941	18.895	17.512	27.006	25.515	28.412	25.430	201.130	36.814
Left	5.167	4.405	16.056	14.589	22.214	20.274	25.645	22.080	31.425	28.334	208.490	38.645
Center	5.167	4.965	17.060	15.970	18.344	17.470	28.036	25.765	26.329	25.437	199.292	35.643
Center	5.124	4.907	17.725	17.225	20.366	18.843	29.666	27.693	31.669	28.704	220.866	37.641
Right	5.202	4.926	17.540	16.163	21.194	19.410	28.011	26.062	33.863	29.966	222.660	39.824
Left	5.202	5.042	16.883	16.351	18.278	17.443	25.806	24.713	26.221	25.443	195.233	36.242
Left	5.132	4.978	15.925	15.865	21.634	18.933	25.500	24.573	35.010	29.351	217.419	39.824
Center	5.135	4.933	17.111	16.315	17.965	17.401	25.437	24.300	26.040	25.448	193.758	36.154
Right	5.287	4.029	15.868	13.377	20.026	18.744	29.748	27.338	29.730	27.236	211.301	37.894
Right	5.265	5.124	17.011	15.802	20.137	18.755	26.362	24.716	27.063	25.519	199.973	36.477
Left	5.243	5.056	16.787	15.736	18.202	17.421	25.605	23.432	28.089	24.995	195.085	37.018
Center	5.323	5.075	16.850	15.974	17.978	17.440	26.271	24.142	29.722	26.866	202.088	38.196
Center	5.323	5.195	17.247	15.974	19.959	18.813	26.357	21.839	25.900	23.786	192.270	34.846
Right	5.325	5.203	17.272	15.982	20.770	19.380	28.118	24.863	33.963	29.406	220.077	39.562
Left	5.268	4.370	15.810	14.509	17.913	17.413	25.926	23.660	25.494	25.262	190.418	37.215
Left	5.273	5.197	17.255	15.827	17.753	17.517	26.099	19.190	27.563	25.476	190.543	35.646
Center	5.245	5.073	16.842	15.742	21.540	19.517	25.432	23.464	29.730	27.236	205.779	37.989
Right	5.298	5.119	16.994	15.901	20.026	18.744	30.039	25.792	30.722	28.758	216.134	38.303
Center	5.290	5.170	17.163	15.876	20.137	18.755	26.541	25.091	32.851	30.023	216.396	40.543
Right	5.243	5.056	16.787	15.736	18.202	17.421	27.084	24.622	30.274	30.111	209.383	39.938
Left	5.329	4.998	16.343	15.939	18.094	17.375	30.001	29.750	29.485	28.743	215.587	37.010

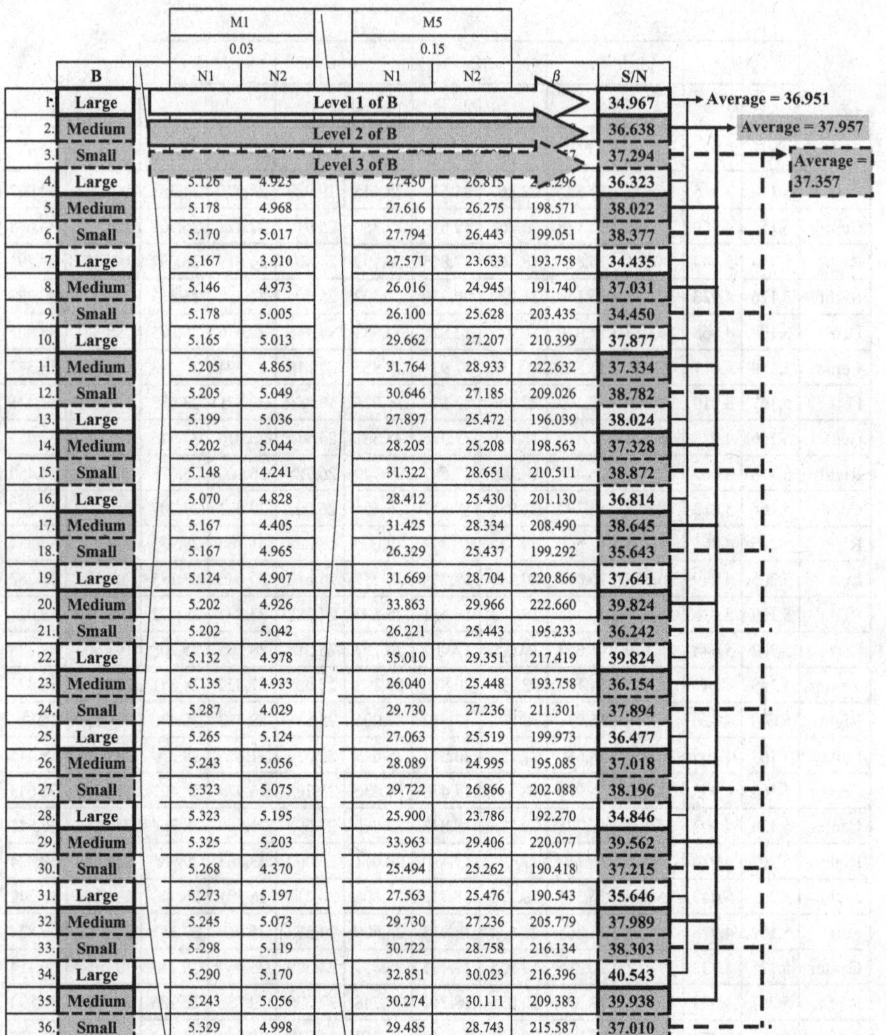

	B	M1 0.03 N1	M1 0.03 N2	M5 0.15 N1	M5 0.15 N2	β	S/N
1.	Large	[illegible]	[illegible]	[illegible]	[illegible]	[illegible]	34.967
2.	Medium	[illegible]	[illegible]	[illegible]	[illegible]	[illegible]	36.638
3.	Small	[illegible]	[illegible]	[illegible]	[illegible]	[illegible]	37.294
4.	Large	5.126	4.923	27.450	26.815	[illegible]	36.323
5.	Medium	5.178	4.968	27.616	26.275	198.571	38.022
6.	Small	5.170	5.017	27.794	26.443	199.051	38.377
7.	Large	5.167	3.910	27.571	23.633	193.758	34.435
8.	Medium	5.146	4.973	26.016	24.945	191.740	37.031
9.	Small	5.178	5.005	26.100	25.628	203.435	34.450
10.	Large	5.165	5.013	29.662	27.207	210.399	37.877
11.	Medium	5.205	4.865	31.764	28.933	222.632	37.334
12.	Small	5.205	5.049	30.646	27.185	209.026	38.782
13.	Large	5.199	5.036	27.897	25.472	196.039	38.024
14.	Medium	5.202	5.044	28.205	25.208	198.563	37.328
15.	Small	5.148	4.241	31.322	28.651	210.511	38.872
16.	Large	5.070	4.828	28.412	25.430	201.130	36.814
17.	Medium	5.167	4.405	31.425	28.334	208.490	38.645
18.	Small	5.167	4.965	26.329	25.437	199.292	35.643
19.	Large	5.124	4.907	31.669	28.704	220.866	37.641
20.	Medium	5.202	4.926	33.863	29.966	222.660	39.824
21.	Small	5.202	5.042	26.221	25.443	195.233	36.242
22.	Large	5.132	4.978	35.010	29.351	217.419	39.824
23.	Medium	5.135	4.933	26.040	25.448	193.758	36.154
24.	Small	5.287	4.029	29.730	27.236	211.301	37.894
25.	Large	5.265	5.124	27.063	25.519	199.973	36.477
26.	Medium	5.243	5.056	28.089	24.995	195.085	37.018
27.	Small	5.323	5.075	29.722	26.866	202.088	38.196
28.	Large	5.323	5.195	25.900	23.786	192.270	34.846
29.	Medium	5.325	5.203	33.963	29.406	220.077	39.562
30.	Small	5.268	4.370	25.494	25.262	190.418	37.215
31.	Large	5.273	5.197	27.563	25.476	190.543	35.646
32.	Medium	5.245	5.073	29.730	27.236	205.779	37.989
33.	Small	5.298	5.119	30.722	28.758	216.134	38.303
34.	Large	5.290	5.170	32.851	30.023	216.396	40.543
35.	Medium	5.243	5.056	30.274	30.111	209.383	39.938
36.	Small	5.329	4.998	29.485	28.743	215.587	37.010

Figure 6.20. Effects of Factor B on the SN ratio.

- Step 2: Choose the factors that influence only β but not (or less) the SN ratio to adjust the slope to the target value.

At Step 1, we would select A12B2C2D2E3F3G2H1I1J3 as the optimal condition: A = (Y2 + X1) on right side, B = medium, C = 20, D = sparse, E = thin, F = thin, G = medium, H = left, I = left, and J = right. In order to confirm that the mechanism design can reduce functional variability, we need to predict the SN ratio gain of the mechanism design and compare it with the gain calculated from actual data. If the actual SN ratio gain is within the ±30%

TABLE 6.20. Response Table for the SN Ratio

	A	B	C	D	E	F	G	H	I	J
Level 1	36.300	36.951	37.230	37.315	37.120	37.452	37.277	37.622	37.636	37.094
Level 2	37.574	37.957	37.671	37.634	37.165	36.913	37.654	37.360	37.124	37.484
Level 3	35.305	37.357	37.364		37.980	37.899	37.334	37.283	37.505	37.686
Level 4	37.998									
Level 5	38.075									
Level 6	37.034									
Level 7	37.902									
Level 8	37.957									
Level 9	37.230									
Level 10	37.207									
Level 11	37.313									
Level 12	39.163									
Range	3.858	1.006	0.441	0.318	0.860	0.986	0.377	0.340	0.512	0.592
Rank	1	2	7	10	4	3	8	9	6	5

range of the predicted SN ratio gain, it means that the factorial effects are confirmed.

Since the sum of response values includes the element of "the total average of 36 SN ratios" which is repeated computed, while we compute the predicted value of SN ratio by the response table, the element should be subtracted. In this case, because there are 10 parameters, the above-mentioned average (which is 37.422) is repeatedly computed nine times. The predicted SN ratio of the Step 1 condition is calculated as follows:

$$\begin{bmatrix}(39.163)+(37.957)+(37.671)+(37.634)+(37.980)+(37.899)+\\(37.654)+(37.622)+(37.636)+(37.686)\end{bmatrix}$$
$$-[(10-1)\times(37.422)]=42.109$$

Applying the same calculation approach, the predicted SN ratio of initial design is 37.741. Hence, the predicted SN ratio gain is 4.368.

Table 6.22 shows the collected data in the confirmation experiment.

Based on the actual data, we compute the SN ratio of the Step 1 condition as follows:

$$S_T=(6.476)^2+(5.928)^2+\cdots+(33.485)^2+(31.761)^2=4533.04921$$
$$r=(0.03)^2+(0.06)^2+(0.09)^2+(0.12)^2+(0.15)^2=0.04950$$

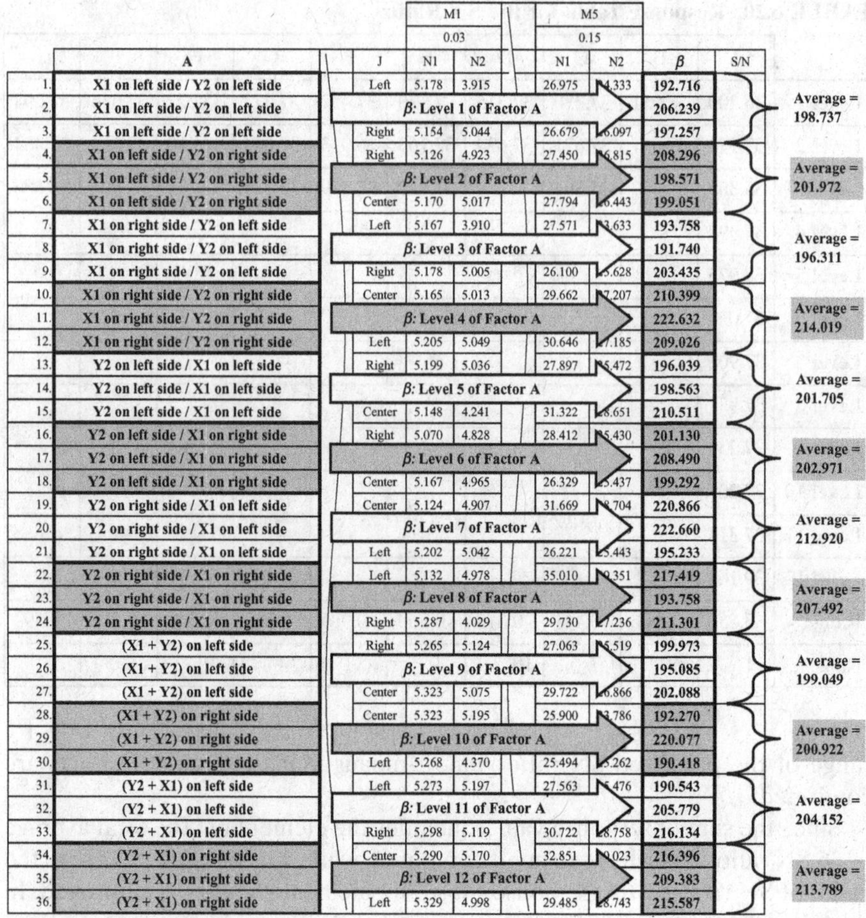

	A	J	M1 (0.03) N1	M1 (0.03) N2	M5 (0.15) N1	M5 (0.15) N2	β	S/N
1.	X1 on left side / Y2 on left side	Left	5.178	3.915	26.975	4.333	192.716	
2.	X1 on left side / Y2 on left side	β: Level 1 of Factor A					206.239	Average = 198.737
3.	X1 on left side / Y2 on left side	Right	5.154	5.044	26.679	6.097	197.257	
4.	X1 on left side / Y2 on right side	Right	5.126	4.923	27.450	6.815	208.296	
5.	X1 on left side / Y2 on right side	β: Level 2 of Factor A					198.571	Average = 201.972
6.	X1 on left side / Y2 on right side	Center	5.170	5.017	27.794	6.443	199.051	
7.	X1 on right side / Y2 on left side	Left	5.167	3.910	27.571	3.633	193.758	
8.	X1 on right side / Y2 on left side	β: Level 3 of Factor A					191.740	Average = 196.311
9.	X1 on right side / Y2 on left side	Right	5.178	5.005	26.100	5.628	203.435	
10.	X1 on right side / Y2 on right side	Center	5.165	5.013	29.662	7.207	210.399	
11.	X1 on right side / Y2 on right side	β: Level 4 of Factor A					222.632	Average = 214.019
12.	X1 on right side / Y2 on right side	Left	5.205	5.049	30.646	7.185	209.026	
13.	Y2 on left side / X1 on left side	Right	5.199	5.036	27.897	5.472	196.039	
14.	Y2 on left side / X1 on left side	β: Level 5 of Factor A					198.563	Average = 201.705
15.	Y2 on left side / X1 on left side	Center	5.148	4.241	31.322	8.651	210.511	
16.	Y2 on left side / X1 on right side	Right	5.070	4.828	28.412	5.430	201.130	
17.	Y2 on left side / X1 on right side	β: Level 6 of Factor A					208.490	Average = 202.971
18.	Y2 on left side / X1 on right side	Center	5.167	4.965	26.329	5.437	199.292	
19.	Y2 on right side / X1 on left side	Center	5.124	4.907	31.669	8.704	220.866	
20.	Y2 on right side / X1 on left side	β: Level 7 of Factor A					222.660	Average = 212.920
21.	Y2 on right side / X1 on left side	Left	5.202	5.042	26.221	5.443	195.233	
22.	Y2 on right side / X1 on right side	Left	5.132	4.978	35.010	9.351	217.419	
23.	Y2 on right side / X1 on right side	β: Level 8 of Factor A					193.758	Average = 207.492
24.	Y2 on right side / X1 on right side	Right	5.287	4.029	29.730	7.236	211.301	
25.	(X1 + Y2) on left side	Right	5.265	5.124	27.063	5.519	199.973	
26.	(X1 + Y2) on left side	β: Level 9 of Factor A					195.085	Average = 199.049
27.	(X1 + Y2) on left side	Center	5.323	5.075	29.722	6.866	202.088	
28.	(X1 + Y2) on right side	Center	5.323	5.195	25.900	3.786	192.270	
29.	(X1 + Y2) on right side	β: Level 10 of Factor A					220.077	Average = 200.922
30.	(X1 + Y2) on right side	Left	5.268	4.370	25.494	5.262	190.418	
31.	(Y2 + X1) on left side	Left	5.273	5.197	27.563	5.476	190.543	
32.	(Y2 + X1) on left side	β: Level 11 of Factor A					205.779	Average = 204.152
33.	(Y2 + X1) on left side	Right	5.298	5.119	30.722	8.758	216.134	
34.	(Y2 + X1) on right side	Center	5.290	5.170	32.851	9.023	216.396	
35.	(Y2 + X1) on right side	β: Level 12 of Factor A					209.383	Average = 213.789
36.	(Y2 + X1) on right side	Left	5.329	4.998	29.485	8.743	215.587	

Figure 6.21. Effects of Factor A on β.

$$L_1 = (6.476 \times 0.03) + (12.838 \times 0.06) + \cdots + (26.964 \times 0.12) + (33.485 \times 0.15) \\ = 10.97934$$

$$L_2 = (5.928 \times 0.03) + (11.680 \times 0.06) + \cdots + (24.996 \times 0.12) + (31.761 \times 0.15) \\ = 10.18032$$

r_0 is 2 in this case.

$$S_\beta = \frac{(10.97934 + 10.18032)^2}{0.04950 \times 2} = 4522.53749$$

$$S_{\beta \times N} = \frac{(10.97934)^2 + (10.18032)^2}{0.04950} - 4522.53749 = 6.44882$$

TABLE 6.21. Response Table for β

	A	B	C	D	E	F	G	H	I	J
Level 1	198.737	203.317	204.319	203.620	202.107	202.505	206.523	203.508	205.769	200.451
Level 2	201.972	206.081	206.572	206.270	203.730	204.401	205.461	207.883	203.224	204.032
Level 3	196.311	204.111	202.619		207.673	206.604	201.526	202.118	204.517	209.026
Level 4	214.019									
Level 5	201.705									
Level 6	202.971									
Level 7	212.920									
Level 8	207.492									
Level 9	199.049									
Level 10	200.922									
Level 11	204.152									
Level 12	213.789									
Range	17.708	2.764	3.953	2.650	5.566	4.099	4.997	5.765	2.545	8.576
Rank	1	8	7	9	4	6	5	3	10	2

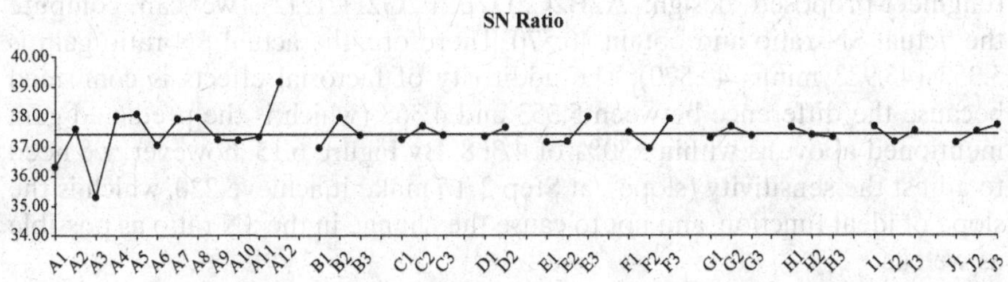

Figure 6.22. Response graph for the SN ratio.

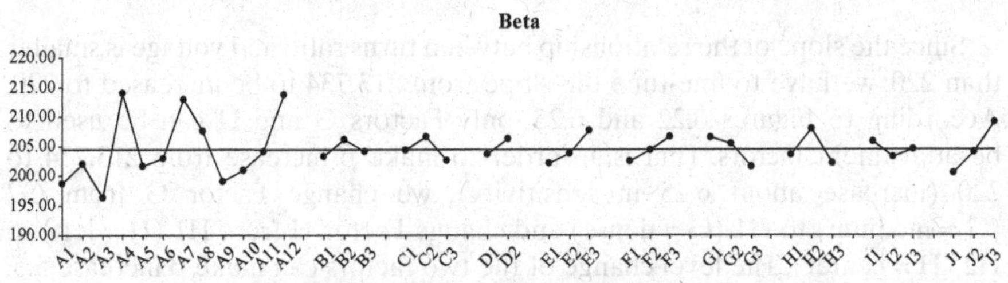

Figure 6.23. Response graph for β.

TABLE 6.22. Collected Data in the Confirmation Experiment

M1		M2		M3		M4		M5	
0.03		0.06		0.09		0.12		0.15	
N1	N2	N1	N2	N1	N2	N1	N2	N1	N2
6.476	5.928	12.838	11.68	19.515	17.089	26.964	24.996	33.485	31.761

$$S_e = 4533.04921 - (4522.53749 + 6.44882) = 4.06290$$

$$V_e = \frac{4.06290}{10-2} = 0.50786$$

$$V_N = \frac{6.44882 + 4.06290}{10-1} = 1.16797$$

Thus,

$$\mathrm{SN} = 10\log\left[\frac{\frac{1}{0.04950\times 2}(4522.53749 - 0.50786)}{1.16797}\right] = 45.923$$

Since we have the actual data (as shown in Table 6.23) of initial design (engineer-proposed design: A2B2C2D2E2F2G2H2I2J2), we can compute the actual SN ratio and obtain 40.570. Therefore, the actual SN ratio gain is 5.353 (45.923 minus 40.570). The additivity of factorial effects is confirmed because the difference between 5.353 and 4.368 (which is the predicted gain mentioned above)is within ±30% of 4.368. By Figure 6.15, however, we need to adjust the sensitivity (slope), at Step 2, to make it achieve 220, which is the slope of ideal function, and not to cause the change in the SN ratio as possible as we can.

We can use the data in Table 6.22 to calculate β as follows:

$$\beta = \frac{10.97934 + 10.18032}{0.04950\times 2} = 213.734$$

Since the slope of the relationship between turns ratio and voltage is smaller than 220, we have to fine-tune the slope from 213.734 to be increased to 220. According to Figures 6.22 and 6.23, only Factors G and H can be used to be adjustment factors. That is, in order to make β increase from 213.734 to 220 (increase about 6.25 in sensitivity), we change Factor G from G2 (G = medium) to G1 (G = dense) and change Factor H from H1 (H = left) to H2 (H = center). The level change of the two factors can make β increase 5.5, which is very close to 6.25 we expected, but the effect on the SN ratio is very

TABLE 6.23. Data Collection for Initial Versus Optimal Design

	M1		M2		M3		M4		M5	
	0.03		0.06		0.09		0.12		0.15	
	N_1	N_2	N_1	N_2	N_1	N_2	N_1	N_2	N_1	N_2
Initial design	7.603	5.828	15.007	11.783	21.863	18.282	28.791	24.063	36.093	30.722
Optimal design	6.753	6.134	13.662	12.304	20.635	18.797	26.896	24.645	35.182	31.205

small. Therefore, we finally select A12B2C2D2E3F3G1H2I1J3 as the optimal condition: A = (Y2 + X1) on right side, B = medium, C = 20, D = sparse, E = thin, F = thin, G = dense, H = center, I = left, and J = right.

We need to collect data again for the sake of confirming that functional variability and sensitivity can be simultaneously improved by optimal condition, and we need to compare the results with the initial design (engineer-proposed design: A2B2C2D2E2F2G2H2I2J2).

Table 6.23 shows the data of voltage under different turns ratio and noise conditions.

Based on the equations for the SN ratio and β, we can obtain the values of initial design and optimal design respectively:

- *Initial design:* SN ratio = 40.570 and $\beta = 222.102$.
- *Optimal design:* SN ratio = 44.853 and $\beta = 218.550$.

Improving SN ratio by 6 dB means to reduce the functional variability by 50%. Thus, we can estimate that, from the improvement of 4.283 dB on the SN ratio, the functional variability of the new component design under noise conditions is smaller than the initial design by 39.03%. In the aspect of sensitivity, the slope of the input/output relationship is closer to 220. Figure 6.24 shows such difference.

7. Standardization and Future Plans. Regarding the design of the component, there are still some parameters that have influence on output voltage and can be considered—for example, the stack status among layers, and layers of material in components, and so on. In addition, the variability of material and the variability in manufacturing process can be considered as noise conditions—for example, the part-to-part variability of iron core, and so on.

If the design of experiment takes customer usage conditions, part-to-part variability, and process variability into consideration, R&D engineers can realize the concept of "concurrent engineering" concretely. In other words,

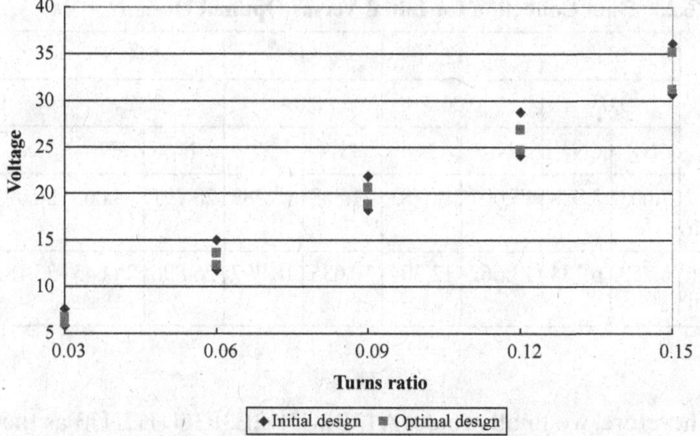

Figure 6.24. Initial design versus optimal design.

R&D engineers don't need to wait until the stage of trial run to know if there is any problem in product design to determine whether they should implement design changes, instead they would implement design for manufacturability simultaneously at the design stage. Therefore, applying the RE method is a very effective and efficient approach to realize concurrent engineering.

6.5.3 Accuracy Engineering of a Measurement System

The development of measurement capability precedes the development of technology. For example, when the energy transformation related to some physical or chemical characteristics can be precisely measured, R&D engineers have the chance to implement some technical operations and controls on this energy transformation, so as to successfully break the technological bottleneck. Thus, the upgrade of measurement capability has an important influence on technology development and breakthrough.

We can describe the input/output relationship of a measurement system as follows: The true value of an object to be measured is the input, and the result of measurement is the output. The common method to improve system capability is to analyze and identify the causes of making the deviations between input values (true values) and output values (measured values) to explore how to minimize the deviations by the changes of system design to enhance measurement accuracy. The analytical steps of this approach are as follows:

- In the situations of different input values, collect the deviation values (or error rate) between them and the corresponding output values.
- Analyze which distribution or pattern is formed by the deviations.
- Analyze the patterns of deviation in various operating conditions.

- Observe and analyze if these patterns have the same trends or different trends, and propose the hypotheses about the root causes of the deviations.
- Confirm the hypotheses and analyze the contribution of system parameters to deviation reduction.
- Modify the design parameters (such as settings, layouts, circuitry, etc.) with larger contribution or to increase numerical compensation mechanisms to fix that measurement system.

Different from the above-mentioned approach in the perspective of RE, a more effective improvement approach is to conduct *direct system design* with design of experiments to realize the ideal function and improve the accuracy of a measurement system. We explain the case study thereinafter:

1. Identify the System to Be Optimized. In a certain electrical circuit, there is a measurement mechanism that is very important and is used to detect the abnormality of functional performance and provide real-time signals to make another protection circuit activate. In this case, the measurement accuracy of the system needs to be optimized.

2. Define Ideal Function and Its Input Signal and Output Response. For a measurement system, the true value of an object to be measured is the input, and the result of measurement is the output. In a good system, the output value must be proportional to the input value. Also, a good measurement system must also be sensitive to different inputs. The input/output relationship can be described by using the terminology of RE: Output value = $\beta \times$ Input value, as shown in Figure 6.25.

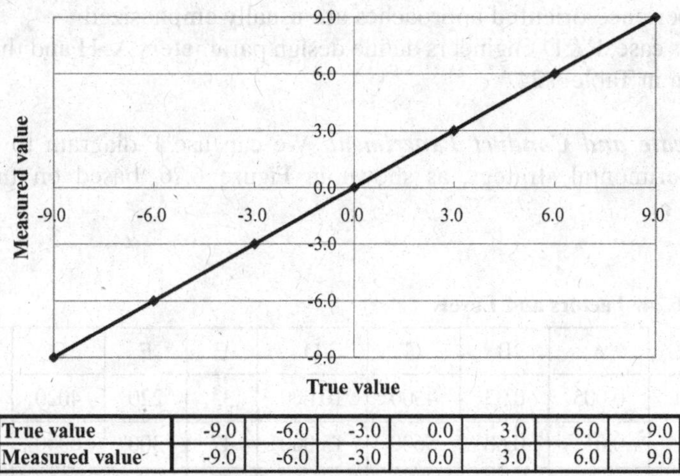

True value	-9.0	-6.0	-3.0	0.0	3.0	6.0	9.0
Measured value	-9.0	-6.0	-3.0	0.0	3.0	6.0	9.0

Figure 6.25. Ideal function.

3. Identify Sources of Variability and Formulate Experimental Strategy. In the design of measurement system, some key components are sensitive to temperature. However, we can't control the conditions under which users use the system, this is because the range of product application is too wide, and the temperature change of usage environment could be very drastic. The temperature change would cause functional variability, so how R&D engineers efficiently design a robust system that can realize measurement accuracy under various different operating temperatures is a key issue.

Despite the fact that several key components are sensitive to temperature, we can, by means of a design (i.e., the combination of component values on the electrical circuit and system settings in the system design), develop a circuit that is "insensitive" to temperature. In other words, design itself can be regarded as a countermeasure for solving problems. R&D engineers define temperature as noise factor whose two levels are:

- N_1 condition: Low temperature.
- N_2 condition: High temperature.

We expect to, through the experimental design approach, make the measurement system have almost the same good measurement accuracy under the conditions of low temperature and high temperature.

4. Define Design Parameters and Their Levels. In designing this system, there are eight design parameters, A–H, whose component values and system setting values can be freely selected by R&D engineers. These parameters have two or three levels. It would be very inefficient, without applying RE method, to find the factor-level combination that can optimize measurement accuracy from all the possible combinations of the parameters. It's impossible for R&D engineers to find the optimum from the whole design space. That's why experience-oriented approaches are usually emphasized.

In this case, R&D engineers define design parameters A–H and their levels as shown in Table 6.24.

5. Allocate and Conduct Experiment. We can use P-diagram to illustrate our experimental strategy, as shown in Figure 6.26, based on the above discussion.

TABLE 6.24. Factors and Levels

	A	B	C	D	E	F	G	H
Level 1	0.005	0.33	4300	10,000	33	220	4020	15,400
Level 2	0.01	0.68	5600	15,000	47	300	6040	17,800
Level 3		0.91	6340	20,000	68	390	7680	19,600

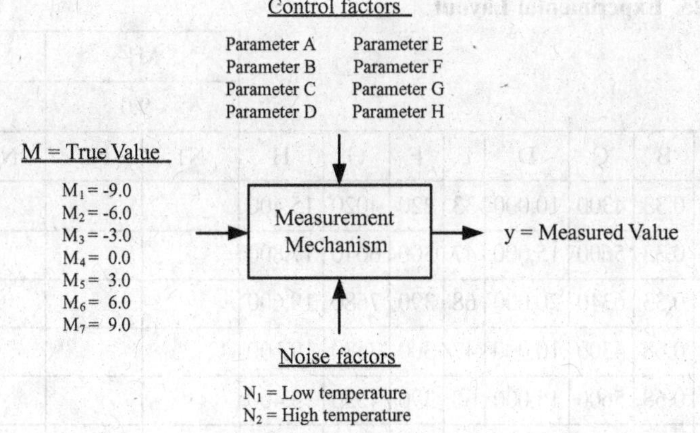

Figure 6.26. P-diagram.

L_{18} orthogonal array can be used for this experiment, and the assignment of factors is shown as the inner array in Table 6.25. R&D engineers need to collect the data of output values under N_1 and N_2 while changing different input values: M_1, M_2, M_3, M_4, M_5, M_6, and M_7. The overall experimental layout (inner array and outer array) is shown as Table 6.25.

Table 6.26 shows all the collected data of 18 runs.

6. Analyze Data and Conduct Two-Step Optimization and Confirmation Experiment. Take the first and second runs of 18 runs as examples; we use the equations explained in Section 5.4.2 to calculate the SN ratio and β values. The calculation of the first run is as follows:

$$S_T = (-7.364)^2 + (-8.556)^2 + \cdots + (14.922)^2 + (14.352)^2 = 1053.086$$

$$r = (-9.0)^2 + (-6.0)^2 + (-3.0)^2 + (0.0)^2 + (3.0)^2 + (6.0)^2 + (9.0)^2 = 252$$

$$L_1 = (-7.364 \times -9.0) + (-4.456 \times -6.0) + \cdots + (13.592 \times 6.0) + (14.922 \times 9.0) = 333.951$$

$$L_2 = (-8.556 \times -9.0) + (-5.564 \times -6.0) + \cdots + (12.892 \times 6.0) + (14.352 \times 9.0) = 341.637$$

r_0 is the number of linear contrasts for noise conditions. It is 2 in this case.

$$S_\beta = \frac{(333.951 + 341.637)^2}{252 \times 2} = 905.595$$

$$S_{\beta \times N} = \frac{(333.951)^2 + (341.637)^2}{252} - 905.595 = 0.117$$

$$S_e = 1053.086 - (905.595 + 0.117) = 147.374$$

TABLE 6.25. Experimental Layout

									M1		M2	
									−9.0		−6.0	
	A	B	C	D	E	F	G	H	N1	N2	N1	N2
1.	0.005	0.33	4300	10,000	33	220	4020	15,400				
2.	0.005	0.33	5600	15,000	47	300	6040	17,800				
3.	0.005	0.33	6340	20,000	68	390	7680	19,600				
4.	0.005	0.68	4300	10,000	47	300	7680	19,600				
5.	0.005	0.68	5600	15,000	68	390	4020	15,400				
6.	0.005	0.68	6340	20,000	33	220	6040	17,800				
7.	0.005	0.91	4300	15,000	33	390	6040	19,600				
8.	0.005	0.91	5600	20,000	47	220	7680	15,400				
9.	0.005	0.91	6340	10,000	68	300	4020	17,800				
10.	0.01	0.33	4300	20,000	68	300	6040	15,400				
11.	0.01	0.33	5600	10,000	33	390	7680	17,800				
12.	0.01	0.33	6340	15,000	47	220	4020	19,600				
13.	0.01	0.68	4300	15,000	68	220	7680	17,800				
14.	0.01	0.68	5600	20,000	33	300	4020	19,600				
15.	0.01	0.68	6340	10,000	47	390	6040	15,400				
16.	0.01	0.91	4300	20,000	47	390	4020	17,800				
17.	0.01	0.91	5600	10,000	68	220	6040	19,600				
18.	0.01	0.91	6340	15,000	33	300	7680	15,400				

$$V_e = \frac{147.374}{14-2} = 12.281$$

$$V_N = \frac{0.117 + 147.374}{14-1} = 11.345$$

Thus,

$$\beta = \frac{333.951 + 341.637}{252 \times 2} = 1.340$$

M3		M4		M5		M6		M7	
−3.0		0.0		3.0		6.0		9.0	
N1	N2	N1	N2	N1	N2	N1	N2	N1	N2

and

$$\mathrm{SN} = 10\log\left[\frac{\frac{1}{252\times 2}(905.595-12.281)}{11.345}\right] = -8.062$$

The calculation of the second run is as follows:

$$S_T = (-10.462)^2 + (-11.521)^2 + \cdots + (10.344)^2 + (9.629)^2 = 675.897$$

TABLE 6.26. Data Collection

									M1		M2	
									−9.0		−6.0	
	A	B	C	D	E	F	G	H	N1	N2	N1	N2
1.	0.005	0.33	4300	10,000	33	220	4020	15,400	−7.364	−8.556	−4.456	−5.564
2.	0.005	0.33	5600	15,000	47	300	6040	17,800	−10.462	−11.521	−6.360	−7.621
3.	0.005	0.33	6340	20,000	68	390	7680	19,600	−7.833	−8.932	−4.728	−5.992
4.	0.005	0.68	4300	10,000	47	300	7680	19,600	−10.692	−12.808	−6.797	−8.132
5.	0.005	0.68	5600	15,000	68	390	4020	15,400	−8.667	−9.948	−6.848	−6.994
6.	0.005	0.68	6340	20,000	33	220	6040	17,800	−8.396	−10.438	−5.387	−7.548
7.	0.005	0.91	4300	15,000	33	390	6040	19,600	−9.047	−9.112	−6.104	−6.247
8.	0.005	0.91	5600	20,000	47	220	7680	15,400	−7.885	−8.159	−5.057	−5.235
9.	0.005	0.91	6340	10,000	68	300	4020	17,800	−5.321	−6.530	−5.217	−6.540
10.	0.01	0.33	4300	20,000	68	300	6040	15,400	−11.372	−12.723	−9.716	−9.812
11.	0.01	0.33	5600	10,000	33	390	7680	17,800	−9.141	−10.606	−6.597	−6.977
12.	0.01	0.33	6340	15,000	47	220	4020	19,600	−8.036	−10.186	−5.973	−6.866
13.	0.01	0.68	4300	15,000	68	220	7680	17,800	−7.662	−9.074	−4.762	−5.184
14.	0.01	0.68	5600	20,000	33	300	4020	19,600	−8.468	−9.645	−5.542	−6.668
15.	0.01	0.68	6340	10,000	47	390	6040	15,400	−8.558	−10.847	−6.446	−7.878
16.	0.01	0.91	4300	20,000	47	390	4020	17,800	−9.802	−10.629	−6.578	−7.719
17.	0.01	0.91	5600	10,000	68	220	6040	19,600	−6.049	−6.273	−3.959	−4.353
18.	0.01	0.91	6340	15,000	33	300	7680	15,400	−9.131	−9.213	−6.146	−6.251

$$r = (-9.0)^2 + (-6.0)^2 + (-3.0)^2 + (0.0)^2 + (3.0)^2 + (6.0)^2 + (9.0)^2 = 252$$

$$L_1 = (-10.462 \times -9.0) + (-6.360 \times -6.0) + \cdots + (7.064 \times 6.0) + (10.344 \times 9.0)$$
$$= 286.622$$

$$L_2 = (-11.521 \times -9.0) + (-7.621 \times -6.0) + \cdots + (5.910 \times 6.0) + (9.629 \times 9.0)$$
$$= 292.065$$

r_0 is 2 in this case.

M3		M4		M5		M6		M7	
−3.0		0.0		3.0		6.0		9.0	
N1	N2	N1	N2	N1	N2	N1	N2	N1	N2
−1.339	−2.963	2.034	1.822	7.024	5.280	13.592	12.892	14.922	14.352
−2.535	−3.750	−0.145	−2.564	3.739	3.093	7.064	5.910	10.344	9.629
−2.111	−3.963	1.923	1.794	8.735	6.457	12.757	11.795	16.242	14.882
−5.723	−6.969	−2.189	−3.617	0.640	0.020	4.785	3.956	8.456	7.125
−4.161	−4.427	1.198	−0.445	5.022	4.171	10.565	9.369	13.215	11.679
−3.542	−3.980	−0.452	−0.947	4.015	2.629	8.347	6.610	11.127	10.040
−5.572	−5.709	−1.054	−1.275	−2.379	−2.687	−0.527	−0.927	2.943	2.143
−2.901	−3.733	0.290	0.150	−1.245	−2.915	6.364	5.149	9.344	8.971
−2.850	−3.613	0.858	0.459	7.934	6.511	8.913	7.132	11.004	10.862
−7.182	−7.290	−2.062	−2.251	4.076	3.807	7.764	7.394	10.796	9.256
−3.051	−4.233	0.169	0.024	1.768	0.216	5.251	3.483	8.823	6.791
−1.898	−2.239	0.588	0.493	5.346	5.006	8.512	7.981	11.182	10.251
−4.546	−4.953	−2.947	−3.781	3.044	2.504	6.184	5.434	10.214	8.964
−3.180	−3.841	−0.320	−0.752	2.240	1.795	6.386	5.779	9.339	9.286
−3.390	−4.723	1.938	1.217	6.937	4.908	80.29	7.377	10.722	9.068
−5.639	−5.787	−3.823	−4.073	3.763	3.119	6.541	6.068	9.878	8.341
0.049	−0.526	3.189	2.083	5.386	4.175	7.653	7.015	10.253	9.765
−3.010	−3.110	0.020	0.009	2.601	2.385	6.027	5.664	8.764	8.407

$$S_\beta = \frac{(286.622 + 292.065)^2}{252 \times 2} = 664.443$$

$$S_{\beta \times N} = \frac{(286.622)^2 + (292.065)^2}{252} - 664.443 = 0.059$$

$$S_e = 675.897 - (664.443 + 0.059) = 11.395$$

$$V_e = \frac{11.395}{14 - 2} = 0.950$$

TABLE 6.27. Data Analysis

									M1		M2	
									−9.0		−6.0	
	A	B	C	D	E	F	G	H	N1	N2	N1	N2
1.	0.005	0.33	4300	10,000	33	220	4020	15,400	−7.364	−8.556	−4.456	−5.564
2.	0.005	0.33	5600	15,000	47	300	6040	17,800	−10.462	−11.521	−6.360	−7.621
3.	0.005	0.33	6340	20,000	68	390	7680	19,600	−7.833	−8.932	−4.728	−5.992
4.	0.005	0.68	4300	10,000	47	300	7680	19,600	−10.692	−12.808	−6.797	−8.132
5.	0.005	0.68	5600	15,000	68	390	4020	15,400	−8.667	−9.948	−6.848	−6.994
6.	0.005	0.68	6340	20,000	33	220	6040	17,800	−8.396	−10.438	−5.387	−7.548
7.	0.005	0.91	4300	15,000	33	390	6040	19,600	−9.047	−9.112	−6.104	−6.247
8.	0.005	0.91	5600	20,000	47	220	7680	15,400	−7.885	−8.159	−5.057	−5.235
9.	0.005	0.91	6340	10,000	68	300	4020	17,800	−5.321	−6.530	−5.217	−6.540
10.	0.01	0.33	4300	20,000	68	300	6040	15,400	−11.372	−12.723	−9.716	−9.812
11.	0.01	0.33	5600	10,000	33	390	7680	17,800	−9.141	−10.606	−6.597	−6.977
12.	0.01	0.33	6340	15,000	47	220	4020	19,600	−8.036	−10.186	−5.973	−6.866
13.	0.01	0.68	4300	15,000	68	220	7680	17,800	−7.662	−9.074	−4.762	−5.184
14.	0.01	0.68	5600	20,000	33	300	4020	19,600	−8.468	−9.645	−5.542	−6.668
15.	0.01	0.68	6340	10,000	47	390	6040	15,400	−8.558	−10.847	−6.446	−7.878
16.	0.01	0.91	4300	20,000	47	390	4020	17,800	−9.802	−10.629	−6.578	−7.719
17.	0.01	0.91	5600	10,000	68	220	6040	19,600	−6.049	−6.273	−3.959	−4.353
18.	0.01	0.91	6340	15,000	33	300	7680	15,400	−9.131	−9.213	−6.146	−6.251

$$V_N = \frac{0.059 + 11.395}{14 - 1} = 0.881$$

Thus,

$$\beta = \frac{286.622 + 292.065}{252 \times 2} = 1.148$$

M3		M4		M5		M6		M7			
−3.0		0.0		3.0		6.0		9.0			
N1	N2	N1	N2	N1	N2	N1	N2	N1	N2	β	S/N
−1.339	−2.963	2.034	1.822	7.024	5.280	13.592	12.892	14.922	14.352	1.340	−8.062
−2.535	−3.750	−0.145	−2.564	3.739	3.093	7.064	5.910	10.344	9.629	1.148	1.744
−2.111	−3.963	1.923	1.794	8.735	6.457	12.757	11.795	16.242	14.882	1.402	−7.485
−5.723	−6.969	−2.189	−3.617	0.640	0.020	4.785	3.956	8.456	7.125	1.059	−7.749
−4.161	−4.427	1.198	−0.445	5.022	4.171	10.565	9.369	13.215	11.679	1.285	−1.382
−3.542	−3.980	−0.452	−0.947	4.015	2.629	8.347	6.610	11.127	10.040	1.131	1.552
−5.572	−5.709	−1.054	−1.275	−2.379	−2.687	−0.527	−0.927	2.943	2.143	0.582	−16.081
−2.901	−3.733	0.290	0.150	−1.245	−2.915	6.364	5.149	9.344	8.971	0.888	−7.077
−2.850	−3.613	0.858	0.459	7.934	6.511	8.913	7.132	11.004	10.862	1.058	−7.124
−7.182	−7.290	−2.062	−2.251	4.076	3.807	7.764	7.394	10.796	9.256	1.334	−3.082
−3.051	−4.233	0.169	0.024	1.768	0.216	5.251	3.483	8.823	6.971	0.955	−3.216
−1.898	−2.239	0.588	0.493	5.346	5.006	8.512	7.981	11.182	10.251	1.144	−0.964
−4.546	−4.953	−2.947	−3.781	3.044	2.504	6.184	5.434	10.214	8.964	0.988	−4.457
−3.180	−3.841	−0.320	−0.752	2.240	1.795	6.386	5.779	9.339	9.286	1.012	4.153
−3.390	−4.723	1.938	1.217	6.937	4.908	8.029	7.377	10.722	9.068	1.173	−2.085
−5.639	−5.787	−3.823	−4.073	3.763	3.119	6.541	6.608	9.878	8.341	1.119	−4.668
0.049	−0.526	3.189	2.083	5.386	4.175	7.653	7.015	10.253	9.765	0.911	−7.645
−3.010	−3.110	0.020	0.009	2.601	2.385	6.207	5.664	8.764	8.407	0.989	10.248

and

$$\mathrm{SN} = 10\log\left[\frac{\frac{1}{252\times 2}(664.443-0.950)}{0.881}\right] = 1.744$$

Likewise, we can obtain the individual β and SN ratio of 18 runs, as shown in Table 6.27.

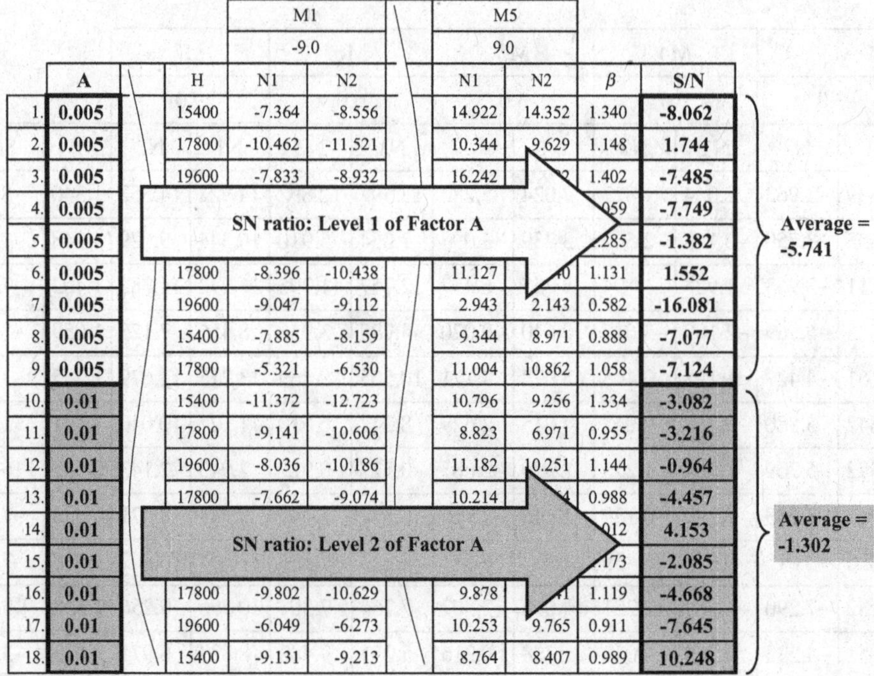

	A	H	M1 (-9.0) N1	M1 (-9.0) N2	M5 (9.0) N1	M5 (9.0) N2	β	S/N
1.	**0.005**	15400	-7.364	-8.556	14.922	14.352	1.340	**-8.062**
2.	**0.005**	17800	-10.462	-11.521	10.344	9.629	1.148	**1.744**
3.	**0.005**	19600	-7.833	-8.932	16.242		1.402	**-7.485**
4.	**0.005**							**-7.749**
5.	**0.005**							**-1.382**
6.	**0.005**	17800	-8.396	-10.438	11.127		1.131	**1.552**
7.	**0.005**	19600	-9.047	-9.112	2.943	2.143	0.582	**-16.081**
8.	**0.005**	15400	-7.885	-8.159	9.344	8.971	0.888	**-7.077**
9.	**0.005**	17800	-5.321	-6.530	11.004	10.862	1.058	**-7.124**
10.	**0.01**	15400	-11.372	-12.723	10.796	9.256	1.334	**-3.082**
11.	**0.01**	17800	-9.141	-10.606	8.823	6.971	0.955	**-3.216**
12.	**0.01**	19600	-8.036	-10.186	11.182	10.251	1.144	**-0.964**
13.	**0.01**	17800	-7.662	-9.074	10.214		0.988	**-4.457**
14.	**0.01**							**4.153**
15.	**0.01**							**-2.085**
16.	**0.01**	17800	-9.802	-10.629	9.878		1.119	**-4.668**
17.	**0.01**	19600	-6.049	-6.273	10.253	9.765	0.911	**-7.645**
18.	**0.01**	15400	-9.131	-9.213	8.764	8.407	0.989	**10.248**

Figure 6.27. Effects of Factor A on the SN ratio.

The analyzed β's and SN ratios shown in Table 6.27 need to be further analyzed to assess the effects of each parameter. Figures 6.27 and 6.28 illustrate how we analyze the effects of Factors A and B on the SN ratio.

Applying the same calculation method, we can yield the effects of each factor on the SN ratio; and we can know, by their ranking, which factors' influences on the SN ratio are more significant and which factors' influences on the SN ratio are insignificant. Table 6.28 shows the information described above; and we can know that the effect of Factor C on the SN ratio is the largest one and that the effect of Factor G on the SN ratio is the smallest.

The approach of calculating factorial effects on β is the same as that of calculating factorial effects on the SN ratio. Figure 6.29 illustrates how we analyze the effects of Factor A on β.

Using the same calculation approach, we can yield the effects of each factor on β. Table 6.29 shows the ranking of the effects of each factor on β: The effect of Factor B on β is the largest one and the effect of Factor A on β is the smallest.

According to the response tables for the SN ratio and β, we can draw the response graphs for the SN ratio and β, as shown in Figures 6.30 and 6.31.

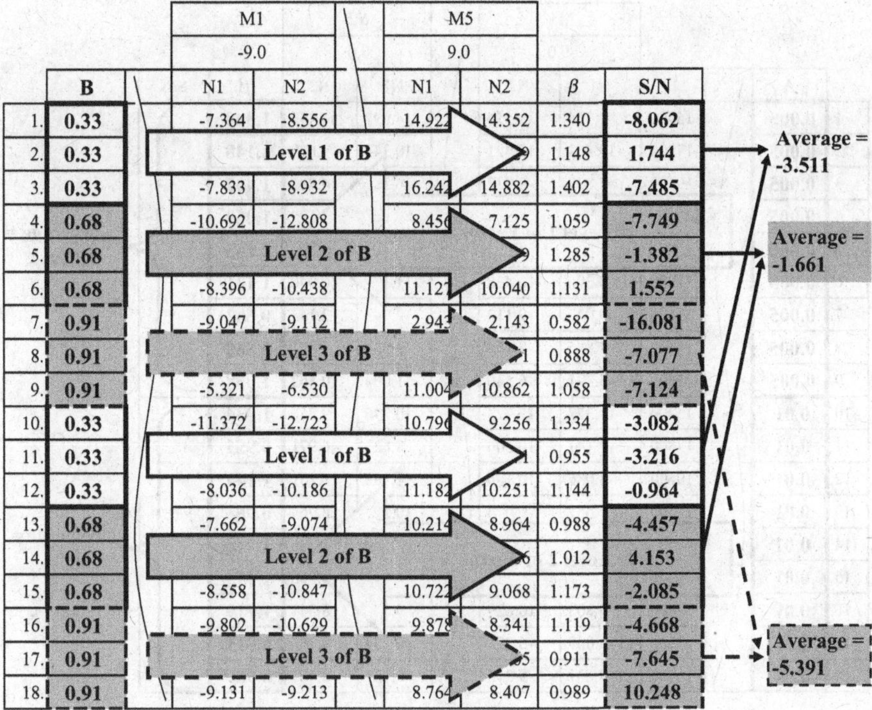

Figure 6.28. Effects of Factor B on the SN ratio.

TABLE 6.28. Response Table for the SN Ratio

	A	B	C	D	E	F	G	H
Level 1	−5.741	−3.511	−7.350	−5.980	−1.901	−4.442	−3.008	−1.907
Level 2	−1.302	−1.661	−2.237	−1.816	−3.467	−0.302	−4.266	−2.695
Level 3		−5.391	−0.977	−2.768	−5.196	−5.820	−3.290	−5.962
Range	4.439	3.730	6.373	4.165	3.295	5.518	1.259	4.055
Rank	3	6	1	4	7	2	8	5

Once we have the information about factorial effects, we can easily identify the optimal condition by the process of two-step optimization:

Step 1. Choose the level with the largest SN ratio to reduce the functional variability under noise conditions.

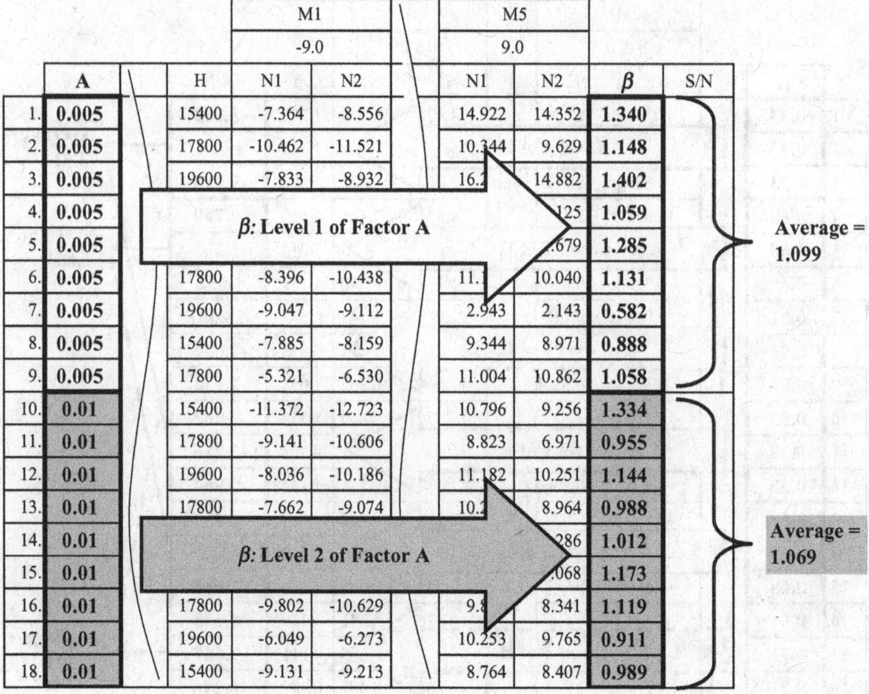

Figure 6.29. Effects of Factor A on β.

TABLE 6.29. Response Table for β

	A	B	C	D	E	F	G	H
Level 1	1.099	1.221	1.070	1.083	1.002	1.067	1.160	1.168
Level 2	1.069	1.108	1.033	1.023	1.088	1.100	1.046	1.066
Level 3		0.924	1.149	1.148	1.163	1.086	1.047	1.018
Range	0.030	0.296	0.116	0.125	0.161	0.033	0.113	0.150
Rank	8	1	5	4	2	7	6	3

Step 2. Choose the factors that influence only β but not (or less) the SN ratio as "adjustment factors" to fine-tune the slope to be placed on the target.

At Step 1, we choose A2B2C3D2E1F2G1H1 (A = 0.01, B = 0.68, C = 6340, D = 15,000, E = 33, F = 300, G = 4020, and H = 15,400) as the factor-level combination that can minimize functional variability. In order to confirm that

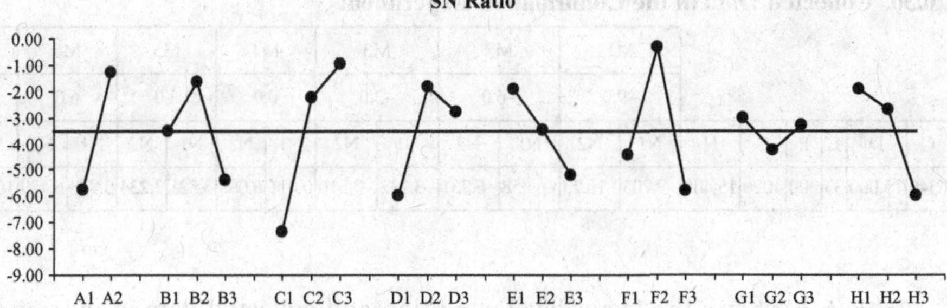

Figure 6.30. Response graph for the SN ratio.

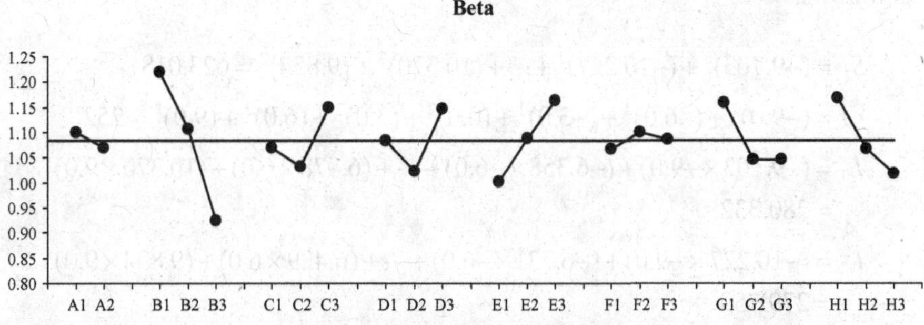

Figure 6.31. Response graph for β.

the functional variability under noise conditions can be reduced as we predicted, we need to predict the SN ratio of the system design and then must collect the actual result of measurement under various conditions of input values and noises. If the actual SN ratio gain computed from the actual data is within the ±30% range of the predicted SN ratio gain, it means that the factorial effects are confirmed.

One point needs to be indicated: since the element of the total average of 18 SN ratios which is repeatedly computed is included in the sum of response values, while computing the predicted value of the SN ratio, the element should be subtracted. In this case, because there are eight factors, the above-mentioned average (which is −3.521) is repeatedly computed seven times. The predicted SN ratio of Step 1 condition is computed as follows:

$$\begin{bmatrix}(-1.302)+(-1.661)+(-0.977)+(-1.816)+(-1.901)+(-0.302)+\\(-3.008)+(-1.907)\end{bmatrix}$$
$$-[(8-1)\times(-3.521)]=11.776$$

Using the same calculation approach, the predicted SN ratio of the initial design is 6.903. Hence, the predicted SN ratio gain is 4.873.

TABLE 6.30. Collected Data in the Confirmation Experiment

								M1		M2		M3		M4		M5		M6		M7	
								−9.0		−6.0		−3.0		0.0		3.0		6.0		9.0	
A	B	C	D	E	F	G	H	N1	N2	N1	N2	N1	N2	N1	N2	N1	N2	N1	N2	N1	N2
0.01	0.68	6340	15,000	33	300	4020	15,400	−9.703	−10.23	−6.358	−6.731	−3.232	−3.516	0.011	0.009	3.521	3.234	6.878	6.429	10.37	9.834

Table 6.30 shows the factor-level condition and collected data in the confirmation experiment.

We compute the SN ratio of the Step 1 conditions, based on the actual data, as follows:

$$S_T = (-9.703)^2 + (-10.227)^2 + \cdots + (10.370)^2 + (9.834)^2 = 623.018$$

$$r = (-9.0)^2 + (-6.0)^2 + (-3.0)^2 + (0.0)^2 + (3.0)^2 + (6.0)^2 + (9.0)^2 = 252$$

$$L_1 = (-9.703 \times -9.0) + (-6.358 \times -6.0) + \cdots + (6.878 \times 6.0) + (10.370 \times 9.0) = 280.332$$

$$L_2 = (-10.227 \times -9.0) + (-6.731 \times -6.0) + \cdots + (6.429 \times 6.0) + (9.834 \times 9.0) = 279.759$$

r_0 is 2 in this case.

$$S_\beta = \frac{(280.332 + 279.759)^2}{252 \times 2} = 622.424$$

$$S_{\beta \times N} = \frac{(280.332)^2 + (279.759)^2}{252} - 622.424 = 0.001$$

$$S_e = 623.018 - (622.424 + 0.001) = 0.593$$

$$V_e = \frac{0.593}{14 - 2} = 0.049$$

$$V_N = \frac{0.001 + 0.593}{14 - 1} = 0.046$$

Thus,

$$\mathrm{SN} = 10 \log \left[\frac{\frac{1}{252 \times 2}(622.424 - 0.049)}{0.046} \right] = 14.320$$

TABLE 6.31. Data Collection for Initial Versus Optimal Design

	M1		M2		M3		M4		M5		M6		M7	
	−9.0		−6.0		−3.0		0.0		3.0		6.0		9.0	
	N1	N2	N1	N2	N1	N2	N1	N2	N1	N2	N1	N2	N1	N2
Initial design	−9.654	−11.147	−6.539	−7.687	−3.276	−3.710	0.143	0.075	3.793	3.340	7.748	6.599	11.381	9.862
Optimal design	−8.579	−9.325	−5.639	−6.326	−2.808	−3.243	0.006	0.004	3.317	2.845	6.349	5.808	9.491	8.632

Since we have the actual data (as shown in Table 6.31) of initial design (engineer-proposed design: A2B2C2C2E2F2G2H2), we can calculate the actual SN ratio and obtain 6.606. Thus, the actual SN ratio gain is 7.714 (14.320 minus 6.606). The additivity of factorial effects is confirmed very well because the actual gain is greater than +30% of 4.873 (which is the predicted gain mentioned above). Although such result looks very good, what raises the factorial effects still need to be explored when the actual gain is much greater than +30% of the predicted gain. (In this case, we accept the good result without further exploration.) After Step 1 of two-step optimization, the variability is minimized but there is still some deviation in sensitivity (slope) between the result of measurement and ideal relationship. We need to adjust the slope, at Step 2, to be 1, and we do not need to cause the change in the SN ratio.

We can use the data in Table 6.30 to calculate β as follows:

$$\beta = \frac{280.332 + 279.759}{252 \times 2} = 1.111$$

Since the slope of the relationship between input value and output value is 1.111, which is larger than the expected β, we have to fine-tune β to be 1 without affecting the measurement variability. According to Figures 6.30 and 6.31, Factor G can be used to be adjustment factor. At Step 1, we set its level as G1 (G = 4020); and at Step 2, to make β decrease from 1.111 to 1 (decrease 0.111 in sensitivity), we change Factor G from G1 to G3 (G = 7680). This is because, according to Table 6.29, the difference of the effects between G1 and G3 is 0.11, which is very close to the 0.111 we needed, and the effect on the SN ratio is very small (which is estimated, from Table 6.28, as 0.28 by subtracting −3.290 from −3.008).

According to the process of two-step optimization described above, the optimal factor-level combination is finalized as A2B2C3D2E1F2G3H1 (A = 0.01, B = 0.68, C = 6340, D = 15,000, E = 33, F = 300, G = 7680, and H = 15,400). For the sake of confirming that measurement variability and slope can be simultaneously improved by optimal conditions, we need to collect data again and to compare the results with the initial design (engineer-proposed design: A2B2C2D2E2F2G2H2).

Table 6.31 shows the data of output value under different input values and noise conditions.

By using the computation equations of the SN ratio and β, we can obtain their values for the initial design and the optimal design:

- *Initial design:* SN ratio = 6.606 and $\beta = 1.175$.
- *Optimal design:* SN ratio = 10.128 and $\beta = 1.003$.

Since the functional variability is reduced by 50% by every 6-dB improvement of the SN ratio, we can estimate that, from the improvement of 3.522 dB on the SN ratio, the variability of the result of measurement of new system design under different conditions of input values and noises is smaller than the initial design by 33.43%. In the aspect of sensitivity, the slope of the input/output relationship is closer to 1. Figure 6.32 shows such difference.

7. Standardization and Future Plans. The improvement of measurement capability plays a very important role for technology development. This is because only when a technology "can be measured," whether it's good or

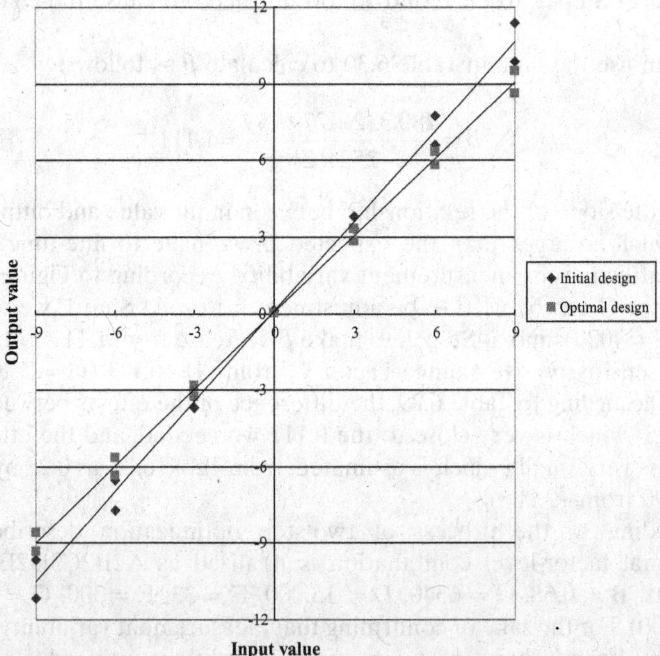

Figure 6.32. Initial design versus optimal design.

bad "can be seen"; and only when a technology is good or bad "can be seen," R&D engineers can collect the data to make the technology "can be optimized."

Based on the study of this case, R&D engineers can, according to the tendency of factorial effects, define the level values on the extension line toward the good tendency of each factor and conduct a new RE experiment to further fine-tune measurement accuracy. In addition, what raises the factorial effects on reducing measurement variability in this case also can be explored for obtaining more accurate engineering knowledge.

The same design logic can be deployed to the field of injection molding in which RE has been widely applied. We just need to change the ideal function used in this case from "Output value = $\beta \times$ Input value" to "Product dimension = $\beta \times$ Mold dimension." Such application shows the power of the RE method.

6.5.4 Stability Engineering of a Cutting Machine

Although this book is focused on exploring the application of the RE method to product design, this mehod, in terms of practical application, is still often bounded to the operations improvement and process design in a factory. The following case study is to explore how to design machine parameters to make the machine stably perform its intended function. Improvement themes such as "designing the parameters or programs of a machine to control machine performance to be stable and not to cause process variability" are frequently seen in the factory. Therefore, as long as production engineers can understand how to apply RE from this case, it would be very helpful for the improvement of similar manufacturing operations.

Next, we illustrate this case step-by-step:

1. Identify the System to Be Optimized. The most obvious difference between design office and factory is that most of the operations conducted in a design office are the engineering activities of confirming product functionality, and therefore the equipment is smaller and the needed pieces are fewer; while in a factory, there are large-scale processing machines and equipment everywhere, and every production line manufacturing the same product has the same equipment. Thus, the universality of the design of machine parameters is very high. In this case, we take the cutting machine which is frequently seen in a factory for instance to illustrate how to use the RE method to optimize the design of parameter settings.

2. Define Ideal Function and Its Input Signal and Output Response. For the cutting machine in this case, the cutting function is controlled by the magnitude of current. In the viewpoint of RE, cutting volume is output response, and the magnitude of current is input signal. The relationship of the two is a proportional relationship: The larger the magnitude of current, the more the

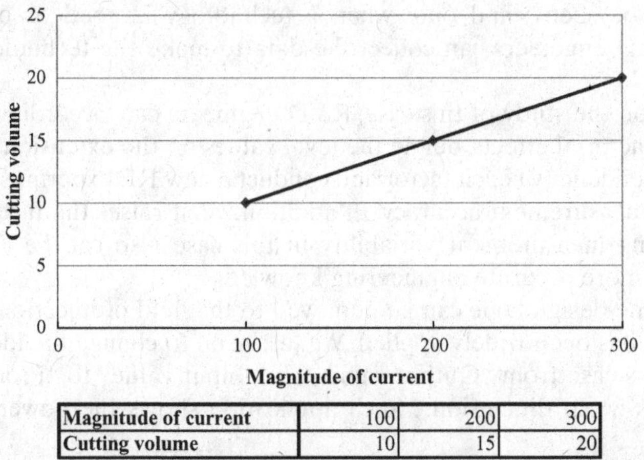

Magnitude of current	100	200	300
Cutting volume	10	15	20

Figure 6.33. Ideal function.

cutting volume; on the contrary, the smaller the magnitude of current, the less the cutting volume.

However, the difference between this case and the above-mentioned cases is that this input/output relationship is a linear relationship that is far away from the origin (0, 0) on axis, as shown in Figure 6.33, because the machine is the cutting machine driven by high current. Thus, production engineers do not treat this relationship between the magnitude of current and cutting volume by using the origin as the starting point, but instead use the magnitude of current over 100 units as the starting point and then manage to realize the ideal proportional relationship between the magnitude of current and cutting volume by parameter design. Therefore, the zero point proportional equation used to calculate the SN ratio in other case studies would not be suitable for this case.

To further explain the above: It's sure that when the magnitude of current is 0, cutting volume is also 0. However, since the application range of current magnitude of this machine is far away from the origin, if we adopt the zero-point proportional equation, it cause us to misjudge that the cutting volume has a larger deviation (i.e., higher variability) than the line through the origin (as shown in Figure 6.34) while evaluating the variability of cutting volume within application range. In fact, if the application range of this machine is far away from the origin, then the cutting function within it would not be affected even if the input/output relationship has a little deviation around the origin. Therefore, we choose to use the reference point proportional equation.

3. Identify Sources of Variability and Formulate Experimental Strategy. When we want to optimize the machining quality of the equipment in the

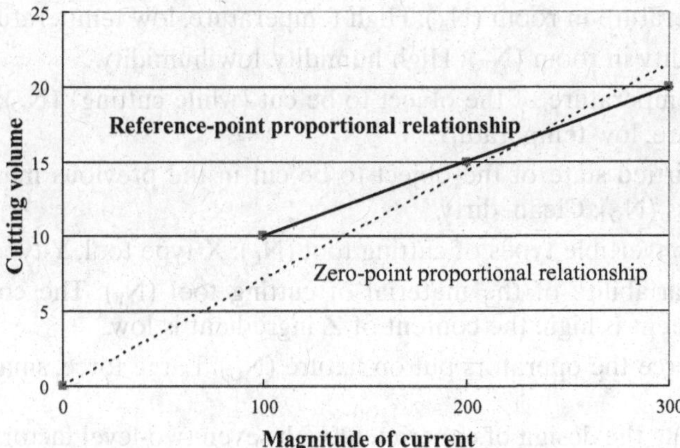

Figure 6.34. Zero point versus reference point proportional relationship.

factory, the major difference between it and product design in the consideration of noise factor is that there is no need to consider customer application conditions in the improvement of machining quality; instead it focuses on the environmental conditions and material variability in the factory. If we have already known the tendencies of the effects of noise factors on machining function, then we can classify noise factors into N_1 and N_2 like the other cases mentioned above:

- N_1: High cutting volume condition.
- N_2: Low cutting volume condition.

In this case, however, for some noise factors, we can know their effect tendencies toward machining quality; but for some other noise factors, we are not sure about their effects on machining function, and therefore we have no way to define clear N_1 and N_2 conditions to collect data in the experimental layout. If there's such situation, we have to implement a preliminary noise experiment first to, through the mean analysis of cutting volume under different conditions, classify N_1 and N_2.

Let's interpret the preliminary analysis method used when the effects of noise factors are unknown. In this situation, our objective is to distinguish which combination of noise factors is N_1 condition and which combination of noise factors is N_2 condition so that we only need a two-level experiment when we conduct preliminary noise experiment. First, we identify the two extreme levels of every noise factor of which we don't know its tendencies of effect (but the two levels must be within the acceptance specifications or the range of machining condition). The noise factors that can influence machining quality, and their two levels are as follows:

- Temperature in room (N_A): High temperature, low temperature.
- Humidity in room (N_B): High humidity, low humidity.
- The temperature of the object to be cut (while cutting) (N_C): High temperature, low temperature.
- The dirtied state of the object to be cut in the previous manufacturing process (N_D): Clean, dirty.
- The convertible types of cutting tool (N_E): X-type tool, Y-type tool.
- The variability of the material of cutting tool (N_F): The content of Z ingredient is high; the content of Z ingredient is low.
- The force the operators put on fixture (N_G): Large force, small force.

Regarding the design of experiment with seven two-level factors, we would choose the orthogonal array of L_8 (2^7) (see Appendix B). We expect to, based on the result (cutting volume) of the eight runs, analyze which level of which factors makes the mean of cutting volume higher and which level of which factors makes the mean of cutting volume lower. The former's factor-level combination is the N_1 condition, and the latter's factor-level combination is the N_2 condition.

To correctly assess the tendencies of noise effects, production engineers have to, while conducting the eight experimental runs, fix the factors other than the above-mentioned seven factors to be the same condition (including fixing the magnitude of current to be 200 units) lest should the experimental results are influenced by the variables other than these seven factors. Table 6.32 shows the layout and collected data of preliminary noise experiment.

Next we conduct mean analysis, as shown in Figure 6.35. Based on such analysis, we can have response table and response graph which are shown in Table 6.33 and Figure 6.36.

TABLE 6.32. Experimental Layout and Data

	N_A	N_B	N_C	N_D	N_E	N_F	N_G	Cutting Volume
1.	High	High	High	Clean	*X*	High	Large	12.8
2.	High	High	High	Dirty	*Y*	Low	Small	15.3
3.	High	Low	Low	Clean	*X*	Low	Small	14.8
4.	High	Low	Low	Dirty	*Y*	High	Large	12.3
5.	Low	High	Low	Clean	*Y*	High	Small	18.7
6.	Low	High	Low	Dirty	*X*	Low	Large	13.5
7.	Low	Low	High	Clean	*Y*	Low	Large	15.6
8.	Low	Low	High	Dirty	*X*	High	Small	10.4

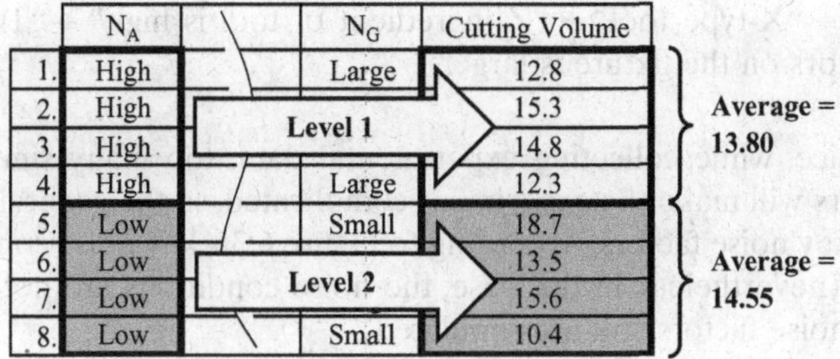

Figure 6.35. Mean analysis.

TABLE 6.33. Response Table

	N_A	N_B	N_C	N_D	N_E	N_F	N_G
Level 1	13.80	15.08	13.53	15.48	12.88	13.55	13.55
Level 2	14.55	13.28	14.83	12.88	15.48	14.80	14.80

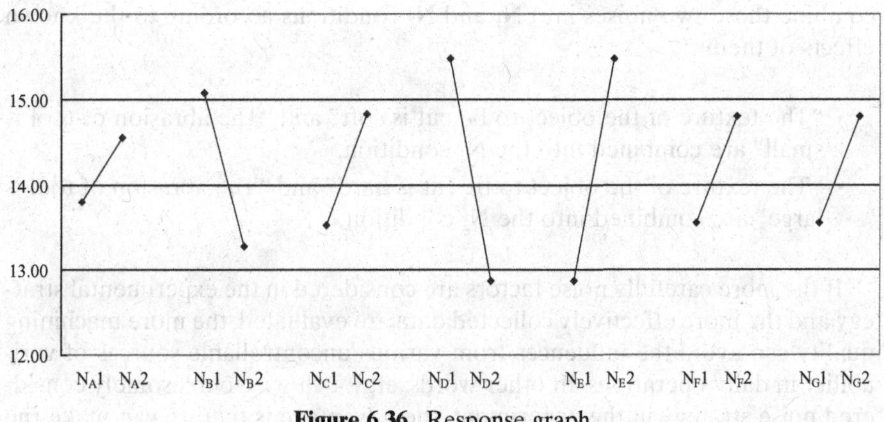

Figure 6.36. Response graph.

Therefore, we can clearly classify noise factors, according to data analysis, as N_1 and N_2:

- **N_1 condition:** "Temperature in room is low" + "Humidity in room is high" + "Temperature of the cut object is low" + "The cut object is clean" + "Y-type tool" + "Z ingredient of tool is low" + "The force of operators on the fixture is small."
- **N_2 condition:** "Temperature in room is high" + "Humidity in room is low" + "Temperature of the cut object is high" + "The cut object is

dirty" + "X-type tool" + "Z ingredient of tool is high" + "The force of operators on the fixture is large."

In practice, while collecting experimental data, too many simulations of noise factors will make the experiment complicated, so the general strategy is to choose key noise factors. According to Figure 6.36, key noise conditions are as follows (nevertheless, in this case, the noise conditions are not simplified, that is, all noise factors still are simulated):

- **N_1 condition:** "Humidity in room is high" + "The cut object is clean" + "Y-type tool."
- **N_2 condition:** "Humidity in room is low" + "The cut object is dirty" + "X-type tool."

When we don't know the effects of noise factors, the above-mentioned method is the analysis method used to deal with this situation. In this case, production engineers actually considered nine noises; however, the effects of two noises are already known so that they were not included in the above-mentioned preliminary noise experiment. Production engineers only need to combine these two noises into N_1 and N_2 conditions according to the known effects of them:

- "The texture of the object to be cut is soft" and "The abrasion of tool is small" are combined into the N_1 condition.
- "The texture of the object to be cut is hard" and "The abrasion of tool is large" are combined into the N_2 condition.

If the more carefully noise factors are considered in the experimental strategy and the more effectively collected data are evaluated, the more machining quality can avoid the influences from various uncontrollable sources of variability in daily operations. In other words, although we troublesomely considered noise strategy in the experiment, the advantage is that we can make the machining quality more insensitive to noises by means of the design of machine parameters. Even if N_1 and N_2, the most extreme conditions, happen, machining quality remains in the good state. In order to achieve this objective, "how to simulate N_1 and N_2 conditions in experiments" to make the data of cutting volume under N_1 and N_2 collectable is very important.

4. Define Design Parameters and Their Levels. In order to make machining quality good, there are eight machine parameters, A–H, that can be designed by production engineers. Since these parameters have two or three levels, if we don't identify the optimal condition through an efficient design method

TABLE 6.34. Machine Parameters and Levels

	A	B	C	D	E	F	G	H
Level 1	Enable	35	Large	1000	2	200	Left	On
Level 2	Disable	40	Medium	1600	4	350	Center	Off
Level 3		45	Small	2200	6	500	Right	Switch

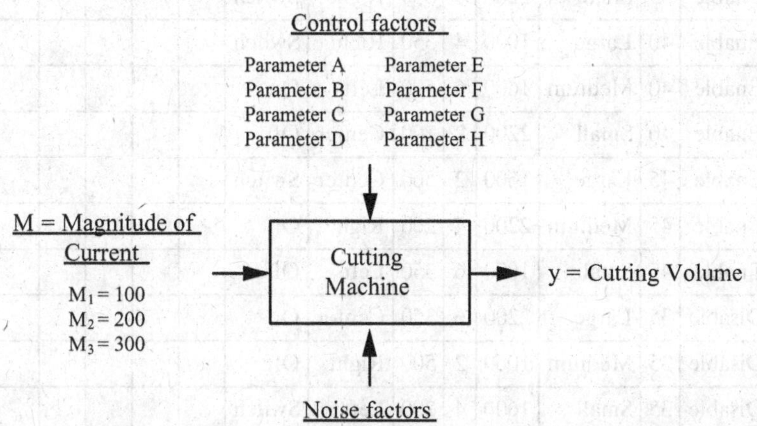

N_1 = Temperature in room is low + Humidity in room is high + Temperature of the cut object is low + The cut object is clean + Y type tool + Z ingredient of tool is very low + The force of operators on the fixture is large + The texture of the cut object is soft + The abrasion of tool is small

N_2 = Temperature in room is high + Humidity in room is low + Temperature of the cut object is high + The cut object is unclean + X type tool + Z ingredient of tool is very high + The force of operators on the fixture is small + The texture of the cut object is hard + The abrasion of tool is large

Figure 6.37. P-diagram.

and depend only on the trial-and-error method, there might be more losses because of ineffectiveness and inefficiency—for example, the time loss from machine adjustment, rework loss, scrap loss, production output loss, and so on.

These machine parameters and their levels are shown in Table 6.34.

5. Allocate and Conduct Experiment. We illustrate our experimental strategy by using a P-diagram in Figure 6.37.

L_{18} orthogonal array can be used for this experiment, and the assignment of factors is shown as the inner array in Table 6.35. Production engineers need to collect the data of cutting volume under N_1 and N_2 while changing different

TABLE 6.35. Experimental Layout

									M1		M2		M3	
									100		200		300	
	A	B	C	D	E	F	G	H	N1	N2	N1	N2	N1	N2
1.	Enable	35	Large	1000	2	200	Left	On						
2.	Enable	35	Medium	1600	4	350	Center	Off						
3.	Enable	35	Small	2200	6	500	Right	Switch						
4.	Enable	40	Large	1000	4	350	Right	Switch						
5.	Enable	40	Medium	1600	6	500	Left	On						
6.	Enable	40	Small	2200	2	200	Center	Off						
7.	Enable	45	Large	1600	2	500	Center	Switch						
8.	Enable	45	Medium	2200	4	200	Right	On						
9.	Enable	45	Small	1000	6	350	Left	Off						
10.	Disable	35	Large	2200	6	350	Center	On						
11.	Disable	35	Medium	1000	2	500	Right	Off						
12.	Disable	35	Small	1600	4	200	Left	Switch						
13.	Disable	40	Large	1600	6	200	Right	Off						
14.	Disable	40	Medium	2200	2	350	Left	Switch						
15.	Disable	40	Small	1000	4	500	Center	On						
16.	Disable	45	Large	2200	4	500	Left	Off						
17.	Disable	45	Medium	1000	6	200	Center	Switch						
18.	Disable	45	Small	1600	2	350	Right	On						

input current: M_1, M_2, and M_3. The overall experimental layout (inner array and outer array) is shown as Table 6.35.

Table 6.36 shows all the collected data of 18 runs.

6. Analyze Data and Conduct Two-Step Optimization and Confirmation Experiment. As described above, since, in this case, the analysis made by the reference point proportional equation is more appropriate than by the zero point proportional equation used in other case studies in this section. Therefore, comparing with the zero point analysis approach, to apply the reference point approach we need to conduct two more things before analyzing the data:

TABLE 6.36. Data Collection

									M1		M2		M3	
									100		200		300	
	A	B	C	D	E	F	G	H	N1	N2	N1	N2	N1	N2
1.	Enable	35	Large	1000	2	200	Left	On	13.4	12.1	17.5	15.5	22.9	20.7
2.	Enable	35	Medium	1600	4	350	Center	Off	13.2	11.9	16.3	14.3	21.3	17.7
3.	Enable	35	Small	2200	6	500	Right	Switch	13.3	12.3	17.5	15.4	22.9	20.7
4.	Enable	40	Large	1000	4	350	Right	Switch	13.6	12.2	15.8	13.8	19.1	16.2
5.	Enable	40	Medium	1600	6	500	Left	On	13.1	11.8	16.8	14.8	22.3	19.2
6.	Enable	40	Small	2200	2	200	Center	Off	13.8	12.4	16.6	14.6	20.8	17.9
7.	Enable	45	Large	1600	2	500	Center	Switch	13.5	12.2	14.9	13.2	17.1	14.4
8.	Enable	45	Medium	2200	4	200	Right	On	14.0	12.6	15.3	13.4	17.5	14.7
9.	Enable	45	Small	1000	6	350	Left	Off	13.5	12.0	24.6	22.1	41.2	36.9
10.	Disable	35	Large	2200	6	350	Center	On	13.3	11.8	14.4	12.6	16.3	13.5
11.	Disable	35	Medium	1000	2	500	Right	Off	13.2	11.9	15.7	13.8	19.4	16.4
12.	Disable	35	Small	1600	4	200	Left	Switch	13.5	12.1	19.2	17.2	27.9	24.7
13.	Disable	40	Large	1600	6	200	Right	Off	14.1	12.6	15.4	13.5	17.9	15.0
14.	Disable	40	Medium	2200	2	350	Left	Switch	13.3	11.9	16.3	14.2	20.4	17.5
15.	Disable	40	Small	1000	4	500	Center	On	13.2	11.8	17.7	15.6	24.4	21.2
16.	Disable	45	Large	2200	4	500	Left	Off	13.3	11.3	15.1	13.1	17.6	14.8
17.	Disable	45	Medium	1000	6	200	Center	Switch	14.5	13.1	20.7	18.4	30.4	26.7
18.	Disable	45	Small	1600	2	350	Right	On	13.6	12.2	15.2	13.3	17.9	15.1

- Select which level of signal factor to be the reference point.
- Transform all the experimental data according to the calculated reference values.

Since M_2 is the magnitude of current needed by most products while cutting, we choose M_2 as the reference point in this case. At this reference point, we compute the average value of the data under N_1 and N_2 conditions of each run in L_{18} and set these values respectively to be the reference value of each run. According to this, all the data values in Table 6.36 must minus that run's reference value according to its run number (M_1 and M_3 must minus the current value, 200, of M_2). After finishing data transformation, we can utilize the zero point approach to analyze the SN ratio and β of each run. The reference-point analysis approach is illustrated in Figure 6.38.

Take the first run of transformed experimental data as an example; the calculation for SN ratio and β is as follows:

$$S_T = (-3.1)^2 + (-4.4)^2 + (1.0)^2 + (-1.0)^2 + (6.4)^2 + (4.2)^2 = 89.57$$

$$r = (-100)^2 + (0)^2 + (100)^2 = 20{,}000$$

$$L_1 = (-3.1 \times -100) + (1.0 \times 0) + (6.4 \times 100) = 950$$

$$L_2 = (-4.4 \times -100) + (-1.0 \times 0) + (4.2 \times 100) = 860$$

r_0 is 2 in this case.

$$S_\beta = \frac{(950+860)^2}{20{,}000 \times 2} = 81.9025$$

$$S_{\beta \times N} = \frac{(950)^2 + (860)^2}{20{,}000} - 81.9025 = 0.2025$$

$$S_e = 89.57 - (81.9025 + 0.2025) = 7.465$$

$$V_e = \frac{7.4650}{6-2} = 1.8663$$

$$V_N = \frac{0.2025 + 7.4650}{6-1} = 1.5335$$

Thus,

$$\beta = \frac{950+860}{20{,}000 \times 2} = 0.0453$$

and

$$\mathrm{SN} = 10\log\left[\frac{\frac{1}{20{,}000 \times 2}(81.9025 - 1.8663)}{1.5335}\right] = -28.8446$$

Applying this computation approach, we can get the individual β and SN ratio of 18 runs, as shown in Table 6.37.

In order to predict the optimal parameter combination, we have to assess the effects of each parameter. The approach of assessing the effects is illustrated in Figure 6.39 by taking the effect analysis of Factor A on the SN ratio as an example.

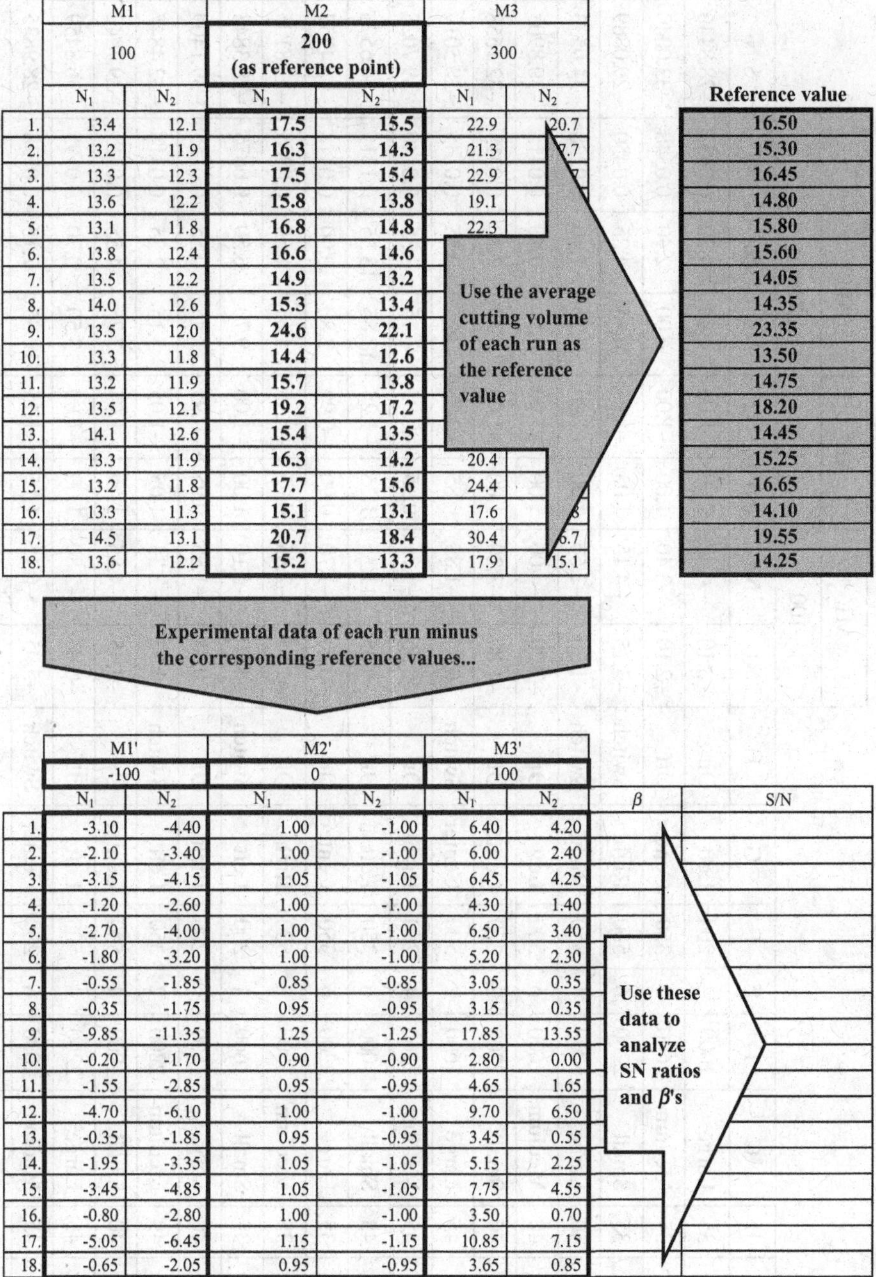

	M1		M2		M3	
	100		**200 (as reference point)**		300	
	N_1	N_2	N_1	N_2	N_1	N_2
1.	13.4	12.1	**17.5**	**15.5**	22.9	20.7
2.	13.2	11.9	**16.3**	**14.3**	21.3	7.7
3.	13.3	12.3	**17.5**	**15.4**	22.9	7
4.	13.6	12.2	**15.8**	**13.8**	19.1	
5.	13.1	11.8	**16.8**	**14.8**	22.3	
6.	13.8	12.4	**16.6**	**14.6**		
7.	13.5	12.2	**14.9**	**13.2**		
8.	14.0	12.6	**15.3**	**13.4**		
9.	13.5	12.0	**24.6**	**22.1**		
10.	13.3	11.8	**14.4**	**12.6**		
11.	13.2	11.9	**15.7**	**13.8**		
12.	13.5	12.1	**19.2**	**17.2**		
13.	14.1	12.6	**15.4**	**13.5**		
14.	13.3	11.9	**16.3**	**14.2**	20.4	
15.	13.2	11.8	**17.7**	**15.6**	24.4	
16.	13.3	11.3	**15.1**	**13.1**	17.6	8
17.	14.5	13.1	**20.7**	**18.4**	30.4	6.7
18.	13.6	12.2	**15.2**	**13.3**	17.9	15.1

Reference value
16.50
15.30
16.45
14.80
15.80
15.60
14.05
14.35
23.35
13.50
14.75
18.20
14.45
15.25
16.65
14.10
19.55
14.25

	M1'		M2'		M3'			
	-100		0		100			
	N_1	N_2	N_1	N_2	N_1	N_2	β	S/N
1.	-3.10	-4.40	1.00	-1.00	6.40	4.20		
2.	-2.10	-3.40	1.00	-1.00	6.00	2.40		
3.	-3.15	-4.15	1.05	-1.05	6.45	4.25		
4.	-1.20	-2.60	1.00	-1.00	4.30	1.40		
5.	-2.70	-4.00	1.00	-1.00	6.50	3.40		
6.	-1.80	-3.20	1.00	-1.00	5.20	2.30		
7.	-0.55	-1.85	0.85	-0.85	3.05	0.35		
8.	-0.35	-1.75	0.95	-0.95	3.15	0.35		
9.	-9.85	-11.35	1.25	-1.25	17.85	13.55		
10.	-0.20	-1.70	0.90	-0.90	2.80	0.00		
11.	-1.55	-2.85	0.95	-0.95	4.65	1.65		
12.	-4.70	-6.10	1.00	-1.00	9.70	6.50		
13.	-0.35	-1.85	0.95	-0.95	3.45	0.55		
14.	-1.95	-3.35	1.05	-1.05	5.15	2.25		
15.	-3.45	-4.85	1.05	-1.05	7.75	4.55		
16.	-0.80	-2.80	1.00	-1.00	3.50	0.70		
17.	-5.05	-6.45	1.15	-1.15	10.85	7.15		
18.	-0.65	-2.05	0.95	-0.95	3.65	0.85		

Figure 6.38. Data transformation in the reference-point analysis approach.

TABLE 6.37. Data Analysis

									M1′		M2′		M3′			
									−100		0		100			
	A	B	C	D	E	F	G	H	N1	N2	N1	N2	N1	N2	β	S/N
1.	Enable	35	Large	1000	2	200	Left	On	−3.10	−4.40	1.00	−1.00	6.40	4.20	0.0453	−28.8446
2.	Enable	35	Medium	1600	4	350	Center	Off	−2.10	−3.40	1.00	−1.00	6.00	2.40	0.0348	−33.0040
3.	Enable	35	Small	2200	6	500	Right	Switch	−3.15	−4.15	1.05	−1.05	6.45	4.25	0.0450	−29.0889
4.	Enable	40	Large	1000	4	350	Right	Switch	−1.20	−2.60	1.00	−1.00	4.30	1.40	0.0238	−34.9533
5.	Enable	40	Medium	1600	6	500	Left	On	−2.70	−4.00	1.00	−1.00	6.50	3.40	0.0415	−30.8903
6.	Enable	40	Small	2200	2	200	Center	Off	−1.80	−3.20	1.00	−1.00	5.20	2.30	0.0313	−32.7658
7.	Enable	45	Large	1600	2	500	Center	Switch	−0.55	−1.85	0.85	−0.85	3.05	0.35	0.0145	−38.5019
8.	Enable	45	Medium	2200	4	200	Right	On	−0.35	−1.75	0.95	−0.95	3.15	0.35	0.0140	−39.7028
9.	Enable	45	Small	1000	6	350	Left	Off	−9.85	−11.35	1.25	−1.25	17.85	13.55	0.1315	−26.6576
10.	Disable	35	Large	2200	6	350	Center	On	−0.20	−1.70	0.90	−0.90	2.80	0.00	0.0118	−41.4757
11.	Disable	35	Medium	1000	2	500	Right	Off	−1.55	−2.85	0.95	−0.95	4.65	1.65	0.0268	−33.8105
12.	Disable	35	Small	1600	4	200	Left	Switch	−4.70	−6.10	1.00	−1.00	9.70	6.50	0.0675	−28.3843
13.	Disable	40	Large	1600	6	200	Right	Off	−0.35	−1.85	0.95	−0.95	3.45	0.55	0.0155	−39.1409
14.	Disable	40	Medium	2200	2	350	Left	Switch	−1.95	−3.35	1.05	−1.05	5.15	2.25	0.0318	−32.4847
15.	Disable	40	Small	1000	4	500	Center	On	−3.45	−4.85	1.05	−1.05	7.75	4.55	0.0515	−29.7942
16.	Disable	45	Large	2200	4	500	Left	Off	−0.80	−2.80	1.00	−1.00	3.50	0.70	0.0195	−36.8459
17.	Disable	45	Medium	1000	6	200	Center	Switch	−5.05	−6.45	1.15	−1.15	10.85	7.15	0.0738	−28.9835
18.	Disable	45	Small	1600	2	350	Right	On	−0.65	−2.05	0.95	−0.95	3.65	0.85	0.0180	−37.2966

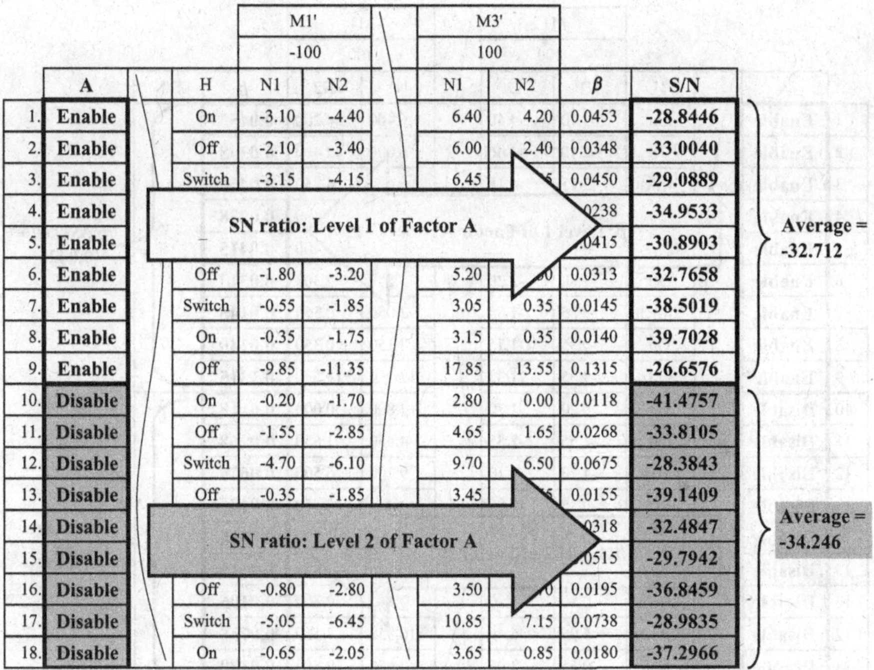

	A	H	M1' (−100) N1	M1' (−100) N2	M3' (100) N1	M3' (100) N2	β	S/N
1.	Enable	On	-3.10	-4.40	6.40	4.20	0.0453	-28.8446
2.	Enable	Off	-2.10	-3.40	6.00	2.40	0.0348	-33.0040
3.	Enable	Switch	-3.15	-4.15	6.45		0.0450	-29.0889
4.	Enable							-34.9533
5.	Enable							-30.8903
6.	Enable	Off	-1.80	-3.20	5.20		0.0313	-32.7658
7.	Enable	Switch	-0.55	-1.85	3.05	0.35	0.0145	-38.5019
8.	Enable	On	-0.35	-1.75	3.15	0.35	0.0140	-39.7028
9.	Enable	Off	-9.85	-11.35	17.85	13.55	0.1315	-26.6576
10.	Disable	On	-0.20	-1.70	2.80	0.00	0.0118	-41.4757
11.	Disable	Off	-1.55	-2.85	4.65	1.65	0.0268	-33.8105
12.	Disable	Switch	-4.70	-6.10	9.70	6.50	0.0675	-28.3843
13.	Disable	Off	-0.35	-1.85	3.45		0.0155	-39.1409
14.	Disable							-32.4847
15.	Disable							-29.7942
16.	Disable	Off	-0.80	-2.80	3.50		0.0195	-36.8459
17.	Disable	Switch	-5.05	-6.45	10.85	7.15	0.0738	-28.9835
18.	Disable	On	-0.65	-2.05	3.65	0.85	0.0180	-37.2966

Figure 6.39. Effects of Factor A on the SN ratio.

TABLE 6.38. Response Table for the SN Ratio

	A	B	C	D	E	F	G	H
Level 1	−32.712	−32.435	−36.627	−30.507	−33.951	−32.970	−30.685	−34.667
Level 2	−34.246	−33.338	−33.146	−34.536	−33.781	−34.312	−34.088	−33.704
Level 3		−34.665	−30.665	−35.394	−32.706	−33.155	−35.665	−32.066
Range	1.534	2.230	5.962	4.887	1.244	1.342	4.981	2.601
Rank	6	5	1	3	8	7	2	4

Based on the same computation approach, we can get the effects of each factor on the SN ratio, as shown in Table 6.38.

Figure 6.40 illustrates how we use the same method to analyze the effects of Factor A on β.

Table 6.39 shows the effects of each factor on β.

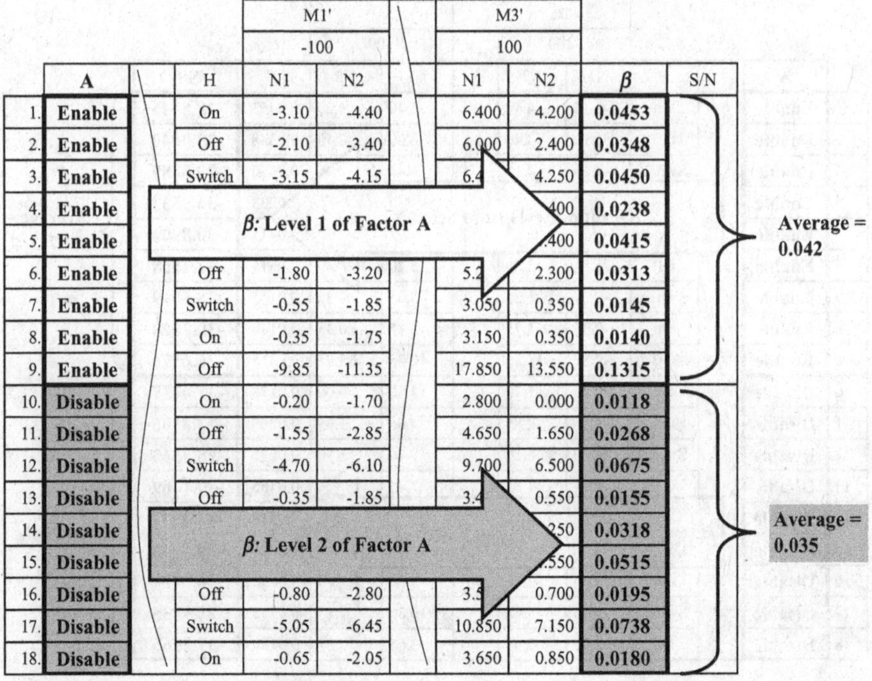

Figure 6.40. Effects of Factor A on β.

TABLE 6.39. Response Table for β

	A	B	C	D	E	F	G	H
Level 1	0.042	0.039	0.022	0.059	0.028	0.041	0.056	0.030
Level 2	0.035	0.033	0.037	0.032	0.035	0.042	0.036	0.043
Level 3		0.045	0.057	0.026	0.053	0.033	0.024	0.043
Range	0.007	0.013	0.036	0.033	0.025	0.009	0.032	0.013
Rank	8	6	1	2	4	7	3	5

Based on the response tables, we can have the response graphs for the SN ratio and β, as shown in Figures 6.41 and 6.42.

At Step 1 of two-step optimization, we select the factor-level combination, A1B1C3D1E3F1G1H3 (A = enable, $B = 35$, $C =$ small, D = 1000, E = 6, F = 200, $G =$ left and H = switch), as the optimal condition that can optimize functionality. In order to make confirmation, we need to check, under this parameter condition, if the SN ratio gain calculated by actual data is within the ±30%

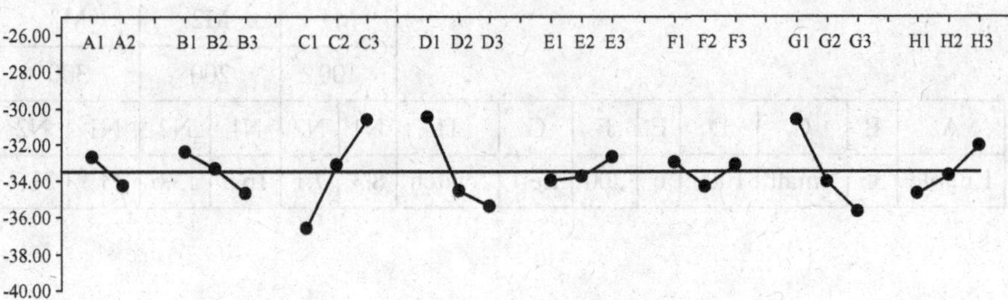

Figure 6.41. Response graph for the SN ratio.

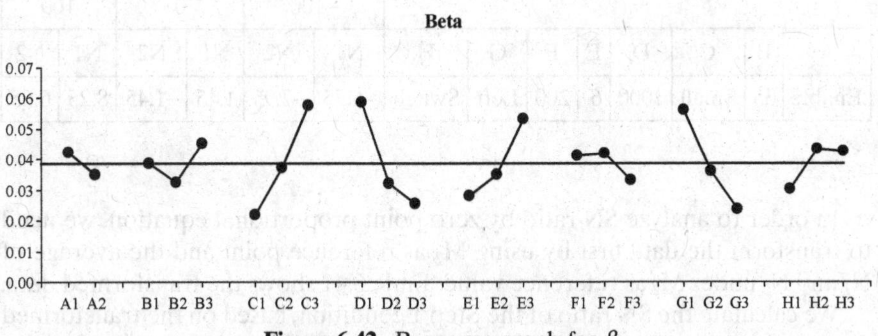

Figure 6.42. Response graph for β.

range of the SN ratio gain predicted by the additive model of factorial effects. If yes, it's confirmed that this parameter condition can reduce functional variability under noise conditions.

Since the element of the total average of 18 SN ratios is repeatedly computed seven times (because of eight factors) while computing the predicted value of the SN ratio, the repeated computed part should be subtracted. In this case, the average is −33.479; hence the predicted SN ratio of Step 1 condition is calculated as follows:

$$\begin{bmatrix}(-32.712)+(-32.435)+(-30.665)+(-30.507)+(-32.706)+\\(-32.970)+(-30.685)+(-32.066)\end{bmatrix}$$
$$-[(8-1)\times(-33.479)]=-20.392$$

Applying the same calculation method, the predicted SN ratio of the initial design is −36.797. Thus, the predicted SN ratio gain is 16.4053.

In order to make confirmation, R&D engineers, based on the selected factor-level combination, organize an experimental run to collect data as shown in Table 6.40.

TABLE 6.40. Collected Data in the Confirmation Experiment

								M1		M2		M3	
								100		200		300	
A	B	C	D	E	F	G	H	N1	N2	N1	N2	N1	N2
Enable	35	Small	1000	6	200	Left	Switch	8.3	7.1	16.5	13.6	23.3	21.7

TABLE 6.41. Transformed Data

								M1′		M2′		M3′	
								−100		0		100	
A	B	C	D	E	F	G	H	N1	N2	N1	N2	N1	N2
Enable	35	Small	1000	6	200	Left	Switch	−6.75	−7.95	1.45	−1.45	8.25	6.65

In order to analyze SN ratio by zero point proportional equation, we need to transform the data first by using M_2 as reference point and the average of N_1 and N_2 under M_2 as reference value. Table 6.41 shows the transformed data.

We calculate the SN ratio of the Step 1 condition, based on the transformed data, as follows:

$$S_T = (-6.75)^2 + (-7.95)^2 + \cdots + (8.25)^2 + (6.65)^2 = 225.255$$
$$r = (-100)^2 + (0)^2 + (100)^2 = 20000$$
$$L_1 = (-6.75\times -100) + (1.45\times 0) + (8.25\times 100) = 1500$$
$$L_2 = (-7.95\times -100) + (-1.45\times 0) + (6.65\times 100) = 1460$$

r_0 is 2 in this case.

$$S_\beta = \frac{(1500+1460)^2}{20,000\times 2} = 219.04$$

$$S_{\beta\times N} = \frac{(1500)^2 + (1460)^2}{20,000} - 219.04 = 0.04$$

$$S_e = 225.255 - (219.04 + 0.04) = 6.175$$

$$V_e = \frac{6.175}{6-2} = 1.5438$$

$$V_N = \frac{0.04 + 6.175}{6-1} = 1.243$$

TABLE 6.42. Data Collection for Initial Versus Optimal Design

	M1		M2		M3	
	100		200		300	
	N_1	N_2	N_1	N_2	N_1	N_2
Initial design	11.9	7.4	16.8	13.8	24.0	18.4
Optimal design	10.5	8.7	15.8	14.2	21.7	19.8

Thus,

$$SN = 10\log\left[\frac{\frac{1}{20{,}000\times 2}(219.04 - 1.5438)}{1.243}\right] = -23.5908$$

Based on the actual data (as shown in Table 6.42) of initial design (engineer-proposed design: A2B2C2D2E2F2G2H2) we had on hand, we can make data trasformation by using M_2 as reference point and the average of N_1 and N_2 under M_2 as reference value to derive the actual SN ratio, −32.8552 from the zero point proportional equation. According to this, we can determine that the actual SN ratio gain is 9.2644 (−23.5908 minus −32.8552). Sine the difference between the actual gain (9.2644) and the predicted gain (16.4053) is not within ±30% range of 16.4053, the additive model of factorial effects is not well confirmed even though the SN ratio is improved (we will comment on this later at the end of this case study). Nevertheless, considering that (1) we still have SN ratio gain when compared with the initial design and (2) the actual SN ratio value is close to the predicted value, we still go to Step 2 and check the final result. Therefore, we calculate β as follow:

$$\beta = \frac{1500 + 1460}{20{,}000 \times 2} = 0.074$$

Since there is a difference between 0.074 and the ideal slope (which is 0.05), we need to conduct adjustment by Step 2 (of two-step optimization). According to Figures 6.41 and 6.42, Factor E can be used to be an adjustment factor. We set the level of Factor E as E3 (E = 6) at Step 1, and we change it from E3 to E2 (E = 4). This adjustment can make the linearity and sensitivity of the technology closer to ideal slope and make the effect on SN ratio (i.e., variability) very small.

According to the above, the optimal factor-level combination is finalized as A1B1C3D1E2F1G1H3 (A = enable, B = 35, C = small, D = 1000, E = 4, F = 200, G = left and H = switch). Table 6.42 shows the data comparison

TABLE 6.43. Transformed Data

	M1′		M2′		M3′	
	–100		0		100	
	N_1	N_2	N_1	N_2	N_1	N_2
Initial design	–3.4	–7.9	1.5	–1.5	8.7	3.1
Optimal design	–4.5	–6.3	0.8	–0.8	6.7	4.8

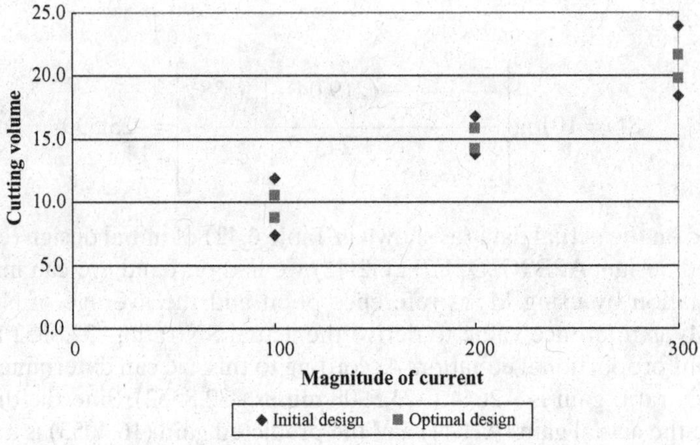

Figure 6.43. Initial design versus optimal design.

between optimal design and initial design (engineer-proposed design: A2B2C2D2E2F2G2H2).

After transforming the data by using M_2 as a reference point and using the average of N_1 and N_2 under M_2 as a reference value (as shown in Table 6.43), we can calculate the SN ratio and β based on the zero point proportional equation:

- **Initial design:** SN ratio = –32.8552 and $\beta = 0.0578$.
- **Optimal design:** SN ratio = –24.9650 and $\beta = 0.0558$.

A 6-dB improvement of SN ratio means 50% reduction of functional variability. Thus, we can estimate that, from the improvement of 7.8902 dB on the SN ratio, the variability of cutting volume of optimal design under different magnitude of current and noise conditions is smaller than the initial design by 59.81%. Moreover, the slope of the input/output relationship is closer to 0.05. Figure 6.43 shows the functional relationship of the two.

7. Standardization and Future Plans. Concerning product functionality, major noise conditions (sources of causing functional variability) are usually customer usage conditions and environmental conditions; while concerning process functionality, major noise conditions are part-to-part variability and operation-to-operation variability. No matter which noise conditions, if engineers can consider them in the experimental design, it's not difficult to find an optimal factor-level combination that is "insensitive" to noises. Seeking such insensitivity (also known as robustness) can more effectively ensure design quality or production quality than the approach of controlling or compensating for the effects of noises.

In factories, it is often seen that the same types of equipment are used on many production lines to execute the same operations in the manufacturing processes. Therefore, the results of this case study can be deployed to all the cutting machines in factory; that is, engineers can implement the standardization of parameter settings to all of the same type of cutting machines. Since we consider many noise conditions in the experimental design, the standardized parameter settings can make operating performance insensitive to machine-to-machine variability (i.e., the relationship between magnitude of current and cutting volume is stable). Regarding plant management for maximizing productivity, such kind of benefit is great. In this case, the marginal confirmation of the additivity of factorial effects means that utilizing the relationship between "magnitude of curent" and "cutting volume" is not a good strategy for optimizing the functionality of this technology. Regarding future palns, R&D engineers should try to confirm the factorial effects well by redefining the ideal function. In practice, the unique and creative formulation of ideal function is regarded as confidentiality. Such process of finding out the best definition of ideal function for optimizing a technology is actually the sketch of technology development activities. There are two views on the result of this case: (1) Form the perspective of improving machine stability, the stability is improved during this project by nearly 60% reduction of variability; however, (2) from the perspective of robust technology development, the experimental strategy in terms of ideal function used in this case study is not recommended. Engineers need to re-optimize the cutting stability with a new experimental planning.

In addition, if engineers already know that the cutting machine that consume higher input current will be adopted in the manufacturing process in the coming future, then it's recommended to consider larger level-values of signal factor (i.e., magnitude of current) in experimental design in advance. As a result, we can simultaneously optimize the current and future equipment, in the same experiment, to realize technology readiness.

6.5.5 Summary

Technology development and product design are complicated jobs because they not only need specialized knowledge, but also involve with the

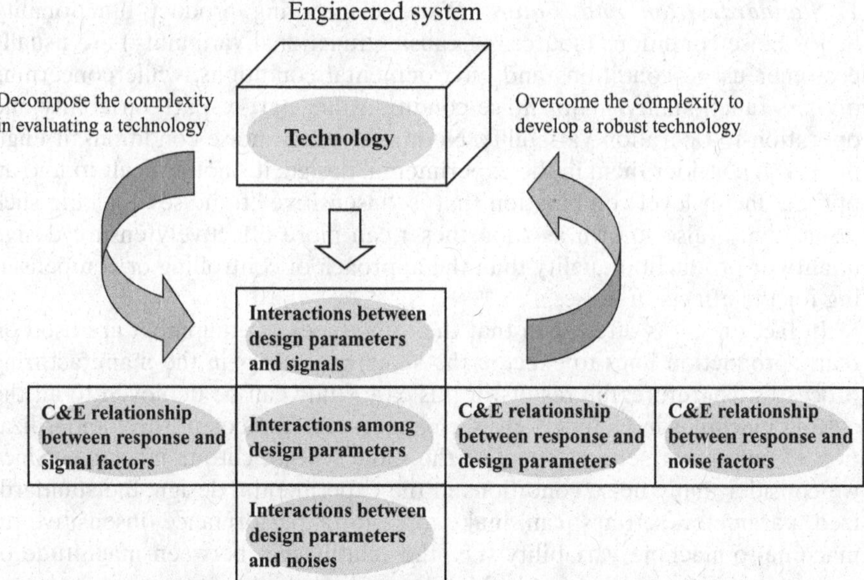

Figure 6.44. Open the black box of a technology.

interactions and cause-and-effect (C&E) relationships among many unknown factors. Hence, R&D engineers often fall into the cycle of "Design–Test–Fix." If we regard the systems or technologies to be optimized in the above-mentioned cases as black boxes, when we open these black boxes and see the configuration stored in those technical systems, we can know how the complexity (as shown in Figure 6.44) of the involved technologies causes the difficulty to R&D engineers while optimizing technologies and makes them difficult to efficiently conduct design optimization.

Next, we interpret the technological complexity that can be seen while opening the black box of a technology:

1. *Interactions among Design Parameters.* The so-called interactions means that the effects of one factor on response value would be influenced to become more or less by the setting of another factor. Interactions determine that there is no additivity among the factorial effects of design parameters, and it's hard to predict the influence of design parameters on response value. Thus, this difficulty makes R&D engineers often fall into a debug cycle.
2. *Cause-and-Effect Relationship between Response and Design Parameters.* The most commonly seen statistical method used to analyze the cause-and-effect relationship between output variables (response variables) and input variables (predictor variables) and model it is regression analysis. However, in practical applications, the coefficient of

determination, R^2 (which is an indicator used in regression analysis), often fails to be statistically significant to make that model helpless to R&D engineers in system (circuits, mechanisms, algorithms, etc.) design. Even because of using such an inappropriate model to predict the effects of design parameters on response value, more Design–Test–Fix activities are caused. Another commonly seen situation is that some engineers implement parameter design according to the analyzed correlation relationship among variables, but this might cause a harder situation for the engineers to efficiently identify the optimal factor-level combination. This is because of the major difference between regression analysis and correlation analysis: The variables that have a correlation relationship do not necessarily have a causal relationship (which should be confirmed by using the experimental design method). If R&D engineers do not apply these statistical methods carefully, the endeavor to analyze the C&E relationship between response and design parameters to make design work efficient might cause more debug cycles instead.

3. *Interactions among Design Parameters and Signal Factors.* In order to make the output response in the operating range of a function stable, R&D engineers need to explore the interaction relationship between signal factors and design parameters, so as to utilize it to minimize the variability around the expected output values under each signal level. Nevertheless, the interaction relationship is difficult to be analyzed by experience-oriented engineering judgment (which may be short of systematic exploration).
4. *Cause-and-Effect Relationship between Response and Signal Factors.* It's difficult in statistical analysis to analyze the influence of the change of the levels of signal factors on response. The main difficulty is that the number of design prototypes is small so that the statistical methods that need a bigger quantity cannot be applied effectively. If the analyzed model is unreliable, engineers also would not refer this causal relationship to implement system (circuits, mechanisms, etc.) design.
5. *Interactions among Design Parameters and Noise Factors.* If a system's actual usage conditions are the same as the conditions in the laboratory, then the functional variability can be controlled easily. However, various noise factors in the real world always cause quality problems. Thus, while developing systems, R&D engineers have to fully consider possible usage conditions to implement various verification tests related to function, reliability, and life at the earlier stages. In this way, the product-development time and cost would be increased. Moreover, even though the engineers try to fully consider noise factors, they might be suffering from the Design–Test–Fix cycle due to the complexity between design parameters and noise factors.
6. *Cause-and-Effect Relationship between Response and Noise Factors.* In practice, the influences of the change of the levels of noise factors on

response usually can only be assessed through tests, thus R&D engineers know the defects in design only when quality problems are discovered by tests. And only when problems are known, design changes can be conducted according to the cause-and-effect relationship between design parameters and that test condition. If, after changing design, new problems occur under another test condition, R&D engineers would implement design changes again according to the cause-and-effect relationship between design parameters and the test condition. The number of such trials may be very few or very many until the system design can pass all the test conditions.

Therefore, how R&D engineers effectively deal with the above-mentioned complexities involving technology or product development is the key to competitiveness. The RE method is a revolutionary approach, based on experimental design and unique analysis methods, to effectively and efficiently deal with these issues. To explain further, we sum up the application cases in this section to characterize the generic model for system (circuits, mechanisms, algorithms, etc.) design without applying RE method as follows:

1. With regard to the design of a new system, R&D engineers usually find a reference model first, and then they modify its design to achieve the design target of the new system. Generally speaking, the senior engineers are fully aware of the influence of each design parameter on response value so that they can quickly judge which parameters should be researched to modify the design, so as to make the response value of the new design reach design target.
2. Reliability and other technical tests are the approaches for inspecting and discovering quality problems while R&D engineers develop systems. When the test result on reliability of new design does not meet the requirements, R&D engineers must conduct design changes.
3. R&D engineers investigate the test failure of the system under certain condition to find out the problem points and reasons causing the design fail in that test, and then they use the explored cause-and-effect relationship to adjust parameter values to make new design pass those test conditions.
4. Regarding the situation necessary to control excessive variability, R&D engineers usually choose higher-grade components or parts to control the variability to an acceptable range.

Nevertheless, before applying the RE method to these practical cases, R&D engineers face several problem points caused by the above-mentioned approach:

1. What the senior engineers understand is only if there is influence of design parameters on the mean value (or slope) of response. The

influence of each parameter on variability can be known only through downstream reliability tests.

2. When the variability problems of a design cause the test result of reliability not meet the requirements, R&D engineers need to implement design changes. However, since engineers fail to clarify which parameter can be used to reduce variability and not influence mean value, they might fall into the debug cycle—that is, spend much time on repeated "test-and-fix" activities until the mean value and variability of the new design get into an acceptable range.
3. To evaluate whether quality is good or bad, R&D engineers develop multiple quality characteristics and measure them; and then they judge, based on the measured values of quality characteristics, if they need to modify the design to improve quality. However, the approach of modifying design according to the response analysis (usually known as mean analysis) to pass test conditions is the design logic of hitting design target first and then controlling variability, which is the reason causing a lot of debug operations in R& D process.
4. According to the response analysis of a system under all the determined test conditions, R&D engineers know how to modify system parameters to make the design pass all test conditions. However, regarding the conditions other than those test conditions, this approach cannot ensure that the system performance can be good under these unknown conditions. Under such circumstance, R&D engineers only can base on the feedback even complaints addressed by customers over system performance to implement design changes.
5. Observed from extending point 4, when R&D engineers want to simultaneously improve multiple quality characteristics, they usually have to make a trade-off; but if they just manage to maintain a balance state among the improvement of certain characteristic value and the deterioration of the others, it's usually hard for them to find optimal parameter design.
6. If the variability of system performance is still large, R&D engineers usually control the variability to an acceptable range by choosing higher-grade components or parts; but the cost is increased in this way.

In these cases, how do we apply the RE method to effectively deal with these issues? We interpret it item by item corresponding to the above-mentioned numbers 1-6 as follows:

1. In the case studies shown above, the reason we take noise factors into consideration when designing an experiment is to compulsorily introduce noise factors to influence experimental results. For instance, we generate an extremely high response and an extremely low response through N_1 and N_2 conditions respectively; and then, according to the SN

ratio analysis, we grasp the factorial effects that can minimize the variability in response caused by these noises. Once the parameter combination of a system can minimize the difference between high response and low response, the functional variability of the system must be very small under the usual conditions other than such extreme conditions, N_1 and N_2. If the engineers do not use this kind of approach, they can only wait for the occurrence of random variation by implementing reliability tests. Only when variation happens, the engineers can judge if there are quality problems by specifications and can then solve problems by firefighting approach. The biggest difference between waiting for random variation and compulsorily making variability is as follows: The occurrence of random variation needs many samples and longer time of test. However, it's hard to answer the questions such as How many samples might be considered as many enough? and How long a peirod of time might be considered as long enough? This is because, now that variation occurs randomly, we have no way to predict how many samples or how long the time we need to discover the variation problems. Therefore, we have to design the noise strategy, so as to discover as early as possible if various design parameters have influence on the mean (or slope) and variability of response, and we don't need to wait until the downstream test stage.

2. If R&D engineers can know factorial effects on variability and mean (or slope) from response tables and graphs, they can quickly find optimal design without repeated test-and-fix. They can know that changing which parameters' levels can reduce variability and that modifying which parameters is helpful to the fine-tuning of the mean (or slope).

3. Each of the electrical, measurement, and machining technologies discussed in the cases shown above has its own criteria used to inspect if the functional performance is good or not. What are quality problems? Usually, R&D engineers use the test specifications or inspection specifications (rather than the deviation from ideality) as the basis of judgment: As long as the functional performance does not exceed the specification limits, there is no problem; if it does, there are problems. In case that all the inspection criteria cannot be simultaneously satisfied, there are problems. Nevertheless, if R&D engineers do not know how to evaluate the degree of the deviation from ideality at the beginning of design, such behavior model of "obtaining quality based on only downstream tests and inspections, along with specification judgment," continues. What we expect is that R&D engineers have an indicator (i.e., SN ratio) to effectively evaluate the degree of the deviation from ideality (ideal function), and they also have a set of procedures that can make functionality (functional variability) close to ideality as early as possible to eliminate various downstream quality problems simultaneously. In this way, engineers have the opportunity to escape from the above-mentioned behavior model.

The unique design approach and analysis techniques of RE are used for this objective.

4. In these case studies, we did not analyze why a system design caused failures under some test conditions so that we could know how to modify the design according to cause-and-effect relationship. This shows that we did not evaluate whether the design is good or bad based on if it passes the test or not; instead what we did is to collect data completely according to the experimental layout of orthogonal array (even if the design performance reflected by the data is poor, we do not need to analyze the failure causes) and then analyze the variability of data to directly identify the optimal parameter combination.
5. In these cases, we collected and analyzed the data related to ideal function, not quality characteristics. Why is such an approach better than the one analyzing quality performance first and then changing the design according to cause-and-effect relationship? This is because inspecting and changing bad design to be good is regarded as tuning or improvement, but is not considered as true design work. If we adopt various corrective actions aiming at inconformity to quality specifications based on the quality criteria or specifications the system design must satisfy, we can easily to fall into the Design–Inspect–Fix cycle. This is because the optimization of a quality characteristic often causes the deterioration of another quality characteristic. A true design work is to study how to make the system design insensitive to the effects brought by various sources of variability (noise factors)—that is, to achieve functional robustness. Only when breaking the framework of downstream quality do R&D engineers have the opportunity to directly improve quality at the upstream by the design logic different from the old ones and from the point of view of functionality.
6. When R&D engineers can know the factorial effects on variability and mean (or slope) from response tables and graphs, they know how to precisely select the level of each design parameter. And when engineers can use the direct parameter design approach to escape from the Design–Test–Fix cycle (debug cycle), they can escape from quality-cost trade-off (which usually means to improve quality by upgrading component and increasing cost) and conduct design optimization in the most cost-effective way. Before grasping the information, it's very hard for R&D engineers to evaluate the influence of any design change on system design and cost.

Although the said concepts have been mentioned in the foregoing chapters, some key concepts, application steps, and practical implications of RE can be more easily understood only by means of using the illustration of case studies. Before entering into the next chapter, we would like to extend the discussion related to these cases, as well as further explain and highlight the two most important concepts in Part II of this book as follows and as the conclusion of this chapter:

1. As for the following two kinds of design logic used by R&D engineers in developing technologies or designing products, there are big differences between the two and the approaches (such as steps for optimization, experimental layout, data collection, analysis measures and methods, etc.) derived from them:

a. Inspecting whether a design or a technology is good or bad.

b. Directly optimizing a design or a technology to be good.

We explain that difference as follows: Suppose that we take the design process as the manufacturing process; the design activities conducted by R&D engineers are just like the manufacturing operations that must be done on a production line. In the manufacturing process at every factory, there must be test or inspection stations to ensure product quality, and the inspection criteria used in the manufacturing process are just like the quality characteristics used to inspect if a system design is good in the design process:

- When the performances of that design are all good in these quality characteristics, it means that the design is a good one that does not need further modification.
- When the quality performance of that design is poor, it means that the design might have defects so that it needs to be returned to R&D engineers to see how to conduct design changes.

However, a poor design has already not been good before inspections, but we can never know that until we conduct inspections (i.e., confirm the performance of multiple quality characteristics). In view of this, based on this approach, we cannot "obtain" quality through inspections; we can only "contain" or "intercept" the quality deficiency flowing to downstream stages. Moreover, it's probable that the engineers can only finalize the design after several times of Design–Inspect–Fix. In practice, inspecting a design and then fixing it is a commonly seen R&D approach.

To inspect and make bad design become good is considered to be tuning or improvement, but it's not regarded as true design work. The better approach we need should be able to directly optimize a design to be good,—that is, to build quality in the system design so that the design can be ensured good quality even without inspections. Naturally, to be cautious in practice, the necessary inspections wouldn't be canceled; however, before conducting inspections, we can feel assured that the design can pass various inspections, without worrying about whether it is good or bad. Thinking through this point, it's clear why we should set ideal function and not quality characteristics as the experimental objective of applying RE: If we can determine the intended function of the system design, then we can build in quality at the beginning of designing a system (circuits, mechanisms, algorithms, etc.).

2. To get quality, don't measure quality! (This is a well-known slogan in the field of RE.) Point 1 mentioned above is to interpret the difference in nature between the conventional design method and the RE method, and the following is to interpret the key concepts regarding why the RE method can realize "direct design approach."

Take Case 1 in this section for instance, the way R&D engineers judge or inspect whether a circuit design is good or bad is based on whether the quality performances such as voltage overshoot, voltage undershoot, heat, and noise conform to specifications; that is, in order to get quality, engineers have to measure if the performances of various quality characteristics are good enough and if satisfying the specifications means good enough. Nevertheless, the complexity of that circuit design is that R&D engineers are usually not able to easily identify a circuit design satisfying multiple quality characteristics. This is because the optimization of a quality characteristic often causes the deterioration of another quality characteristic so that usually the engineers can only find a design after trade-off. The so-called "trade-off" means that the solution may not be able to make every quality characteristic attain the optimal state, but it can make them satisfy all the specifications. Perhaps there is a better solution in the whole design space, but this can be regarded as a good design in consideration of time and cost spent for a better solution.

As follows, we hypothesize a situation for Case 1 to interpret the inefficiency of the above-mentioned method: With regard to that circuit design, assume that R&D engineers have to ensure five quality characteristics satisfying specifications. So far, after several times of trials and errors, the engineers have already been able to make three quality characteristics satisfy specifications; as long as they can make the other two quality characteristics conform to specifications, the design work is completed. At this time, the engineers are concerned with the problem points related to the two quality characteristics displayed on the equipment used to test the circuit performance. Once R&D engineers can, based on engineering experience and knowledge, judge the root causes and eliminate them with design changes, the quality performance related to those two characteristics should become better. The quality improvement approach (i.e., Design–Build–Test–Fix approach) described above is shown in Figure 6.45 (Case 1 is used as illustration), and the design logic behind it is follows:

- Make quality become good by solving problems.
- Solve problems by analyzing and utilizing cause-and-effect relationships.

Under this design logic, every time that R&D engineers design circuits, varying with the different situations fulfilling quality characteristics (for example, there are two quality characteristics not conforming to the specifications at this time, but another three characteristics not satisfying the specifications next time), the problem points that concern engineers are different, the cause-and-effect relationships engineers need to explore are different, and

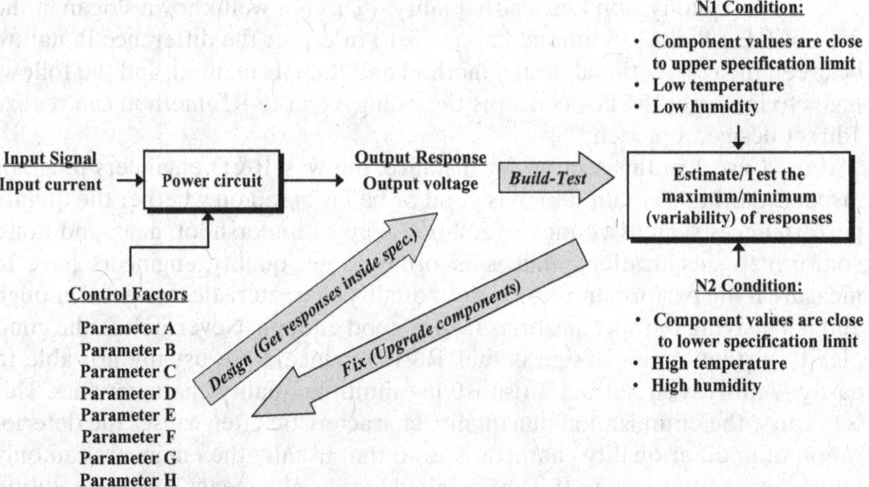

Figure 6.45. Conventional approach: Quality improvement by cause-and-effect relationships.

thereby the design changes needed might also be different. Therefore, the experimental study on individual product needs to be conducted individually. This time's experimental planning and analysis result are not easy to be deployed to the next time's circuit design.

In Case 1, we didn't set multiple quality characteristics as the subjects for data collection and analysis, but focused on the intended function of that circuit. When we can regard the nature of the circuit that converts current to voltage as the subject for data collection and analysis, the design logic used by us is described as follows:

- Now that the intended function of that circuit is conversion between current and voltage, it means that if we can make that circuit perfectly convert current to voltage to realize an ideal circuit, such an ideal circuit should pass the inspections, regardless of the inspection criteria or quality characteristics, and be regarded as an optimized design.
- In order to get quality, if measuring functionality (i.e., evaluating the functional variability of the input/output energy transformation) can be more efficient than measuring quality (i.e., measuring multiple quality characteristics), then functionality improvement can be regarded as a revolutionary design approach, namely killing multiple birds with one stone (i.e., solve multiple quality problems with functionality optimization). The difference in design steps between "directly improving functionality" shown in Figure 6.46 (taking Case 1 for instance) and "eliminating quality problems by cause-and-effect relationships" shown in Figure 6.45 is as follows: In direct design approach, functionality is directly optimized (which means to make functionality insensitive to the effects of noise

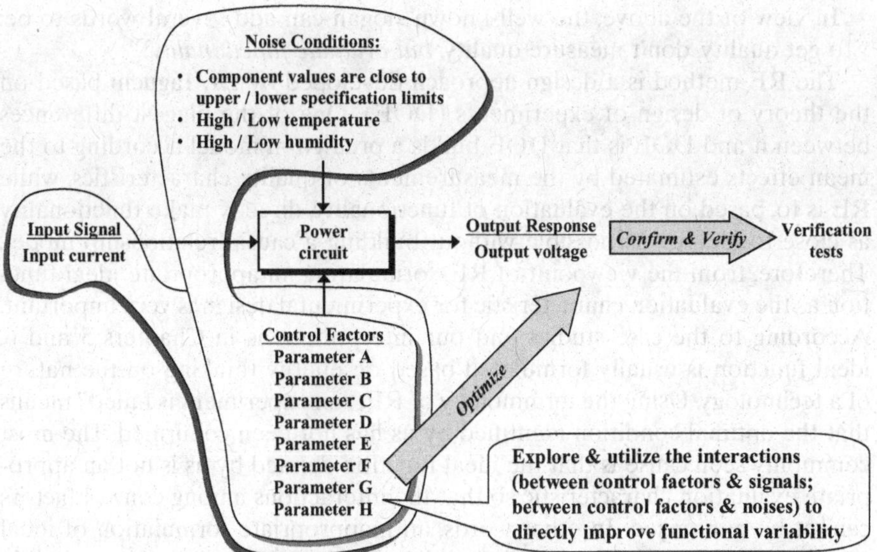

Figure 6.46. Direct design approach: Functionality improvement by minimizing the effects of noise factors.

factors) through experiments by exploring and utilizing the interactions between control factor and signal factors as well as the interactions between control factors and noise factors.

- If we focus on the circuit function of current-to-voltage conversion, what we research and optimize is the technology of how to perfectly convert current to voltage. The optimization of the technology would be reflected by the quality performance of that circuit. Instead, if we focus on the multiple quality characteristics used to check whether a finished circuit design is good or bad, what we do is only to offer corrective countermeasures toward the various quality problems found in the circuit itself (i.e., product itself, not technology itself) whose design is finished. Knowing this subtle difference is the key success factor for deploying the RE method in organizations. This is because technology has universality; that is, the same product family based on the same technology can all apply the result of technology optimization. In the RE application taking technology development as an experimental objective, product-to-product variability can be regarded as "noise factor" to be dealt with. In contrast, if RE application is to reduce the quality problems of every product development with product itself as the subject, then individual optimization is needed when developing different products. However, in view of precedency (i.e., enhancing technology readiness) and universality (i.e., enhancing technology application flexibility), the focus of RE application must be transited from product development to technology development.

In view of the above, the well-known slogan can add several words to be: "To get quality, don't measure quality, *but evaluate functionality*!"

The RE method is a design approach developed by Dr. Taguchi based on the theory of design of experiments (DOE). One of the biggest differences between it and DOE is that DOE builds a prediction model according to the mean effects estimated by the measurements of quality characteristics, while RE is to, based on the evaluation of functionality, directly make functionality as close to ideality as possible without building a causal relationship model. Therefore, from the viewpoint of RE, formulating an appropriate ideal function as the evaluation characteristic for experimental design is very important. According to the case studies and our interpretations in Chapters 5 and 6, ideal function is usually formulated based on energy thinking on the nature of a technology. Using the terminology of RE, "an experiment is failed" means that the optimal condition identified by us has not been confirmed. The most commonly seen cause is that the ideal function defined by us is not an appropriate evaluation characteristic so that the interactions among control factors cannot be minimized. In other words, an inappropriate formulation of ideal function would cause the additive model of factorial effects to be untenable (i.e., no "additivity") so that the optimal condition has no reproducibility. From the viewpoint of RE, the purpose of experiment is to discover the failure of experiments as soon as possible so that we can avoid the situation where by the optimal condition found by experiments with no reproducibility cannot be known until downstream stages. It's too late! The best strategy is to reformulate an ideal function more suitable for evaluating a design or technology to enable the additive model of factorial effects to be established. This is why RE is called an evaluation technology. Once the characteristic (i.e., appropriate ideal function) for evaluating a design or technology is established, we can effectively and efficiently evaluate and improve the functional variability of that design or technology under various conditions of customer application. Figure 6.47 shows the difference between conventional thinking and strategic thinking about Taguchi experiments. From a conventional perspective, the Taguchi experiment is regarded as a tool for design efficiency (that is, if the quality performance is not better by setting the optimal condition found by this tool, it's concluded that this tool is of no help), rather than the method helps to identify the evaluation characteristic for efficientizing technology development and product design.

What we have to remind ourselves is that the RE method mentioned in this book is the so-called dynamic method in the field of Dr. Taguchi's Quality Engineering, not the static method used to analyze quality characteristics (which is viewed as static characteristics). The so-called "static characteristics" are classified as Nominal-the-Best, Larger-the-Better, and Smaller-the-Better according to the expected tendency of improving quality characteristics. Based on the discussion in this section, the analysis method toward static characteristics is broadly used in the RE experiment taking resolving quality problems as the objective. This is also the major contents of most books or articles about

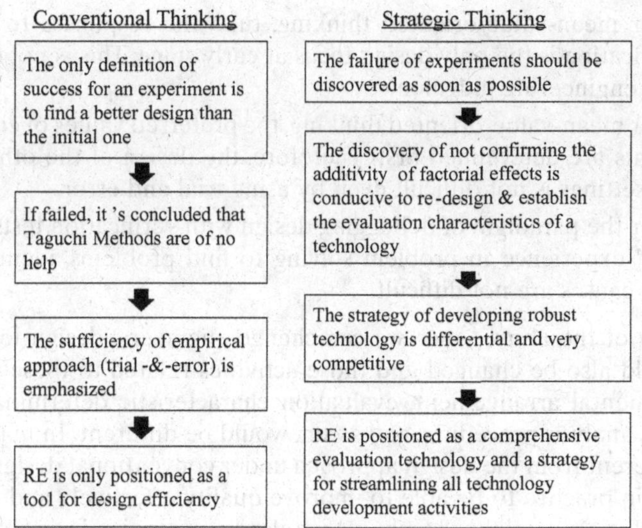

Figure 6.47. Conventional versus strategic thinking about the Taguchi experiments.

RE published in English (because, during the time when the RE method focusing on exploring static analysis was popularized internationally, the dynamic method was just initiated and didn't have enough concrete case studies). However, if R&D engineers want to apply the RE method to technology development and further simultaneously efficientize the product development of the same product family, the dynamic method is the method that can realize one-shot optimization. This is also the evolution of the RE method emphasized by most Japanese books, Japanese articles, and this book.

The improvement of production quality is usually related to the operations seen on the shop floor and the improvement in tangible equipment—for example, to accurately monitor and control every manufacturing process step and machining operation, to invest in new control mechanisms or machining mechanisms, or to introduce automation, and so on. Unlike this, the high leverage of improving design quality is to improve the design logic existing in the brains of R&D engineers. While designing the parameters of a system (circuits, mechanisms, algorithms, etc.), R&D engineers, according to the design steps under that design logic, determine various parameter values that can make quality performance satisfy specifications step by step. The RE method is different from the design logic of the conventional design approach. However, before we really understand how RE is better, based on these differences, than the conventional approach, the most principal reason why R&D engineers don't think they need to use the RE method is relevant to the design logic of mean-value-oriented thinking and Design–Test–Fix described as follows:

- Under mean-value-oriented thinking, tune the responses to get inside specification is the only design focus at early stage. This is not difficult by using engineering judgment.
- Under mean-value-oriented thinking, the preferred values of critical components are determined first. Therefore, the design of the other component settings is not difficult even by using trial and error.
- Under the paradigm of debugging design with verification tests, use engineers' experience in problem solving to find problems, identifying and fixing causes are not difficult.

In view of the above, once we can change engineers' design logic, design steps would also be changed and those activities related to these steps, such as experimental arrangement, evaluation characteristic determination, data collection, analysis procedure, and so on, would be different. In many aspects, RE is different from the design approach under conventional design logic and is proved in practice to be able to improve quality effectively and efficiently.

In conclusion, specialized technology (also known as intrinsic technology—for example, electrical engineering, mechanical engineering, information engineering, etc.) can be used to distinguish various technology fields, while RE is regarded as the generic technology and evaluation technology that can be applied to various technology fields, as shown in Figure 6.48.

RE not only can upgrade design efficiency, but also can make design activities have differentiation and competitiveness compared with the activity model of competitors. In addition, the approach of combining CAE (computer-aided engineering) and RE to develop an "intelligent design system" can be demonstrated in many actual cases at present, hence we can expect that RE can be further powered by such computer-based application.

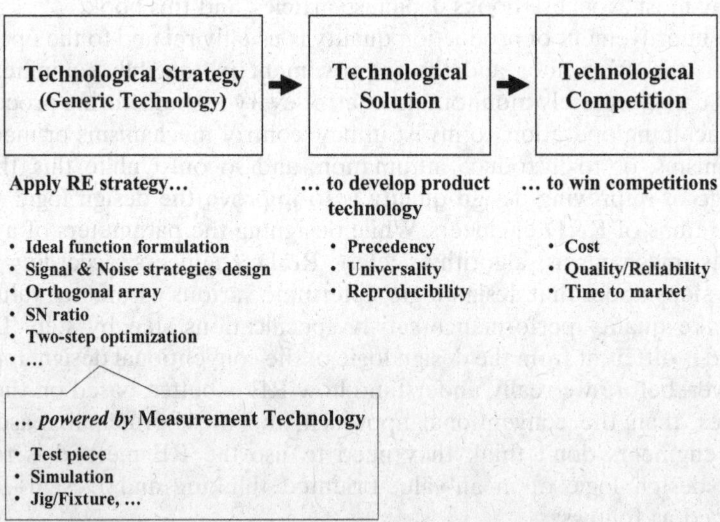

Figure 6.48. RE as generic strategy for developing various technologies.

7 Managing for Paradigm Shift

7.1 WINNING QUALITY-BASED TECHNOLOGY LEADERSHIP

The leading positions in technology and quality are the two key factors for the company to maintain a sustainable competitive advantage. By means of the RE method described in this book, we expect it is helpful for R&D personnel to challenge quality-based technology leadership. The so-called quality-based technology leadership means that a technology, with precedency (readiness) before product planning, is applied to develop the product with competitiveness that is of high quality, low cost, and short time to market, and it has the adjustability satisfying the design targets of products of different generations. RE can be regarded as the quality strategy adopted in the upstream of the company's value chain—that is, applied in the fields of technology and product development. It is used to deal with (a) the technology readiness before product planning (during laboratory research) and (b) technology development during product design; the company, by means of that technology and the products applying it, competes with competitors in the market and gains market share and profit. The so-called technology leadership means that the technology development is competitive in market in both quality and cost.

We can examine the application of RE from two aspects: In the technology development domain, the application strategy of RE is to (a) utilize robust optimization for generic function (related to energy transformation) to make functionality achieve its ideal state even at the stage where no actual products are made and (b) develop the technology that can be used by a group of products with the same generic function. Another angle for examining the application strategy of RE is to look at technology development from harmonizing product and market. The technology development here means to (a) conduct technology deployment for the objective function (related to design intent or customer specification) that the product intended to be realized and (b) identify and break through the technological bottlenecks. In summary, RE application focuses on technical functionality engineering in the technology development field, and it also focuses on product quality designing-in in the product development field.

Quality Strategy for Research and Development, First Edition. Ming-Li Shiu, Jui-Chin Jiang, and Mao-Hsiung Tu.

No matter which examining angle, the company has to effectively develop high-quality technology to protect the current market share or to develop a new market. If the quality management for new product development is the focused quality strategy in late twentith century, then quality-based technology development (called *robust technology development* in this book) is one of the most important quality strategies for the company to gain a competitive advantage in the twenty-first century; at the same time, the company can, by means of this, manage the quality assurance of new product development more effectively. Regarding the RE method described in this book, in micro-viewpoint, it is a methodology of technology optimization; while in macro-viewpoint, it is a strategic innovation which drives the total transformation of paradigm used to dominate the innovations of technology and product.

7.2 KEY SUCCESS FACTORS

For the different purposes of applying RE by managers, their results and influences on organizational behavior will also be different: RE can be just a design optimization method or a paradigm shift accelerator. In this book, applying RE is a strategic thinking that not only is quality strategy and technology development strategy, but also competitive strategy for winning commercial competitions.

Although the concept of quality has evolved from the past "conformance to specification" to "variability within specification," in practice, many organizations still set "zero defects" as their recognition for quality. Under different quality recognition, the organizational behaviors vary, as shown in Table 7.1.

RE is a quality design and engineering method based on the philosophy of variability minimization; hence we would face the challenge brought by the difference of quality recognition while promoting it in an organization. When the managers consider that implementing RE is not only applying a design optimization method but also introducing a paradigm shift accelerator, RE

TABLE 7.1. Zero Defects Versus Variability Minimization

Item	Zero Defects	Variability Minimization
Objective	• Conformance to specification	• Performance of hitting target
Indicator	• Yield rate • Defective rate/failure rate • First pass yield rate, etc.	• Standard deviation • SN ratio, etc.
Measurement	• Quality target easy to be understood, measured, and acceptable	• In the product design and manufacturing process, variability is not easy to be measured, predicted, and improved.

TABLE 7.1. (*Continued*)

Item	Zero Defects	Variability Minimization
Practical behavior	• For the quality characteristic values exceeding the specification, we easily tend to adjust them according to specification or change process-related parameter settings; however, such a way is easy to cause overadjustment and confuse the judgment on *deficiency* and *abnormality*.	• According to the procedure of "two-step optimization": After reducing variability first, quality performance is adjusted to be target value. However, at manufacturing stage, sometimes we might need to try to acquire a beneficial balance between "stabilizing processes (reduce variability)" (which needs to take actions for "abnormal" "good products") and "satisfying specification" (which needs to take actions for "non-abnormal" "defective products")
Quality design and improvement approaches based on this concept	• Easy to be directed to use the debug approach to conduct quality design or improvement; once quality performance enters specification, design or improvement is stopped and downstream activities are executed. However, the quality characteristic values near the limits of specification are easy to exceed specification at downstream stages due to variability. • Easy to be directed to conduct improvement by enlarging test plan (such as increasing test samples, test coverage rate, test time, etc.), enhancing test capability (such as investing on test equipment), or using traditional approaches.	• After identifying the important factors that influence variability, and using them to reduce variability as possible, adjust quality performance to be the target value. If the specification tolerance is wide enough, variability can be enlarged to lower the cost of quality control. • The concept and methods of reducing variability are different from those of achieving specification. In dealing with variability, the importance of reinforcing optimization capability is higher than enhancing test capability; hence we are easy to be directed to seek and apply new (newly adopted by the organization) quality methodologies.
Field feedback	• The shipped products are easy to produce customer loss caused by variability and our cost of poor quality.	• The shipped products are not easy to produce quality loss caused by variability.

implementation becomes a process of managing organizational change. If change management can be conducted under the framework of key success factors (KSF), its success probability will increase a lot and the input of organizational resources can also obtain concrete return. In view of this, this section introduces the KSF of implementing RE and divides them into technical and managerial aspects.

7.2.1 Technical Aspect

We consider that the KSF of technical aspect for realizing the R&D paradigm shift lies on establishing evaluability of quality—that is, establishing the capability to efficiently evaluate quality. The core of Part II of this book is to interpret how to establish such competitive advantage and broadly apply it into various technology fields. The following items are the keys for developing evaluability of quality:

1. Proper Identification of Ideal Function. As described earlier, energy transformation is usually used to express an ideal function. A technological system, in nature, is a tool for making the energy transformation process happen because energy is added to the work piece in order to change its shape, remove material from it, or change its physical properties or function (Taguchi et al., 2005).

Even if the ideal function can be clearly formulated, in practice, the common difficulty is that we might lack measurement capability needed in measuring that ideal function and have to use substitute characteristics. Therefore, the ideal function might have new definitions following new development of measurement technology to enable the optimization of technology to be conducted with a more effective approach. R&D personnel may, through such process, keep evolving their philosophy, tools, and paradigm used to conduct technology development and, further, form the technological competitive advantage that is hard to be imitated.

2. Effective Noise Strategy. In the real world, there many noise factors we can't control, making a function fail to realize its ideality. The common noise includes temperature, humidity, dust, magnetic field, deterioration, aging, wearing, material variability, manufacturing variability, and so on. The more serious these noises intervene functionality, the more the downstream quality problems. As for the consideration about noises by R&D personnel, in practice, the most common way is to, at each stage of development of technology and product, use various verification tests and inspections to ensure the performance of quality characteristics. These tests include functional test, reliability test, lifetime test, thermal test, and other technical tests. These tests are conducted individually and maybe complete the schedule step by step according to the requirements of each stage so that they are not considered simultaneously to examine functionality.

Unlike conventional approach, the RE approach introduces all the important noise factors in experiment compulsorily to study how to make functionality insensitive to these noises. Therefore, how to integrate current verification test methods and equipment to realize the noise strategy for RE experiment is the key to success. Sometimes, current methods and equipment may fail to conduct verification procedures regarding test pieces or conduct tests in simulated environment; in such case, R&D personnel's creativity and experimental ideas would provide useful help.

3. Measurement Capability for Input Signal and Output Response. After ideal function is properly defined, the largest difficulty we may encounter is that we may not be able to measure the input signal and output response we formulated. In fact, for RE application, this is a commonly seen situation. When the situation occurs where we can't measure ideal function, we can choose substitute signal or response or manage to develop measurement capability. The approaches of developing measurement capability include buying related equipment, developing equipment by ourselves, or technological cooperation.

In the long-term perspective, R&D personnel's requirement about measurement capability should not be limited to a one-time experiment only. In fact, many breakthroughs of product and process technologies are from the development of measurement technology. Hence, not only the investment on measurement technology is helpful to RE experiment, but also R&D personnel may conduct various related experiments and technology development based on that capability.

4. SN Ratio Definition and Calculation. The calculation of the dynamic SN ratio has its ordinary formula. For most types of technology research, the dynamic SN ratio in this book can be used to evaluate and design functionality. Under some circumstances, the calculation of the SN ratio needs to be custom-written. For example, for the technology optimization of chemical reactions, we need a method of SN ratio evaluation and functionality design called *dynamic operating window* (Taguchi et al., 2005). This calculation approach of the SN ratio is different from the ordinary formula because of the characteristic of chemical reactions. Another example is electronic design. Since many of electronic characteristics are expressed in complex number form, the input/output relationship also needs to be described and measured using an SN ratio with complex numbers.

RE is a generic technology that can be broadly applied to various technology fields. Thus, sometimes the SN ratio needs custom design according to the characteristics of different technologies.

7.2.2 Managerial Aspect

The following three points are the most important KSF of managerial aspect for realizing R&D paradigm shift:

1. Management Commitment. Management commitment means that managers, in the process of leading R&D paradigm shift, have to declare and participate in this process and display recognition and set themselves as a good example. Application and promotion of any method and tool in an organization can be regarded as the means for managers to realize some intention; hence managing this process of application and promotion to achieve that purpose becomes one of the most important responsibilities for managers.

Regarding RE, it not only is a methodology for technology optimization but also is a mechanism for developing new logic and a new operating model of R&D, as well as organizational quality DNA. When managers can exercise leadership in the process, the organization promotes it; implanting such quality DNA into the organization is very efficient and effective. However, the real situations that managers face are not so easy to deal with; although managers sense that something is important, they take very few actions because they are restricted in many emergent things. R&D personnel have to complete the development of technology and product under various restrictions such as time and cost; although they know that fire prevention is important, they spend most of their time on firefighting to solve various quality problems caused by deficient design. Thus, in order to competitively improve quality and make the organization constantly implement a better R&D operating model, managers have to, with leadership, deal with various difficulties in reality and manage integration effectively.

2. Mechanism Design. Mechanism design is to plan and design the various organizational activities that are helpful to promote new concepts and methods to make organizational members willing to actively participate in the process of paradigm shift and to exercise their role and responsibility and collective learning in the process. Because every organization has a different culture, the mechanism suitable for that organization needs different design.

Mechanism design includes two major activities; one of them is to establish consensus and common language. Regarding RE promotion, the promoters have to establish organizational members' (especially, top managers and R&D managers) cognition and consensus about the necessity of R&D paradigm shift so that promotion of the RE method can be implemented smoothly. Moreover, promoters may, through the activities such as education and trainings, project coaching, and hands-on guidance in various forms urge R&D personnel to set RE language as common language and to be familiar about concept and procedure such as ideal function, SN ratio, two-step optimization, and robust technology development. This situation is just like when Six Sigma becomes the adjective for global companies to describe high quality level. When R&D personnel have a common language, they have a larger possibility to pursue common behavior, which means better R&D operating model. Another major activity of mechanism design is to stimulate R&D personnel's willing to apply RE in daily works and form a standard operating model; this activity corresponds to the deployment needed by management commitment.

The authors would like to highlight the activity of *Management Diagnosis* promoted in some companies of which the spirit is different from Quality Audit. Quality audit is to examine if managers and operational personnel are consistent in speaking, writing, and doing—that is, under existing management system, to examine the operational discipline and execution capability. Nevertheless, Management Diagnosis can be regarded as a horizontal "virtual organization" used to conduct organizational horizontal integration, capability development, and best practice deployment; we intend to, through this activity, effectively reinforce the design of management system and mechanism, letting R&D personnel make good quality in the circumstances. It is human nature to produce good quality; but like Dr. W. Edwards Deming (1982) said, what operational personnel can do about quality is actually limited to the system designed by managers. In other words, in a mechanism having no way to effectively make good quality, R&D personnel have no way to make good quality effectively when depending on their efforts alone. If managers don't have such cognition, no matter how much they require discipline and execution capability, the enhancement of quality is quite limited. The more effective the mechanism that managers can design, the more the operational personnel can do about quality.

3. Coaching by RE Experts. There are many experts specializing in the Taguchi Methods, but the RE experts discussed here are the experts who are specialized and competent to guide robust technology development described in this book. These experts apply dynamic method to technology development, focus on generic function, and are competent to develop robust technologies that can be applied to an entire product family. The way they use RE is different from the commonly seen approach:

- We taking quality characteristics as the object for optimization and using Nominal-the-Best (NTB), Larger-the-Better (LTB), and Smaller-the-Better (STB) subject to the type of quality characteristics.
- Application focus is on the manufacturing processes.
- We use static method to optimize the objective function with fixed target value.
- We determine the optimal factor-level combination according to the trade-off among the performances of different quality characteristic values.

In view of this, the experts specializing in the RE method described in this book play a key role for R&D paradigm shift. Adopting the conventional Taguchi Methods can't realize the R&D paradigm described in this book.

7.3 BENEFIT TO THE ORGANIZATION

Applying RE to develop technology is a competitive strategy, and it is also the invisible management capital used to build organizational capability. Thus, it has influence on the stakeholders of core business processes. These stakeholders include the personnel participating in that process, factories as an internal customer, direct customers, and end users. The benefits of implementing RE strategy for core processes and its stakeholders can be described as follows:

1. *Technology Development Process:* Making robustness be built-in to technologies at the origin stage of development and broadly applying the robust technologies to entire applicable product families to tremendously shorten the time needed by optimizing an individual product, so as to enhance R&D productivity.
2. *Marketing and Product Planning Processes:* Achieving technology readiness effectively before product planning to ensure that product development has no technological problems. When customers raise requests for quotation (RFQ), sales and R&D personnel can quickly respond to customers' demands.
3. *Product Design Process:* Effectively and efficiently optimizing product design technologies and achieving the reproducibility in the mass production stage in the factory.
4. *Production Preparation and Mass Production Processes:* Effectively and efficiently optimizing process design technologies and achieving the reproducibility in the mass production stage.
5. *Supplier Evaluation and Supply Chain Management Processes:* Establishing the competitive advantage that can lower supply chain cost (be able to adopt supply sources with low cost and use the materials/components of lower grade than that used by the competitors) and achieve high-quality technology simultaneously.
6. *After Sales Service Process:* Lowering quality variability and loss caused by poor robustness to reduce the number of complaints and claims from customers.

In summary, implementing RE is helpful for the company to gain benefits in various aspects such as NPD, quality, cost, and delivery, as shown in Table 7.2.

7.4 SLOGAN OR STRATEGY?

In view of current various improvement opportunities in R&D practice, Part II of this book provides actionable strategy and methodology to enhance the efficiency and productivity of R&D itself and acquire the leadership in market

TABLE 7.2. Benefits of RE Implementation

Dimension	Description	Related indicators
New product development	• Effectively enhancing the competitiveness of product quality and cost in market, enhancing sales, and increasing new RFI (request for inquiry) and RFQ (request for quotation) • Effectively shortening the time needed by commercialization of technology	• New product sales ratio • Time to technology commercialization • RFI • RFQ • . . .
Quality	• Effectively improving design quality and its manufacturing quality in factory • Effectively improving the robustness of products under various customer usage conditions (stable state of low variability)	• Project pass rate • Number of design changes • First pass yield rate • First pass yield rate (Direct yield) • Customer line reject rate (LRR) • Number of customer complaints • . . .
Cost	• Effectively reducing material/ component costs and achieving higher quality by lower grades • Effectively reducing the costs of poor quality (appraisal cost, internal failure cost, external failure cost)	• Material cost • Product cost • Quality loss and costs • . . .
Delivery	• Effectively accelerating the development efficiency of new technologies and new products and reducing the time of trials, tests, and inspections • Effectively accelerating the time to market of new products	• Product development cycle time • On-time rate of project activities • Time to market • . . .

in technology, cost, and quality. The authors consider that the RE method developed by Dr. Taguchi is the leading edge of quality technology and it can be broadly applied to technology development, product design, and process design.

To fully interpret RE, this book clearly points out the inefficiency and insufficiency of the current R&D approach; however, maybe because of this, this perspective can't win the recognition from specialized technicians who have profound technological knowledge and development experiences; instead they think that the approach proposed by this book is just slogan. In fact, this book emphasizes that RE is an evaluation technology and evaluation technology that cannot be used to solve technical problems; to solve technical problems still needs specialized technology so that the positioning of the two in R&D can be regarded as two pillars. The common situation is that R&D personnel own profound specialized technological knowledge and development experiences but the evaluation technology used by them lacks competitiveness. We consider that in the organizations which are mutually qualified to be competitors, the "team intelligence" of their R&D personnel is pretty similar so that it's difficult for a company to create distinguishability in team intelligence against the competitors; that is, it's not easy to establish a competitive advantage in this aspect. In other words, among the organizations, there apparently seems to be a competition in products and technologies; actually, it's the competition of the effectiveness of methodologies adopted by R&D personnel. We advocate that if R&D organization applies and promotes the strategy proposed by this book, they can shift its conventional paradigm and, further, establish the advantage that is not easily imitated.

When RE is regarded as a strategy and not a slogan, R&D managers will find that it is effective in totally improving R&D productivity and in predicting quality. R&D managers can take good advantage of the opportunities of applying RE to examine whether RE is really a slogan or a strategy.

PART III
Integration Strategy

8 Structure for Design Activity Integration

8.1 UNIVERSAL ROADMAP AND NINE TOOLS FOR DESIGN ENGINEERING

This book develops a so-called IC^2DV (identify, characterize, concept, design, verify) model as a universal roadmap for design engineering by QFD, as shown in Table 8.1.

At the identify phase, the company first has to define its target market—that is, to clarify who is the customer and to investigate voice of the customer (VOC) toward target customers. VOC information is the demanded items about a product so that the company has to identify which items that are critical to customer (CTC) according to the importance evaluated by the customers as well as to identify which items are critical to business (CTB) according to the product-market strategy determined by the company and then combine the two considerations to identify critical to value (CTV) from VOC, which is the very item that can create value for both customer and the company.

At the characterize phase, R&D personnel implement the characterization of VOC to convert customer language into technical language (i.e., substitute quality characteristics). Quality characteristics will yield characteristic values through measurement systems, so measurement system capability needs to be analyzed and improved to ensure accuracy and precision of the measured values of quality characteristics. The expected design specifications (nominal value and tolerance) of CTV quality characteristics are calculated and developed based on a good measurement system capability.

At the concept phase, R&D personnel develop the product concept that can reach the objective function—that is, to implement a system design to define functional blocks of product and their relationships. R&D personnel then assess its technical feasibility and choose the optimal system from several competitive design concepts.

At the design phase, R&D personnel, setting system requirements and its specifications as the target, implement the deployment of subsystems and component units and map its manufacturing process. R&D personnel then

---

Quality Strategy for Research and Development, First Edition. Ming-Li Shiu, Jui-Chin Jiang, and Mao-Hsiung Tu.

TABLE 8.1. IC^{2}DV Model as a Universal Roadmap for QFD Implementation

Phases		Quality deployment charts	Tasks
I	Identify	Demanded-quality deployment chart	1. Define target market 2. Collect voice of the customer (VOC) 3. Identify critical to value (CTV) from VOC
C	Characterize	Finished product quality characteristics deployment chart	4. Characterize VOC 5. Establish measurement system capability 6. Develop specifications of CTV quality characteristics
C	Concept		7. Develop product concept 8. Assess technical feasibility
D	Design	Subsystems and parts deployment chart Process deployment chart	9. Define product architecture 10. Map process and specify critical to quality (CTQ) and critical to process (CTP) 11. Develop specifications of CTQ and CTP 12. Establish control plan
V	Verify	Product test plans	13. Verify product and process design 14. Manage design changes

specify semi-finished product characteristics that are the key to the quality of finished product (also known as CTQ, which stands for critical to quality), and, according to design of process, specify the process characteristics that are the key to the quality of semi-finished product (also known as CTP which stands for critical to process). After defining CTQ and CTP, R&D personnel have to develop their expected design specifications and establish control plan so that they can minimize process and product variation through design, selection, and implementation of control methods.

At the verify phase, R&D personnel implement the verification of product design and process design through various product test plans and, regarding the items that need design change after the test, ensure the appropriate management in the whole product development cycle.

QFD is the QA system managing NPD and can be regarded as a framework that can integrate various design tools so that QFD may, through its integration with the other design tools, strengthen its effectiveness in NPD. It is worth exploring which tools can be systematically organized to be the essential tool set used in NPD and QFD. Because, if we divide the whole NPD cycle from upstream (market strategy formulation and product portfolio determination) to downstream (operations management for mass production) into the stages of business planning, product planning, product and process design, and production operations management, then we can observe and know: Besides the stage of product and process design, researchers and practitioners, toward different stages, have developed tool sets (as shown in Table 8.2) that are systematically organized to provide for related personnel to use following them so that they can implement planning, designing or analyzing tasks more logically. In view of such, we develop so-called "nine tools for design engineering" (D9) as the essential tools for product and process design. The objectives of tools are described as follows:

1. *Quality Chart (House of Quality, HOQ):* Identifying customer requirements and planned quality and converting them into substitute quality characteristics to determine the design quality of finished product for converting product quality from the world of customer to the world of technology.
2. *TRIZ (Russian acronym for Theory of Inventive Problem Solving):* Solving the inventive problems while designing a system (developing product concept) to develop the optimal system (optimal product concept).
3. *Measurement Systems Analysis (MSA):* Analyzing components of variability of measurement systems used in product design and process design, in order to evaluate accuracy and precision of measurement systems.

TABLE 8.2. Tool Set for NPD Cycle from Marketing Plan Formulation to Production

Stages	Tool Usage	
Business planning	Seven new QC tools (NQC7)	1. Relations diagram 2. KJ method (affinity diagram) 3. Systematic diagram 4. Matrix diagram 5. Matrix data analysis 6. Process decision program chart (PDPC) 7. Arrow diagram
	Seven strategic tools (S7) (Osada, 1998)	1. Environment analysis 2. Product analysis 3. Market analysis 4. Product-market analysis 5. Product portfolio analysis 6. Strategic elements analysis 7. Resource allocation analysis

(*Continued*)

TABLE 8.2. ***(Continued)***

Stages	Tool Usage	
Product planning	Seven new QC tools (NQC7)	1. Relations diagram 2. KJ method (affinity diagram) 3. Systematic diagram 4. Matrix diagram 5. Matrix data analysis 6. PDPC 7. Arrow diagram
	Seven tools for product planning (P7) (Kanda, 1995)	1. Group interview 2. Questionnaire 3. Positioning 4. Creativity checklist 5. Conception diagram 6. Conjoint analysis 7. Quality chart
Product and process design	Essential tools for this stage need to be identified and systematically organized	
Production operations management	Seven QC tools (QC7)	1. Cause-and-effect diagram 2. Pareto chart 3. Checklist 4. Histogram 5. Scatter diagram 6. Stratification 7. Various graphs
	Eight Six-Sigma tools (SS8) (Snee, 2003)	1. Process map 2. Cause and effects matrix 3. Measurement systems analysis 4. Capability study 5. Failure mode and effects analysis 6. Multi-vari analysis 7. Design of experiments 8. Control plan
	Seven industrial engineering tools (IE7)	1. Job/Worksite analysis guide 2. Operation process chart 3. Flow process chart 4. Flow diagram 5. Worker and machine process charts 6. Gang process charts 7. Line balancing

4. *Regression Analysis (RA):* Establishing the transfer functions between finished product characteristics and semi-finished product characteristics as well as between semi-finished product characteristics and process conditions, in order to identify key characteristic items and establish the guidelines for product design and process design.
5. *Robust Engineering (RE):* Developing the optimal specifications (nominal value and tolerance) for semi-finished product characteristics and process conditions to minimize variability of product and process performance.
6. *Fault Tree Analysis (FTA):* Analyzing fault symptoms of product and process, in order to accumulate experiences and knowledge of product design and process design to achieve problem recurrence prevention and problem prevention by prediction.
7. *Failure Mode and Effects Analysis (FMEA):* Early identifying the potential problems in product design and process design and preventing their impact on quality.
8. *Process Performance Analysis (PPA):* Characterizing the performance of a product to satisfy expected design specifications.
9. *Control Plan (CP):* Providing a structured approach for the design, selection, and implementation of value-added control methods for minimizing process and product variation.

Furthermore, this book uses the model of four domains of the design world (Suh, 2005) to be the simple model for describing NPD cycle to help explain the interrelations between nine tools when they are used in NPD. The four domains are customer domain, functional domain, physical domain, and process domain. The customer domain consists of the needs or attributes (customer attributes, CAs) that the customer is looking for in a product. The functional domain consists of functional requirements (FRs), often defined as engineering specifications, and constraints. The physical domain is the domain in which the key design parameters (DPs) are chosen to satisfy the FRs. Finally, the process domain specifies the manufacturing process variables (PVs) that can produce the DPs (Suh, 2005). The NPD cycle could be viewed as a mapping process among the four domains: map CAs into FRs, map FRs into DPs, and map DPs into PVs.

Figure 8.1 shows how D9 tools are systematically organized to link the elements (CAs, FRs, DPs, and PVs) in the four domains of NPD.

The interrelations between nine tools in the four domains of NPD, as shown in Figure 8.1, are described as follows:

1. *House of Quality (HOQ):* HOQ is a chart that converts customer expressed requirements (CAs) into technical and engineering specification (FRs) that can be understood by R&D personnel in the company. In other

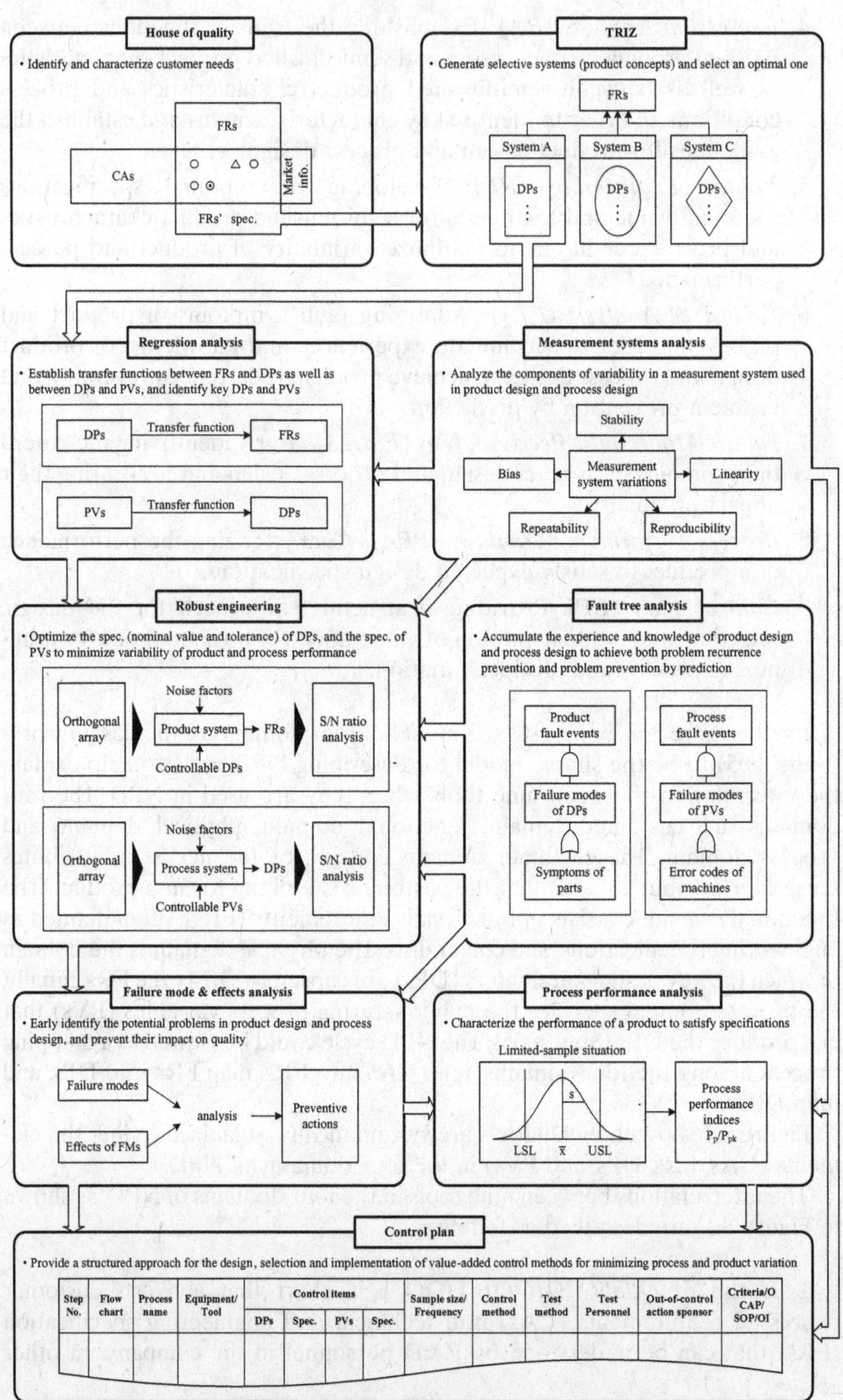

Figure 8.1. Nine tools for design engineering.

words, it's made at product planning and extended to apply to design engineering of product so that it is both the last tool of P7 (Kanda, 1995) and the first tool of D9.

The company can use HOQ to effectively select the quality niche that can comprehensively achieve or support customer satisfaction, corporate value proposition, and competitive advantage; also, it can convert planned quality into world of technology to realize it to achieve the harmonization between product and market.

2. *TRIZ:* In the functional domain, R&D personnel must develop the design concept of the finished product—that is, define the functional blocks that constitute the whole finished product and describe their generic input/output relationships. In the subsequent other domains, R&D personnel's design and engineering efforts toward product and process deployment are to realize the ultimate product concept.

To develop the product concept is a task that deals with invention, the inventive problems it may encounter and the solutions have been generalized to be a complete theory: TRIZ. R&D personnel can use that theory to conduct concept engineering or creativity engineering.

3. *Measurement Systems Analysis (MSA):* In the physical domain and the process domain, R&D personnel need to use measurement systems to conduct design engineering of product and process, and the measured values of various characteristics of final product design also have to meet engineering specifications (FRs) defined in the functional domain.

Therefore, R&D personnel need to, once they enter physical domain and process domain, analyze and improve the components of variability of measurement systems used in design engineering to minimize measurement variation.

4. *Regression Analysis (RA):* In the physical domain, the finished product is made from the semi-finished product, and the quality level achieved by finished product characteristics (FRs) is determined by the combined effect of all semi-finished product characteristics (DPs) so that the relationship between them usually can be described by a transfer function. Regression analysis is the method to establish the transfer function. The regression coefficient determined by it can be deemed as the importance of DPs so that R&D personnel can utilize it to establish product design guidelines. Likewise, in the process domain, transfer functions between semi-finished product characteristics (DPs) and process conditions (PVs) can also be established and be process design guidelines.

While developing a new product, R&D personnel may select the most similar product in specification and performance (FRs) as the baseline product or reference model and then use these design guidelines to extend the past product development experience to achieve efficient design engineering.

5. *Robust Engineering (RE):* In the physical domain, R&D personnel set the engineering specifications (FRs) of the finished product as the objectives

and develop optimal specification of important DPs based on the formula of product design guidelines or the experimental methods of parameter design and tolerance design to minimize the variation of product performance. In the process domain, R&D personnel set the design specifications (DPs) of semi-finished product characteristics as the objectives and develop optimal specification of important PVs based on the formula of process design guidelines or the experimental methods of parameter design and tolerance design to minimize the variation of process performance.

6. *Fault Tree Analysis (FTA):* FTA can be regarded as an effective method to accumulate experience and knowledge of product design and process design. It is used to integrate all fault symptoms of product and process and their relationships. R&D personnel can use FTA to (a) grasp fault symptoms and manufacturing problems that occured in the baseline product and (b) think about how a new product and its process can achieve optimal design to prevent problems and, in addition, enhance the capability for the product design and the process design.

7. *Failure Mode and Effects Analysis (FMEA):* FMEA refers to the analyses for failure mode and effects. In the physical domain, R&D personnel can use product design FMEA to early predict and analyze whether a new product design has potential impact toward quality, so as to prevent problems in the earliest phase of design engineering. In the process domain, R&D personnel can use process design FMEA to analyze whether process design has a potential impact toward quality in order to prevent problems by early prediction.

8. *Process Performance Analysis (PPA):* In the process domain, R&D personnel can use process performance analysis to characterize the performance of product design fulfilling engineering specifications (FRs), in order to provide plant personnel the preliminary confirmation information that product design can be transferred to production operations.

Since the number of engineering samples is limited at this time, R&D personnel can utilize P_p/P_{pk} indices (process performance indices proposed by Automotive Industry Action Group) that are different from C_p/C_{pk} indices (process capability indices) used in mass production to analyze process performance. The difference between the two in calculation is that the estimation method for standard deviation in formula is different.

9. *Control Plan (CP):* Before product and process design will be transferred from process domain to production operations, R&D personnel have to establish a control plan to provide a written summary of contents and standards of shop floor control to ensure that design objectives (PVs, DPs, and FRs) can be achieved in production.

The applications of D9 in the IC^{2}DV model are shown in Table 8.3.

TABLE 8.3. IC²DV Model and D9

Phases		Quality Deployment Charts	Tasks	Tool Usage
I	Identify	Demanded-quality deployment chart	1. Define target market 2. Collect VOC 3. Identify CTV from VOC	• HOQ
C	Characterize	Finished product quality characteristics deployment chart	4. Characterize VOC 5. Establish measurement system capability 6. Develop specifications of CTV quality characteristics	• HOQ • MSA
C	Concept		7. Develop product concept 8. Assess technical feasibility	• TRIZ
D	Design	Subsystems and parts deployment chart Process deployment chart	9. Define product architecture 10. Map process and specify CTQ and CTP 11. Develop specifications of CTQ and CTP 12. Establish control plan	• MSA • RA • RE • FTA • FMEA • PPA • CP
V	Verify	Product test plans	13. Verify product and process design 14. Manage design changes	• MSA • RA • RE • FTA • FMEA • PPA • CP

8.2 INTEGRATION OF QFD AND OTHER BREAKTHROUGH STRATEGIES

This section explains how QFD integrates other relevant breakthrough strategies, so as to make R&D personnel simultaneously implement all effective methods to realize their synergy by a more efficient approach. These breakthrough strategies include DFX, DFSS, and blue ocean strategy (BOS).

8.2.1 Simplified EQFD Model and Its Integration with DFX and DFSS

The methodology of QFD is one of the few technologies that could have a significant quality improvement impact throughout a company's NPD cycle. QFD's evolution is for it to become an integral quality assurance architecture (IQAA) that can be implemented concurrently with the NPD cycle, rather than exist as a mere series of matrices.

This book, based on the concept of broadly defined QFD, develops a so-called expanded QFD (EQFD) system. There are eight major differences between an EQFD system and the original QFD which can be used to reinforce the quality designing-in methods of the original QFD and can effectively support the corporate NPD cycle. In addition, the implementation process of EQFD, which includes four stages, eight phases, and 36 steps, is developed to clarify how to use the NPD approach to establish an EQFD system.

This section summarizes EQFD and its process by a simplified model shown in Figure 8.2, and focuses on illustrating how DFX and DFSS integrate with QFD.

The first step for EQFD implementation is to formulate a market strategy and determine the required product portfolio. And then, a survey is conducted over target customers, and a demanded-quality deployment chart is made. According to market evaluation information such as competitive analysis and

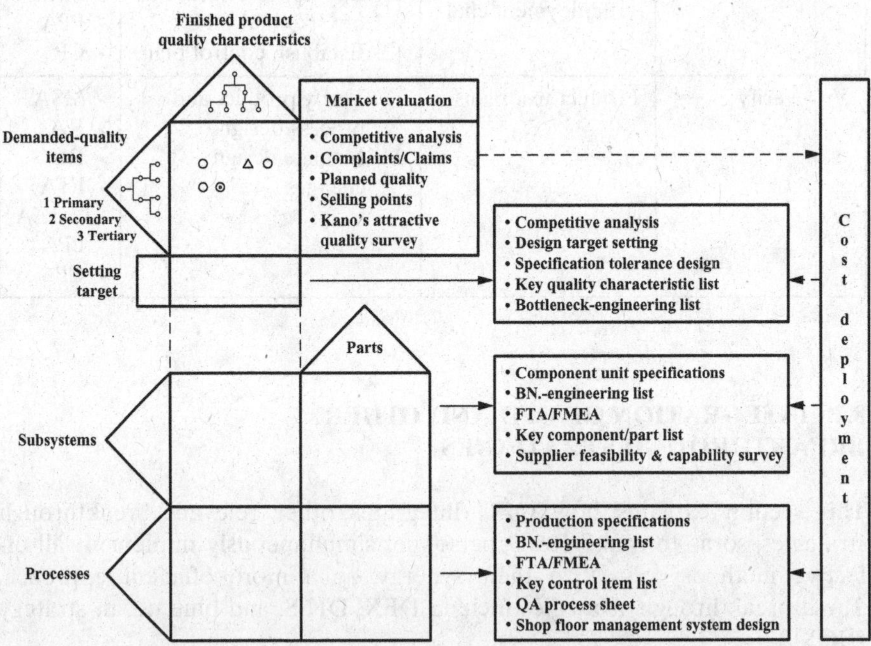

Figure 8.2. Simplified model of EQFD (Jiang et al., 2007a).

claims analysis, the company may conduct quality planning and decide the new product's "individuality" or selling points. Kano's attractive quality survey helps to conduct product planning for attractive quality creation. Demanded qualities are expressed by customers by directly describing and perceiving what product quality is with customer language, which is equivalent to true qualities. However, demanded qualities must be converted into substitute quality characteristics—that is, the technical language used by company's R&D personnel to understand how to achieve them technically. Only in this way is it possible to materialize them through development technology. Therefore, it is necessary to carry out a quality characteristics deployment of the finished product to transform product quality from the world of customer into the world of technology. The company may decide design specification values according to competitive analysis—technically—and extract BNE that is not on the extension line of the past technology and hinders the realization of the design quality.

Even if the fabrication supplier's NPD activities start from accepting the specified specification contract (i.e., deployed quality characteristics), the necessity of deployment for demanded quality is emphasized in EQFD. If the fabrication supplier can, through transforming the customer specifications inversely into demanded quality, actively grasp the demanded significance behind customer specifications, they can evaluate the adequacy of customer specifications in applications to propose possible design changes to customers (e.g., propose design changes due to the overdesign of customer specifications or propose design changes because the company can achieve higher specification level) and not to totally accept the specifications proposed by customers. That way, the company could create the chances to have discussions with the customer and collaboratively develop products, and even influence the customer's orientation of developing future products to create a competitive advantage. After the company's development of a demanded-quality deployment chart, in the future, companies can utilize it together with a quality characteristics deployment chart. For example: Sales personnel can conduct quality planning and decide the new product's selling points based on market evaluation information; while R&D personnel can study how to, in customer specifications, find technical differentiation chances available for supporting the planned quality to set design quality as well as extract BNE that is not on the extension line of the past technology so that they can set it as the direction for future R&D.

When the design quality of the finished product is determined, R&D personnel do not directly design the entire product but make the subsystem or component unit (also known as building block or chunk) of the intermediate layer of the product architecture the design unit. As a result, a subsystems deployment is necessary, in order to allocate the specification tolerance of finished product quality characteristics to relevant component units. Besides, in the design phase the product must effectively prevent the recurrence of the design problems of the existing product as well as the design potential problems of the new product. To that end, FTA and FMEA can be employed. With respect to parts and materials needed for constructing subsystems, deployment is required, and an evaluation of the vendor's feasibility and capability should be conducted.

Products are made through processes and are completed by assembling their semi-finished products. Therefore, a semi-finished product and its specifications defined by subsystems deployment are made by using process deployment and its design and by deciding specifications of process conditions (also known as production specifications). Like product design, process design must effectively prevent the recurrence of the design problems of the existing process and the design potential problems of the new process. This can be done by drawing on equipment FTA and process FMEA. According to process deployment and FMEA-related information, a control plan can be created to provide an overview of information needed for process control. The design of the SFMS needed to execute the control plan, which is part of the process design, is intended to ensure that before a product enters into mass production, adequate preparations are already in place to achieve manufacturing quality assurance.

The foregoing quality deployment that includes technology and reliability can realize customers' demanded qualities and failure-free qualities, yet it may increase the cost as a result. Therefore, by using market evaluation information to determine the target cost of the finished product and by corresponding to the quality deployment process to set up cost targets such as for components and parts, materials, and labor, a balance between quality assurance and cost reduction can be achieved.

The structural integrity of QFD in dealing with NPD's crucial issues (quality, technology, cost, reliability) helps to position and integrate DFX and DFSS, as illustrated in Figure 8.3.

DFX is a design methodology that can integrate with QFD in converting demanded qualities into substitute quality characteristics. In other words, it ensures that in planning all desirable dimensions of demanded quality, a value proposition is developed based on the selection of quality niches (i.e., one X or several Xs); that in the phase of demanded-quality characterization, the design targets that can support the value proposition are set; and that product design for the DFX and its verification can then be conducted downstream.

DFSS is a design methodology that can integrate with QFD in tolerance design and process design of a product. That is, after setting the nominal values of quality characteristics, the reference to specification values, actual performance, and process capability of similar products developed in the past is made, in order to develop a product specification tolerance with six-sigma design quality and to allocate the tolerance to related subsystems, component units, parts, and materials. Moreover, DFSS also includes the design and verification of optimal process conditions and the design of SFMS in order to ensure the manufacture of a semi-finished product that can achieve six-sigma design quality.

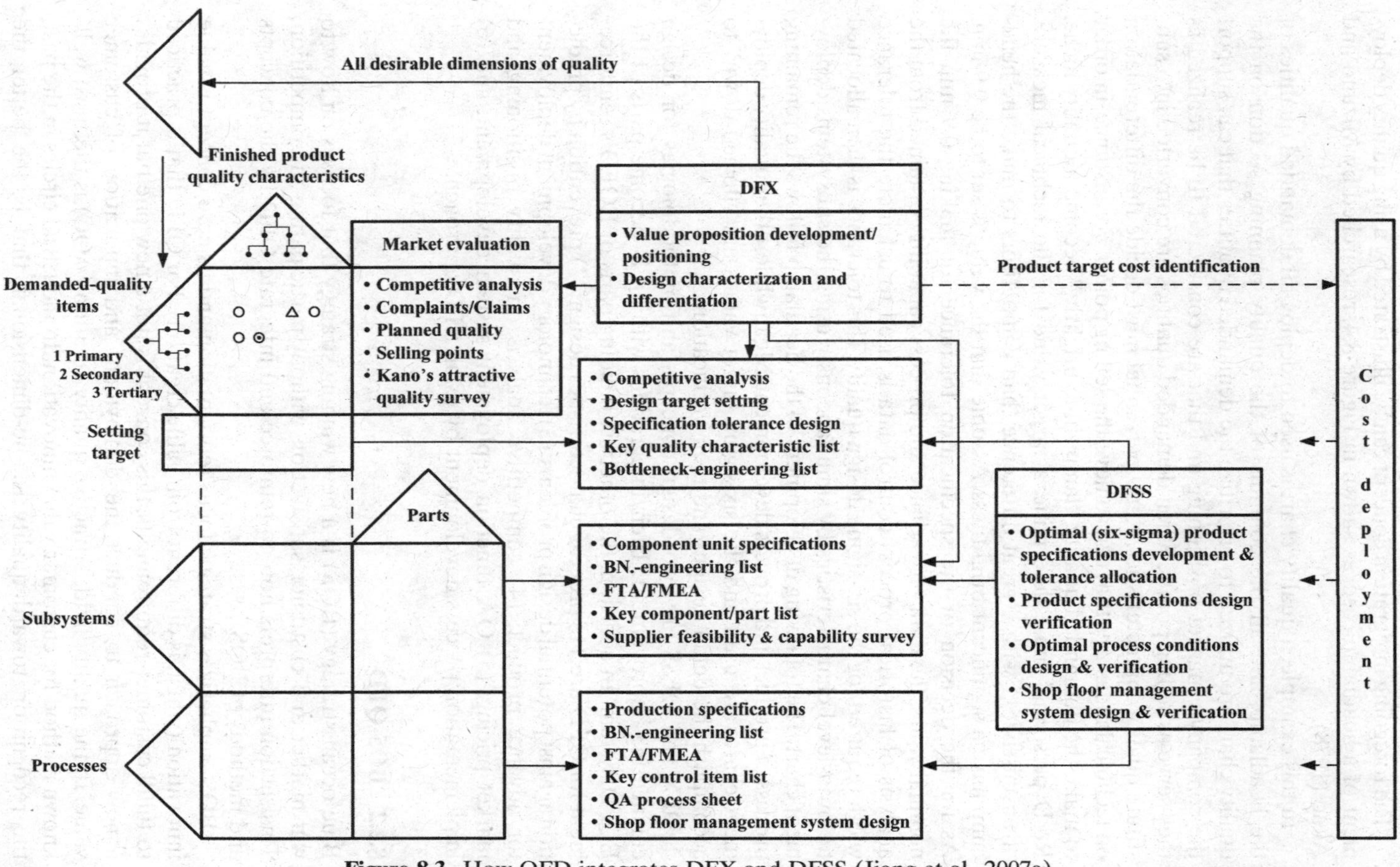

Figure 8.3. How QFD integrates DFX and DFSS (Jiang et al., 2007a).

The foregoing integration concept can be illustrated by using quality deployment of a headlamp case, as shown in Figure 8.4, researched by Mizuno and Akao (1978).

In this example, a quality chart is used to deploy all demanded qualities of the headlamp, and all the columns of the quality planning section on the quality chart are utilized to prioritize the demanded qualities that can support the company's market positioning and that are considered to be realized as they serve as selling points. When demanded qualities are converted into substitute quality characteristics, R&D personnel must decide the differentiation on technical measures that can achieve the selling points. For instance, in order to make the demanded quality, "lamp shine brightly," become a quality niche, R&D personnel must determine a target value for the technical measure, "transmissivity," with a great difference than competitors to make the headlamp have a significant brightness. Assume that the value is set to be 0.9 min. As for the decision of the specification tolerance for making 0.9 min the nominal value, the data on variance and process capability obtained from the analysis of historical process control data is used to calculate the tolerance width required for the six-sigma design quality. This tolerance is then allocated to lower-level characteristics by simultaneously using the subsystems deployment chart. The following development of the headlamp follows the remaining process of product and process development. Meanwhile, other quality deployment charts are used and the six-sigma target values are "flowed down" to accomplish the detail designs and their verifications.

DFX and DFSS are two design concepts and methodologies for design engineering improvement. Their integrations with QFD become parts of the EQFD system. In addition, the other six building parts of EQFD's reinforcement for QFD are: (1) process characteristics design; (2) prevention by prediction against potential design problems; (3) technology development deployment for advance product; (4) competitive analysis on quality positioning and market pricing; (5) QA function deployment system development; and (6) implementation process development by using NPD approach.

8.2.2 BOS-QFD

Blue ocean strategy (BOS) is a new winning strategy that focuses on how to win market and customer satisfaction without participating in competition. This section interprets how to reinforce and integrate QFD by the concepts and methods of BOS.

BOS emphasizes strategic focus reorientation and its core concept is value innovation (which is different from value creation). For QFD that is practiced to fulfill customer requirements, this concept is not a new one but it provides a new approach to conduct the deployment and innovation of customer value. In the quality field, attractive quality creation (AQC) is the most well-known method for customer value innovation. In industrial circles in the past, the recognition toward quality is one-dimensional; that is, the better the

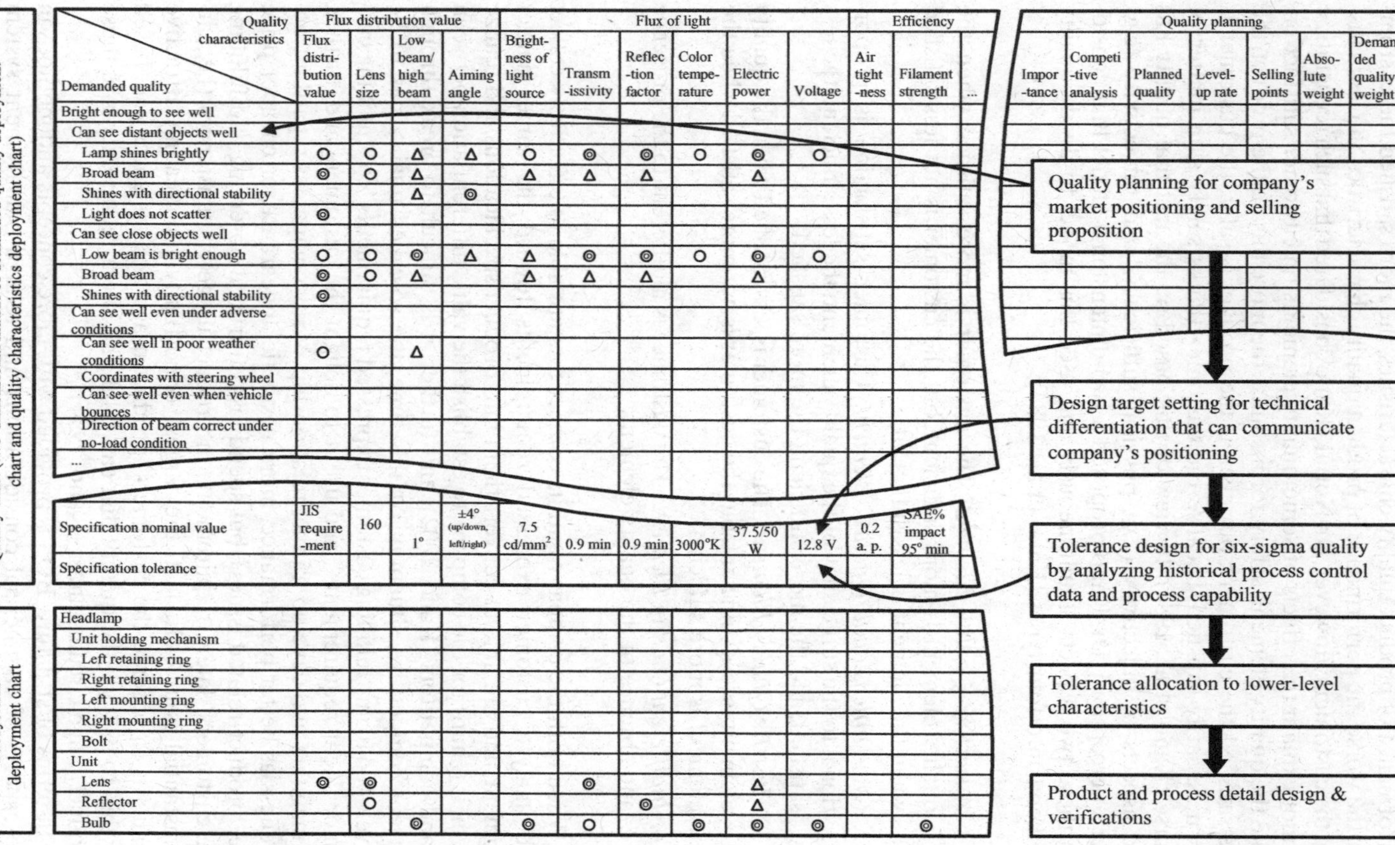

Figure 8.4. How QFD integrates DFX and DFSS: Example of a headlamp (Jiang et al., 2007a).

performance of product quality characteristics, the more satisfied the customers; the worse the performance of product quality characteristics, the less satisfied the customers. However, Dr. Noriaki Kano considered that such recognition cannot explain the effects of some quality elements on customer satisfaction. Take the quality element, safety, for instance, the more unsafe the product, the less satisfied the customers; but even if the product is safer, the customer is not necessarily satisfied, because the customer thinks safety as a matter of course (Kano et al., 1984). In view of this, based on the fact that quality has two aspects of objectivity (e.g., physical sufficiency) and subjectivity (e.g., user's perception), Dr. Kano proposed a two-dimensional model that considers these two aspects simultaneously and used this model to divide quality elements into four types (Kano et al., 1984):

1. *Attractive Quality Element:* The customer is more satisfied when the quality element is more sufficient, but the customer is not less satisfied when the element is less sufficient.
2. *One-Dimensional Quality Element:* Customer satisfaction is proportional to the sufficiency of the quality element; the less sufficient, the less satisfied, and the more sufficient, the more satisfied.
3. *Must-Be Quality Element:* The customer is less satisfied when the quality element is less sufficient, but the customer is not more satisfied when the element is more sufficient.
4. *Indifferent Quality Element:* The customer is indifferent to the presence and absence of the quality element.

When both the company and its competitors are able to effectively collect and analyze customer expressed requirements toward the same market segment, and are able to convert them into product specifications and realize them, it is hard for the company to establish the value differentiation between it and competitors (Kano, 2002); and further, it is difficult to break through current competitive situation. At this time, the so-called quality has to be developed from "fulfilling customer expressed requirements" to "fulfilling customer latent requirements" and use it to delight the customers and, further, to maintain and develop trading relationship. Many companies regard attractiveness as better performance, more aesthetic appearance, or cheaper price (i.e., product-related issues), but spend less efforts on observing and fulfilling the customers' latent requirements in product usage circumstances (i.e., circumstantial issues). In fact, the largest opportunity for creating attractive quality is to observe the customers' product usage circumstances and, based on this, to provide the quality that can fulfill the latent requirements in those product usage circumstances (Kano, 1994).

In terms of structure, BOS is provided with more concepts and methods in the aspects of commercial considerations and related management system development than AQC. In this book, AQC is considered able to reinforce the demanded-quality deployment in QFD system. Likewise, we also consider that BOS can provide added values in several aspects to QFD; these added values can be shown in three aspects:

1. Demanded-Quality Deployment. In BOS, a tool used to capture the current state of play in the known marketplace is called strategy canvas. It is used to understand the factors the industry and competitors compete on, as well as what customers receive from the existing competition in the market (Kim and Mauborgne, 2005). From another perspective, strategy canvas can be viewed as the so-called value curve drawn based on the two-dimensional matrix formed by demanded-quality items and customer perceived value. Strategists then use four actions framework (ERRC) to define the value curves that are different from the competitors'. The new value curve can be regarded as a group of strategic profile; such planning process is to implement quality planning activities and propose the value proposition with differential positioning. The so-called four actions framework (ERRC) means the following (Kim and Mauborgne, 2005):

- *Eliminate:* Which of the factors that the industry takes for granted should be eliminated?
- *Reduce:* Which factors should be reduced well below the industry's standard?
- *Raise:* Which factors should be raised well above the industry's standard?
- *Create:* Which factors should be created that the industry has never offered?

Through ERRC framework, we can simultaneously realize the competitive strategies of differentiation and low cost: ER thinking is helpful for planning how to positioning to establish value differentiation, while RC thinking is helpful for evaluating how to realize the strategic positioning we want to deliver with low cost. In the demanded-quality deployment chart (DQDC), the value curve approach can be used to reinforce DQDC's competitive analysis, and the strategic profile is the planned quality information in DQDC. The core concept of BOS is value innovation which is based on the simultaneous realization of differentiation and low cost and, further, connects customer value with technology innovation to develop new market space. QFD's DQDC and its related deployment charts are systematic realization means.

Just like the attractive quality has a life cycle (Kano, 2002), BOS can't sustain permanently. However, QFD can serve as the communication platform between marketing personnel and R&D personnel and the mechanism of strategy formulation to make competitive strategy continue renewal according to the dynamic priority of customer value and technology innovation.

2. Cost Deployment. Both BOS and QFD's cost deployment emphasize the strategic pricing approach (price-minus costing) of "minusing price with profit to set cost," not "setting price by adding profit on cost." From the viewpoint of QFD, price design can be conducted by the QD_m (quality deployment for market pricing) method, but BOS uses the strategic pricing method to determine the price corridor of the mass and then sets price in that price corridor depending on whether blue ocean ideas are protected by laws and whether they are easily imitated. Also, BOS provides three levers for lowering cost which can be used to reinforce QFD's cost deployment: (1) streamlining operations and introducing cost innovations from manufacturing to distribution, (2) partnering, and (3) changing the pricing model of the industry. Therefore, we can utilize QFD's cost deployment to systematically deploy target cost which is needed to be achieved and its cost bottleneck to realize profit, and then we can use those three levers to break through cost bottleneck to make quality and cost hit the balance. The value innovation emphasized by BOS pushes the profit forward to utmost through those three levers.

3. Business Model Buildup. BOS emphasizes the importance of designing and establishing a profitable business model to realize "smart strategic move." A business model supporting BOS comes from the combined evaluation and verification of four aspects such as buyer utility, price, cost, and adoption. This approach can be used to reinforce the relationship between QFD implementation and the business model to ensure and accelerate the benefits of QFD on profitability and competitiveness.

In summary, the framework of BOS-QFD is described as Figure 8.5.

8.3 INTEGRATED PROCESS FOR IMPLEMENTING QFD AND RE

As mentioned above, there are two types of quality: customer-driven quality and engineered quality. The former emphasizes fulfilling various expressed or latent customer requirements; the latter focuses on ensuring the functional robustness of products under various customer usage conditions to reduce various quality loss caused by poor functionality. To effectively manage quality, we need strategy and its corresponding methods to achieve the purpose. The authors divide quality strategies as the following four types:

1. *Strategy for "Quality by Tests/Inspections":* Using various test or inspection approaches to contain and screen quality deficiencies to make defective products not delivered to the customers. These defective products may be scrap and may be delivered again after rework.
2. *Strategy for "Quality by Production Operations Management":* Ensuring that the production line not manufacture the products with defective quality by designing and executing the shop floor management systems. If quality abnormalities occur, shop floor personnel can also implement

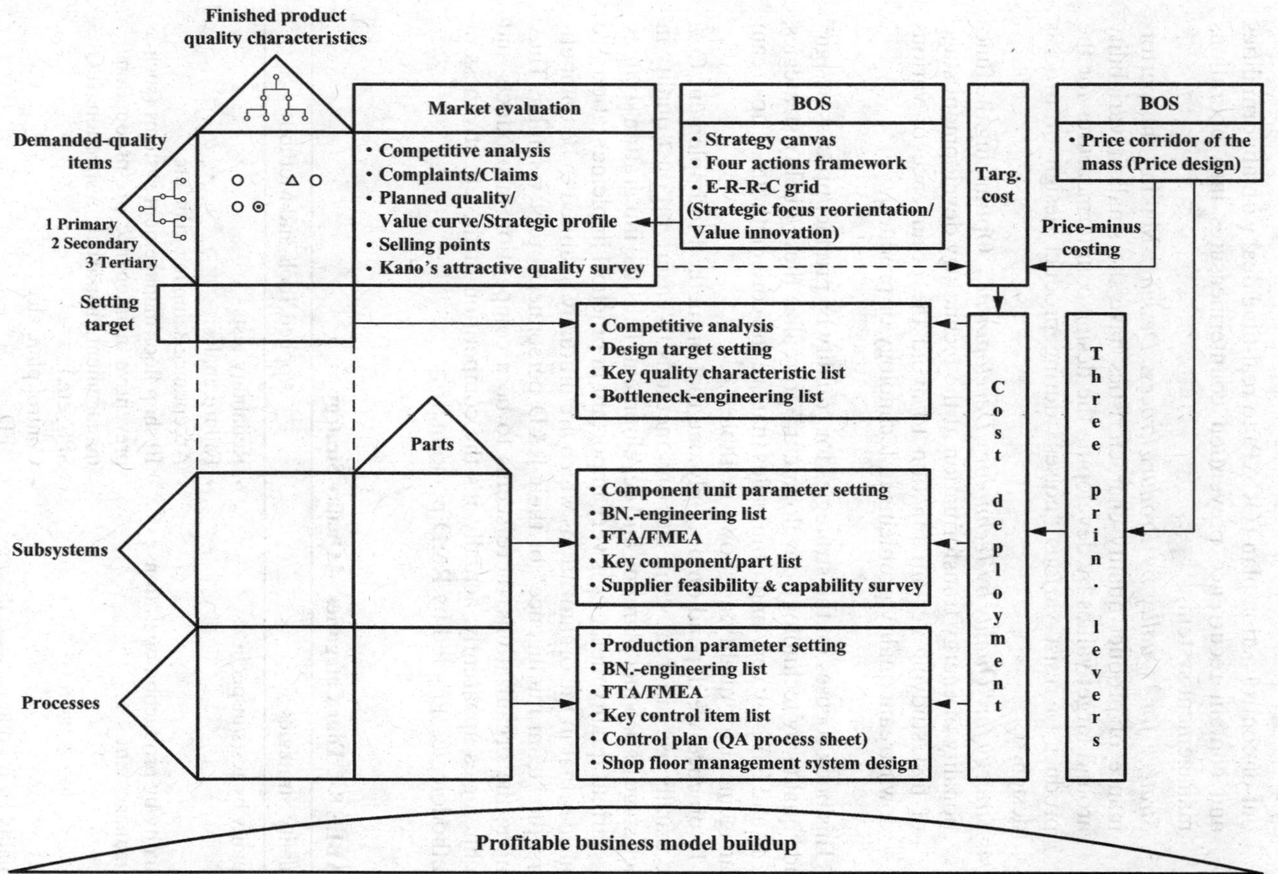

Figure 8.5. BOS-QFD.

out-of-control action plan (OCAP) to real-time deal with abnormalities and contain recurrence prevention countermeasures into operations management system.

3. *Strategy for "Quality by Product/Process Design":* Making the performance of product quality characteristics have the smallest variability around target values by developing the design countermeasures for the product to resist various "noises" during product design or process design.
4. *Strategy for "Quality by Technology Development":* Optimizing the functionality of energy transformation at the technology development stage of first studying certain function to avoid the occurrence of various downstream quality problems in technology or product.

This book focuses on the strategies for "quality by product/process design" and "quality by technology development." To realize these quality strategies, QFD and RE play very important roles that are the core methods of upstream management for quality, as shown in Table 8.4.

In practice, we depend on not only one tool or method to implement QA; we usually use different strategies and methods to ensure product quality in each stage of product planning and development. Thus, the integration strategy for various quality methods is very important in practical implementation. We consider that in the organizations which are mutually qualified to be competitors, the "team intelligence" of their R&D personnel is pretty similar. Thus, among the organizations, there seems to be a competition in products and technologies apparently; actually, it's the competition of the effectiveness of methodologies adopted by R&D personnel.

TABLE 8.4. Four Categories of Quality Strategy

Quality Strategies	Applied Tools and Methods
Quality by tests/inspections	• Reliability tests • Failure analysis • Acceptance-sampling plan, etc.
Quality by production operations management	• 10 shop floor management systems (such as preventive maintenance, pre-production management, first-article inspection, IPQC, SPC, etc.) • Control plan, etc.
Quality by product/process design	• QFD • TRIZ • Robust engineering • FMEA, etc.
Quality by technology development	• Robust engineering (Robust technology development)

TABLE 8.5. Integrated Process for Implementing QFD and RE

I. Business Planning

1. Conduct market research and develop STP (segmentation, targeting, positioning) strategy
2. Formulate product portfolio and business plan

II. Product Planning

3. Conduct demanded-quality deployment, and formulate value proposition
4. Plan the marketing mix (product, price, place, promotion, service)
5. Conduct cost deployment

III. Product Design

6. Conduct quality characteristics deployment
7. Conduct system design and assess robustness
8. Conduct subsystems deployment
9. Conduct FMEA and manage design changes
10. Implement design reviews and verification tests

IV. Robust Technology Development

3′. Plan advanced technology
4′. Conduct function deployment and mechanism deployment
5′. Identify the mechanisms/subsystems to be optimized
6′. Identify ideal function and develop measurement capability
7′. Conduct parameter design for functionality and confirm the reproducibility
8′. Implement tolerance design (if necessary) for the quality versus cost trade-off
9′. Conduct technology combination for integrating subsystems

V. Process Design

11. Conduct process deployment
12. Conduct FMEA and manage design changes
13. Optimize process technology
14. Implement design reviews and verifications

VI. Production Preparation

15. Design 10 shop floor management systems
16. Verify control plan

VII. Mass Production

17. Execute 10 shop floor management systems
18. Manage operations by dynamic priority

We think, among so many methods, that the methods applied in upstream stages such as technology and product development are the most powerful; QFD and RE are not only methodology itself, but also they play the role as a framework to integrate various quality tools and design engineering tools. Moreover, viewing from the objectives of these two methods, RE is to optimize functionality, while QFD focuses on function design and its comprehensive quality deployment. The integration of QFD and RE can create the synergy used to enhance the applicability in practice. In view of this, in this section, we propose the integrated implementation process of QFD and RE which can be divided into 7 stages and 18 steps. Among them, steps 3–9 and steps 3′–9′ may be implemented simultaneously; that is, in the stages of product planning and product design, R&D personnel also need to conduct planning and development of technology to enable a new product's technological difficulties to be predicted and solved at the earlier stage of the development process and wouldn't cause the increment of time and cost. Also, the integrated implementation process can be viewed as the quality planning process for advanced or new product development.

Based on the well-known concept of core competence (Prahalad and Hamel, 1990), we illustrate how the integration of QFD and RE is the engine for competitiveness by Figure 8.6: Once the core competencies 1–4 can be effectively and efficiently developed through QFD and RE and embodied to be core products 1 and 2, the business units 1–4 and a wide range of end products 1–12 can be competitively engendered.

8.4 CUSTOMER-FOCUSED, RATIONALIZED AND EFFICIENT R&D

We use a form of house to express the entire framework realizing customer-focused, rationalized, and efficient R&D, as shown in Figure 8.7.

Just like the structure of a house, we divide the consisting elements of that structure into "foundation," "pillars," and "beams" and "roof" used to connect the pillars. They are interpreted as follows:

1. "Foundation" includes three items:
   a. *Specialized/Intrinsic Technology.* This term represents some technological capabilities for the company to enter some industry or to survive in that industry. Without these technological capabilities, the company has no way to enter or survive in that industry. Another description for it is the specialized knowledge and experience accumulation in some technological field.
   b. *Process and Project Management Systems.* Even though the various methods mentioned in this book are not adopted, a company must have technology and product development processes as well as the execution of various improvement projects to maintain its business operations. In other words, current systems of process management and project management can be regarded as an application platform for the methods in this book to obtain the added values making R&D more customer-oriented, rationalized, and efficient.
   c. *Information Technology (IT).* IT is an effective platform used to accelerate the diffusion of methods, knowledge, and experience in organization. On the other hand, because the core business functions

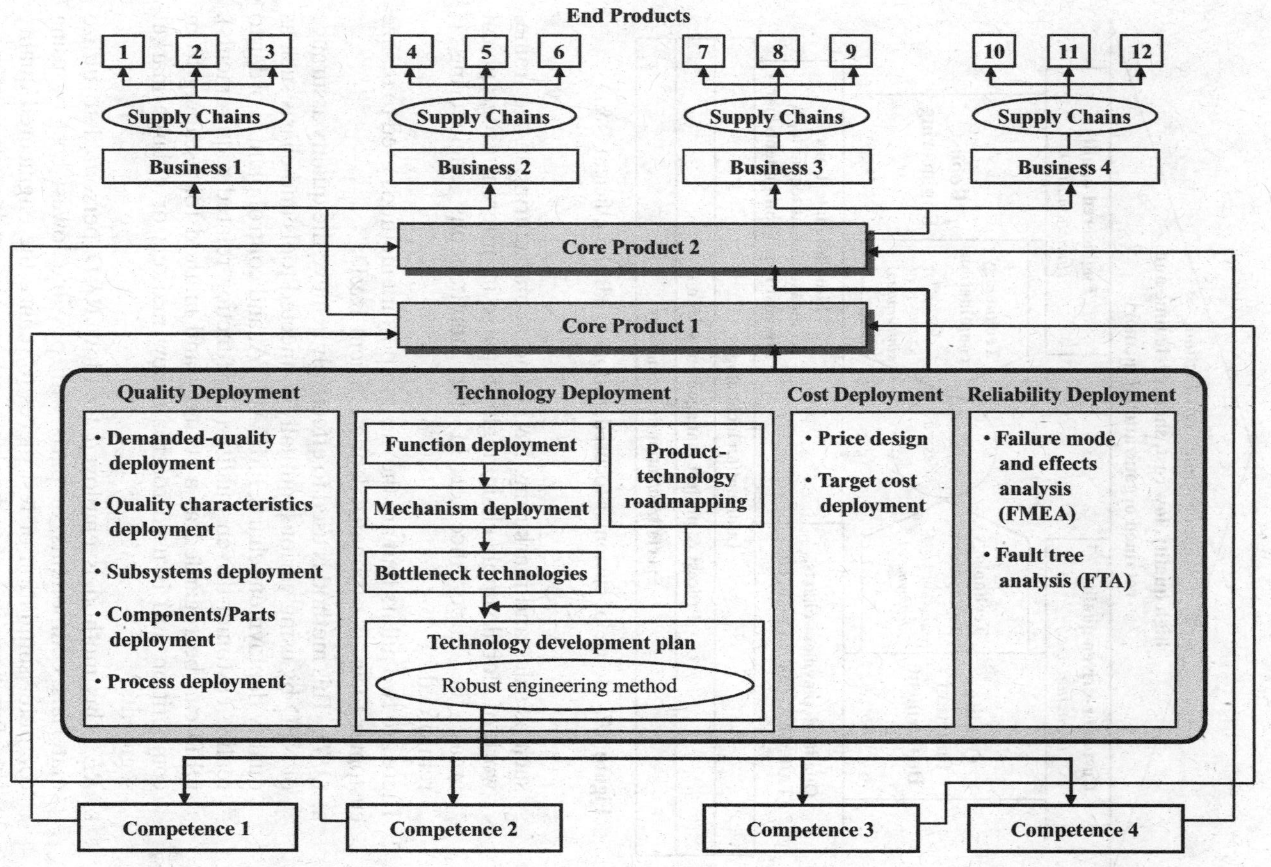

Figure 8.6. Linking up technologies and methodologies with competencies for competitive advantage.

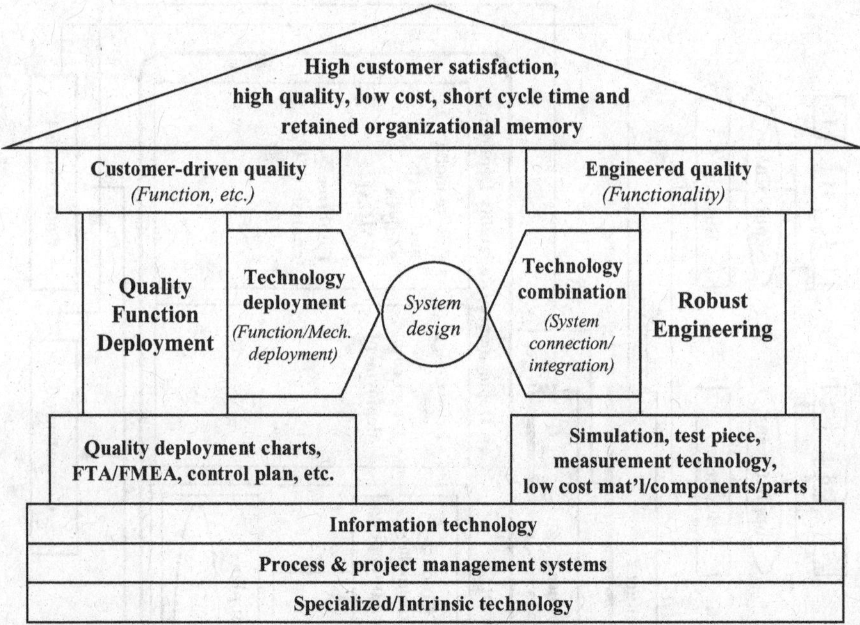

Figure 8.7. House of customer-focused, rationalized, and efficient R&D.

such as sales and marketing, R&D, and manufacturing in the framework of global supply chain might disperse in different geographical areas, the importance of IT for simultaneous engineering is reinforced.

2. There are two pillars that are the most important methods used to realize customer-focused, rationalized, and efficient R&D:
   a. *QFD.* This method is used to effectively achieve the quality assurance of NPD by using various mutually connected tools and charts such as quality deployment charts, FMEA, FTA, and control plan, in order to realize customer-driven quality (i.e., function-related requirements). QFD can be regarded as a strategy and method for escaping from competition and is used to create new markets or enlarge market segments.
   b. *RE.* This method is employed to allow R&D personnel to utilize materials, components, or parts with possible lowest cost by using computer simulation or test pieces to realize the engineered quality (i.e., functionality) of technology and product under sufficient measurement capability (that is, while conducting functionality optimization, measurement systems are capable of measuring the output response defined by us). RE can be regarded as a strategy and method for dealing with competition, and it is used to win market share by creating quality and cost (price) advantages.

3. The beam connecting the two pillars indicates that the technology management activities are conducted according to system design:
   a. *Technology Deployment.* Since technology is invisible, it can be regarded as something that is stored in mechanisms and performs its function through that mechanism. We can, based on system design, utilize function deployment and mechanism deployment methods to analyze the required technologies and bottleneck-engineering (BNE) to conduct related technology management activities as early as possible (such as technology planning, technology sourcing, etc.).
   b. *Technology Combination.* One mechanism (subsystem) is usually taken as a research unit for functionality optimization, thus while the various mechanisms required by a system have been optimized, the activities that R&D personnel need to conduct are to connect and integrate each mechanism and tune performance if necessary. In the perspective of technology, such connection and integration is technology combination, and various technologies have different combinations varying with different system designs.
4. Roof represents the objectives of entire framework: high customer satisfaction, high quality, low cost, short cycle time, and retained organizational memory. Also, these indicators are usually used to evaluate competitive advantage. To gain competitive advantage, the company depends on adopting more effective and efficient methods than what its competitors adopted. This is because a different result comes from a different process, and the process is related to the adopted strategy and method. In Figure 8.7, QFD and RE are the quality strategies and methodologies we suggest to adopt.

The objectives of entire framework expressed in Figure 8.7 are to make all the various activities conducted in R&D "customer-focused," "rationalized," and "efficient" to create competitive advantage. We interpret the significance represented by these items as follows, and we summarize them in Table 8.6:

1. *Customer-Focused.* The function design method mentioned in this book is applied to grasp customer-expressed and latent requirements and to effectively develop the fitness-for-use products and products with attractive quality.
2. *Rationalized.* The frequently seen approach of R&D focuses on quality characteristics and utilizes various validation tests (or verification tests, reliability tests) to confirm if the performance of quality characteristics conforms to specification. Once it exceeds the specification, R&D personnel commence various problem-solving and design-change activities to make quality performance conform to specification. The engineering methods mentioned in this book focus on the optimization of ideal function to (a) ensure that various quality characteristic values measured in

TABLE 8.6. Elements of the Customer-Focused, Rationalized, and Efficient R&D

Item	Description
Customer-focused	• Fitness for use • Attractive quality creation
Rationalized	• Focus on ideal function, not quality characteristics • Focus on optimization, not validation/verification/reliability tests (debug approach) • Focus on prevention for the unknown in the market, not firefighting • Focus on technology readiness, flexibility, and reproducibility, not tuning
Efficient	• Cost efficient • Time efficient • Evaluation efficient

downstream can pass various tests and (b) prevent the quality problems caused by various unknown usage conditions in the market. In the aspect of technology development, the methods in this book are better than the tuning approach, so they can effectively realize technology readiness, flexibility, and reproducibility.

3. *Efficient.* The R&D efficiency improvement attained by applying the methods in this book can be divided into three aspects: cost, time, and functionality evaluation. It's possible for R&D personnel to achieve the product of high robustness with the components of low cost and to realize shorter time to market with more efficient functionality evaluation (or, say, design idea evaluation) approach.

Although pursuing quality is a "sweating task," the function design and functionality design methods provided in this book are the strategies that can reduce sweating and firefighting but can improve quality fundamentally. Ultimately speaking, the authors suggest using this book as an approach to participate in establishing a world where people can enjoy higher quality.

APPENDIX A
Recommended Orthogonal Arrays for R&D

$L_{12}\ (2^{11})$

No.	1	2	3	4	5	6	7	8	9	10	11
1	1	1	1	1	1	1	1	1	1	1	1
2	1	1	1	1	1	2	2	2	2	2	2
3	1	1	2	2	2	1	1	1	2	2	2
4	1	2	1	2	2	1	2	2	1	1	2
5	1	2	2	1	2	2	1	2	1	2	1
6	1	2	2	2	1	2	2	1	2	1	1
7	2	1	2	2	1	1	2	2	1	2	1
8	2	1	2	1	2	2	2	1	1	1	2
9	2	1	1	2	2	2	1	2	2	1	1
10	2	2	2	1	1	1	1	2	2	1	2
11	2	2	1	2	1	2	1	1	1	2	2
12	2	2	1	1	2	1	2	1	2	2	1

Quality Strategy for Research and Development, First Edition. Ming-Li Shiu, Jui-Chin Jiang, and Mao-Hsiung Tu.

L_{18} $(2^1 \times 3^7)$

No.	1	2	3	4	5	6	7	8
1	1	1	1	1	1	1	1	1
2	1	1	2	2	2	2	2	2
3	1	1	3	3	3	3	3	3
4	1	2	1	1	2	2	3	3
5	1	2	2	2	3	3	1	1
6	1	2	3	3	1	1	2	2
7	1	3	1	2	1	3	2	3
8	1	3	2	3	2	1	3	1
9	1	3	3	1	3	2	1	2
10	2	1	1	3	3	2	2	1
11	2	1	2	1	1	3	3	2
12	2	1	3	2	2	1	1	3
13	2	2	1	2	3	1	3	2
14	2	2	2	3	1	2	1	3
15	2	2	3	1	2	3	2	1
16	2	3	1	3	2	3	1	2
17	2	3	2	1	3	1	2	3
18	2	3	3	2	1	2	3	1

L_{18} $(6^1 \times 3^6)$

No.	1	2	3	4	5	6	7
1	1	1	1	1	1	1	1
2	1	2	2	2	2	2	2
3	1	3	3	3	3	3	3
4	2	1	1	2	2	3	3
5	2	2	2	3	3	1	1
6	2	3	3	1	1	2	2
7	3	1	2	1	3	2	3
8	3	2	3	2	1	3	1
9	3	3	1	3	2	1	2
10	4	1	3	3	2	2	1
11	4	2	1	1	3	3	2
12	4	3	2	2	1	1	3
13	5	1	2	3	1	3	2
14	5	2	3	1	2	1	3
15	5	3	1	2	3	2	1
16	6	1	3	2	3	1	2
17	6	2	1	3	1	2	3
18	6	3	2	1	2	3	1

L_{36} ($2^3 \times 3^{13}$)

No.	1	2	3	4	5	6	7	8	9	10	11	12	13	14	15	16
1	1	1	1	1	1	1	1	1	1	1	1	1	1	1	1	1
2	1	1	1	1	2	2	2	2	2	2	2	2	2	2	2	2
3	1	1	1	1	3	3	3	3	3	3	3	3	3	3	3	3
4	1	2	2	1	1	1	1	1	2	2	2	2	3	3	3	3
5	1	2	2	1	2	2	2	2	3	3	3	3	1	1	1	1
6	1	2	2	1	3	3	3	3	1	1	1	1	2	2	2	2
7	2	1	2	1	1	1	2	3	1	2	3	3	1	2	2	3
8	2	1	2	1	2	2	3	1	2	3	1	1	2	3	3	1
9	2	1	2	1	3	3	1	2	3	1	2	2	3	1	1	2
10	2	2	1	1	1	1	3	2	1	3	2	3	2	1	3	2
11	2	2	1	1	2	2	1	3	2	1	3	1	3	2	1	3
12	2	2	1	1	3	3	2	1	3	2	1	2	1	3	2	1
13	1	1	1	2	1	2	3	1	3	2	1	3	3	2	1	2
14	1	1	1	2	2	3	1	2	1	3	2	1	1	3	2	3
15	1	1	1	2	3	1	2	3	2	1	3	2	2	1	3	1
16	1	2	2	2	1	2	3	2	1	1	3	2	3	3	2	1
17	1	2	2	2	2	3	1	3	2	2	1	3	1	1	3	2
18	1	2	2	2	3	1	2	1	3	3	2	1	2	2	1	3
19	2	1	2	2	1	2	1	3	3	3	1	2	2	1	2	3
20	2	1	2	2	2	3	2	1	1	1	2	3	3	2	3	1
21	2	1	2	2	3	1	3	2	2	2	3	1	1	3	1	2
22	2	2	1	2	1	2	2	3	3	1	2	1	1	3	3	2
23	2	2	1	2	2	3	3	1	1	2	3	2	2	1	1	3
24	2	2	1	2	3	1	1	2	2	3	1	3	3	2	2	1
25	1	1	1	3	1	3	2	1	2	3	3	1	3	1	2	2
26	1	1	1	3	2	1	3	2	3	1	1	2	1	2	3	3
27	1	1	1	3	3	2	1	3	1	2	2	3	2	3	1	1
28	1	2	2	3	1	3	2	2	2	1	1	3	2	3	1	3
29	1	2	2	3	2	1	3	3	3	2	2	1	3	1	2	1
30	1	2	2	3	3	2	1	1	1	3	3	2	1	2	3	2
31	2	1	2	3	1	3	3	3	2	3	2	2	1	2	1	1
32	2	1	2	3	2	1	1	1	3	1	3	3	2	3	2	2
33	2	1	2	3	3	2	2	2	1	2	1	1	3	1	3	3
34	2	2	1	3	1	3	1	2	3	2	3	1	2	2	3	1
35	2	2	1	3	2	1	2	3	1	3	1	2	3	3	1	2
36	2	2	1	3	3	2	3	1	2	1	2	3	1	1	2	3

L_{36} $(2^{11} \times 3^{12})$

No.	1	2	3	4	5	6	7	8	9	10	11	12	13	14	15	16	17	18	19	20	21	22	23
1	1	1	1	1	1	1	1	1	1	1	1	1	1	1	1	1	1	1	1	1	1	1	1
2	1	1	1	1	1	1	1	1	1	1	1	2	2	2	2	2	2	2	2	2	2	2	2
3	1	1	1	1	1	1	1	1	1	1	1	3	3	3	3	3	3	3	3	3	3	3	3
4	1	1	1	1	1	2	2	2	2	2	2	1	1	1	1	2	2	2	2	3	3	3	3
5	1	1	1	1	1	2	2	2	2	2	2	2	2	2	2	3	3	3	3	1	1	1	1
6	1	1	1	1	1	2	2	2	2	2	2	3	3	3	3	1	1	1	1	2	2	2	2
7	1	1	2	2	2	1	1	1	2	2	2	1	1	2	3	1	2	3	3	1	2	2	3
8	1	1	2	2	2	1	1	1	2	2	2	2	2	3	1	2	3	1	1	2	3	3	1
9	1	1	2	2	2	1	1	1	2	2	2	3	3	1	2	3	1	2	2	3	1	1	2
10	1	2	1	2	2	1	2	2	1	1	2	1	1	3	2	1	3	2	3	2	1	3	2
11	1	2	1	2	2	1	2	2	1	1	2	2	2	1	3	2	1	3	1	3	2	1	3
12	1	2	1	2	2	1	2	2	1	1	2	3	3	2	1	3	2	1	2	1	3	2	1
13	1	2	2	1	2	2	1	2	1	2	1	1	2	3	1	3	2	1	3	3	2	1	2
14	1	2	2	1	2	2	1	2	1	2	1	2	3	1	2	1	3	2	1	1	3	2	3
15	1	2	2	1	2	2	1	2	1	2	1	3	1	2	3	2	1	3	2	2	1	3	1
16	1	2	2	2	1	2	2	1	2	1	1	1	2	3	2	1	1	3	2	3	3	2	1
17	1	2	2	2	1	2	2	1	2	1	1	2	3	1	3	2	2	1	3	1	1	3	2
18	1	2	2	2	1	2	2	1	2	1	1	3	1	2	1	3	3	2	1	2	2	1	3

19	2	1	2	2	1	1	2	2	1	2	1	1	2	1	3	3	3	1	2	2	1	2	3
20	2	1	2	2	1	1	2	2	1	2	1	2	3	2	1	1	1	2	3	3	2	3	1
21	2	1	2	2	1	1	2	2	1	2	1	3	1	3	2	2	2	3	1	1	3	1	2
22	2	1	2	1	2	2	2	1	1	1	2	1	2	2	3	3	1	2	1	1	3	3	2
23	2	1	2	1	2	2	2	1	1	1	2	2	3	3	1	1	2	3	2	2	1	1	3
24	2	1	2	1	2	2	2	1	1	1	2	3	1	1	2	2	3	1	3	3	2	2	1
25	2	1	1	2	2	2	1	2	2	1	1	1	3	2	1	2	3	3	1	3	1	2	2
26	2	1	1	2	2	2	1	2	2	1	1	2	1	3	2	3	1	1	2	1	2	3	3
27	2	1	1	2	2	2	1	2	2	1	1	3	2	1	3	1	2	2	3	2	3	1	1
28	2	2	2	1	1	1	1	2	2	1	2	1	3	2	2	2	1	1	3	2	3	1	3
29	2	2	2	1	1	1	1	2	2	1	2	2	1	3	3	3	2	2	1	3	1	2	1
30	2	2	2	1	1	1	1	2	2	1	2	3	2	1	1	1	3	3	2	1	2	3	2
31	2	2	1	2	1	2	1	1	1	2	2	1	3	3	3	2	3	2	2	1	2	1	1
32	2	2	1	2	1	2	1	1	1	2	2	2	1	1	1	3	1	3	3	2	3	2	2
33	2	2	1	2	1	2	1	1	1	2	2	3	2	2	2	1	2	1	1	3	1	3	3
34	2	2	1	1	2	1	2	1	2	2	1	1	3	1	2	3	2	3	1	2	2	3	1
35	2	2	1	1	2	1	2	1	2	2	1	2	1	2	3	1	3	1	2	3	3	1	2
36	2	2	1	1	2	1	2	1	2	2	1	3	2	3	1	2	1	2	3	1	1	2	3

$L_{36}\ (6^1 \times 2^2 \times 3^{12})$

No.	1	2	3	4	5	6	7	8	9	10	11	12	13	14	15
1	1	1	1	1	1	1	1	1	1	1	1	1	1	1	1
2	1	1	1	2	2	2	2	2	2	2	2	2	2	2	2
3	1	1	1	3	3	3	3	3	3	3	3	3	3	3	3
4	1	2	2	1	1	1	1	2	2	2	2	3	3	3	3
5	1	2	2	2	2	2	2	3	3	3	3	1	1	1	1
6	1	2	2	3	3	3	3	1	1	1	1	2	2	2	2
7	2	1	2	1	1	2	3	1	2	3	3	1	2	2	3
8	2	1	2	2	2	3	1	2	3	1	1	2	3	3	1
9	2	1	2	3	3	1	2	3	1	2	2	3	1	1	2
10	2	2	1	1	1	3	2	1	3	2	3	2	1	3	2
11	2	2	1	2	2	1	3	2	1	3	1	3	2	1	3
12	2	2	1	3	3	2	1	3	2	1	2	1	3	2	1
13	3	1	1	1	2	3	1	3	2	1	3	3	2	1	2
14	3	1	1	2	3	1	2	1	3	2	1	1	3	2	3
15	3	1	1	3	1	2	3	2	1	3	2	2	1	3	1
16	3	2	2	1	2	3	2	1	1	3	2	3	3	2	1
17	3	2	2	2	3	1	3	2	2	1	3	1	1	3	2
18	3	2	2	3	1	2	1	3	3	2	1	2	2	1	3
19	4	1	2	1	2	1	3	3	3	1	2	2	1	2	3
20	4	1	2	2	3	2	1	1	1	2	3	3	2	3	1
21	4	1	2	3	1	3	2	2	2	3	1	1	3	1	2
22	4	2	1	1	2	2	3	3	1	2	1	1	3	3	2
23	4	2	1	2	3	3	1	1	2	3	2	2	1	1	3
24	4	2	1	3	1	1	2	2	3	1	3	3	2	2	1
25	5	1	1	1	3	2	1	2	3	3	1	3	1	2	2
26	5	1	1	2	1	3	2	3	1	1	2	1	2	3	3
27	5	1	1	3	2	1	3	1	2	2	3	2	3	1	1
28	5	2	2	1	3	2	2	2	1	1	3	2	3	1	3
29	5	2	2	2	1	3	3	3	2	2	1	3	1	2	1
30	5	2	2	3	2	1	1	1	3	3	2	1	2	3	2
31	6	1	2	1	3	3	3	2	3	2	2	1	2	1	1
32	6	1	2	2	1	1	1	3	1	3	3	2	3	2	2
33	6	1	2	3	2	2	2	1	2	1	1	3	1	3	3
34	6	2	1	1	3	1	2	3	2	3	1	2	2	3	1
35	6	2	1	2	1	2	3	1	3	1	2	3	3	1	2
36	6	2	1	3	2	3	1	2	1	2	3	1	1	2	3

L_{54} ($2^1 \times 3^{25}$)

No.	1	2	3	4	5	6	7	8	9	10	11	12	13	14	15	16	17	18	19	20	21	22	23	24	25	26
1	1	1	1	1	1	1	1	1	1	1	1	1	1	1	1	1	1	1	1	1	1	1	1	1	1	1
2	1	1	1	1	1	1	1	1	2	2	2	2	2	2	2	2	2	2	2	2	2	2	2	2	2	2
3	1	1	1	1	1	1	1	1	3	3	3	3	3	3	3	3	3	3	3	3	3	3	3	3	3	3
4	1	1	2	2	2	2	2	2	1	1	1	1	1	1	2	3	2	3	2	3	2	3	2	3	2	3
5	1	1	2	2	2	2	2	2	2	2	2	2	2	2	3	1	3	1	3	1	3	1	3	1	3	1
6	1	1	2	2	2	2	2	2	3	3	3	3	3	3	1	2	1	2	1	2	1	2	1	2	1	2
7	1	1	3	3	3	3	3	3	1	1	1	1	1	1	3	2	3	2	3	2	3	2	3	2	3	2
8	1	1	3	3	3	3	3	3	2	2	2	2	2	2	1	3	1	3	1	3	1	3	1	3	1	3
9	1	1	3	3	3	3	3	3	3	3	3	3	3	3	2	1	2	1	2	1	2	1	2	1	2	1
10	1	2	1	1	2	2	3	3	1	1	2	2	3	3	1	1	1	1	2	3	2	3	3	2	3	2
11	1	2	1	1	2	2	3	3	2	2	3	3	1	1	2	2	2	2	3	1	3	1	1	3	1	3
12	1	2	1	1	2	2	3	3	3	3	1	1	2	2	3	3	3	3	1	2	1	2	2	1	2	1
13	1	2	2	2	3	3	1	1	1	1	2	2	3	3	2	3	2	3	3	2	3	2	1	1	1	1
14	1	2	2	2	3	3	1	1	2	2	3	3	1	1	3	1	3	1	1	3	1	3	2	2	2	2
15	1	2	2	2	3	3	1	1	3	3	1	1	2	2	1	2	1	2	2	1	2	1	3	3	3	3
16	1	2	3	3	1	1	2	2	1	1	2	2	3	3	3	2	3	2	1	1	1	1	2	3	2	3
17	1	2	3	3	1	1	2	2	2	2	3	3	1	1	1	3	1	3	2	2	2	2	3	1	3	1
18	1	2	3	3	1	1	2	2	3	3	1	1	2	2	2	1	2	1	3	3	3	3	1	2	1	2

(*Continued*)

L_{54} $(2^1 \times 3^{25})$ *(Continued)*

No.	1	2	3	4	5	6	7	8	9	10	11	12	13	14	15	16	17	18	19	20	21	22	23	24	25	26
19	1	3	1	2	1	3	2	3	1	2	1	3	2	3	1	1	2	3	1	1	3	2	2	3	3	2
20	1	3	1	2	1	3	2	3	2	3	2	1	3	1	2	2	3	1	2	2	1	3	3	1	1	3
21	1	3	1	2	1	3	2	3	3	1	3	2	1	2	3	3	1	2	3	3	2	1	1	2	2	1
22	1	3	2	3	2	1	3	1	1	2	1	3	2	3	2	3	3	2	2	3	1	1	3	2	1	1
23	1	3	2	3	2	1	3	1	2	3	2	1	3	1	3	1	1	3	3	1	2	2	1	3	2	2
24	1	3	2	3	2	1	3	1	3	1	3	2	1	2	1	2	2	1	1	2	3	3	2	1	3	3
25	1	3	3	1	3	2	1	2	1	2	1	3	2	3	3	2	1	1	3	2	2	3	1	1	2	3
26	1	3	3	1	3	2	1	2	2	3	2	1	3	1	1	3	2	2	1	3	3	1	2	2	3	1
27	1	3	3	1	3	2	1	2	3	1	3	2	1	2	2	1	3	3	2	1	1	2	3	3	1	2
28	2	1	1	3	3	2	2	1	1	3	3	2	2	1	1	1	3	2	3	2	2	3	2	3	1	1
29	2	1	1	3	3	2	2	1	2	1	1	3	3	2	2	2	1	3	1	3	3	1	3	1	2	2
30	2	1	1	3	3	2	2	1	3	2	2	1	1	3	3	3	2	1	2	1	1	2	1	2	3	3
31	2	1	2	1	1	3	3	2	1	3	3	2	2	1	2	3	1	1	1	1	3	2	3	2	2	3
32	2	1	2	1	1	3	3	2	2	1	1	3	3	2	3	1	2	2	2	2	1	3	1	3	3	1
33	2	1	2	1	1	3	3	2	3	2	2	1	1	3	1	2	3	3	3	3	2	1	2	1	1	2
34	2	1	3	2	2	1	1	3	1	3	3	2	2	1	3	2	2	3	2	3	1	1	1	1	3	2
35	2	1	3	2	2	1	1	3	2	1	1	3	3	2	1	3	3	1	3	1	2	2	2	2	1	3
36	2	1	3	2	2	1	1	3	3	2	2	1	1	3	2	1	1	2	1	2	3	3	3	3	2	1

37	2	2	1	2	3	1	3	2	1	2	3	1	3	2	1	1	2	3	3	2	1	1	3	2	2	3
38	2	2	1	2	3	1	3	2	2	3	1	2	1	3	2	2	3	1	1	3	2	2	1	3	3	1
39	2	2	1	2	3	1	3	2	3	1	2	3	2	1	3	3	1	2	2	1	3	3	2	1	1	2
40	2	2	2	3	1	2	1	3	1	2	3	1	3	2	2	3	3	2	1	1	2	3	1	1	3	2
41	2	2	2	3	1	2	1	3	2	3	1	2	1	3	3	1	1	3	2	2	3	1	2	2	1	3
42	2	2	2	3	1	2	1	3	3	1	2	3	2	1	1	2	2	1	3	3	1	2	3	3	2	1
43	2	2	3	1	2	3	2	1	1	2	3	1	3	2	3	2	1	1	2	3	3	2	2	3	1	1
44	2	2	3	1	2	3	2	1	2	3	1	2	1	3	1	3	2	2	3	1	1	3	3	1	2	2
45	2	2	3	1	2	3	2	1	3	1	2	3	2	1	2	1	3	3	1	2	2	1	1	2	3	3
46	2	3	1	3	2	3	1	2	1	3	2	3	1	2	1	1	3	2	2	3	3	2	1	1	2	3
47	2	3	1	3	2	3	1	2	2	1	3	1	2	3	2	2	1	3	3	1	1	3	2	2	3	1
48	2	3	1	3	2	3	1	2	3	2	1	2	3	1	3	3	2	1	1	2	2	1	3	3	1	2
49	2	3	2	1	3	1	2	3	1	3	2	3	1	2	2	3	1	1	3	2	1	1	2	3	3	2
50	2	3	2	1	3	1	2	3	2	1	3	1	2	3	3	1	2	2	1	3	2	2	3	1	1	3
51	2	3	2	1	3	1	2	3	3	2	1	2	3	1	1	2	3	3	2	1	3	3	1	2	2	1
52	2	3	3	2	1	2	3	1	1	3	2	3	1	2	3	2	2	3	1	1	2	3	3	2	1	1
53	2	3	3	2	1	2	3	1	2	1	3	1	2	3	1	3	3	1	2	2	3	1	1	3	2	2
54	2	3	3	2	1	2	3	1	3	2	1	2	3	1	2	1	1	2	3	3	1	2	2	1	3	3

APPENDIX B
General Orthogonal Arrays

$L_8\ (2^7)$

No.	1	2	3	4	5	6	7
1	1	1	1	1	1	1	1
2	1	1	1	2	2	2	2
3	1	2	2	1	1	2	2
4	1	2	2	2	2	1	1
5	2	1	2	1	2	1	2
6	2	1	2	2	1	2	1
7	2	2	1	1	2	2	1
8	2	2	1	2	1	1	2

$L_8\ (4^1 \times 2^4)$

No.	1	2	3	4	5
1	1	1	1	1	1
2	1	2	2	2	2
3	2	1	1	2	2
4	2	2	2	1	1
5	3	1	2	1	2
6	3	2	1	2	1
7	4	1	2	2	1
8	4	2	1	1	2

Quality Strategy for Research and Development, First Edition. Ming-Li Shiu, Jui-Chin Jiang, and Mao-Hsiung Tu.

$L_9\ (3^4)$

No.	1	2	3	4
1	1	1	1	1
2	1	2	2	2
3	1	3	3	3
4	2	1	2	3
5	2	2	3	1
6	2	3	1	2
7	3	1	3	2
8	3	2	1	3
9	3	3	2	1

$L_{16}\ (2^{15})$

No.	1	2	3	4	5	6	7	8	9	10	11	12	13	14	15
1	1	1	1	1	1	1	1	1	1	1	1	1	1	1	1
2	1	1	1	1	1	1	1	2	2	2	2	2	2	2	2
3	1	1	1	2	2	2	2	1	1	1	1	2	2	2	2
4	1	1	1	2	2	2	2	2	2	2	2	1	1	1	1
5	1	2	2	1	1	2	2	1	1	2	2	1	1	2	2
6	1	2	2	1	1	2	2	2	2	1	1	2	2	1	1
7	1	2	2	2	2	1	1	1	1	2	2	2	2	1	1
8	1	2	2	2	2	1	1	2	2	1	1	1	1	2	2
9	2	1	2	1	2	1	2	1	2	1	2	1	2	1	2
10	2	1	2	1	2	1	2	2	1	2	1	2	1	2	1
11	2	1	2	2	1	2	1	1	2	1	2	2	1	2	1
12	2	1	2	2	1	2	1	2	1	2	1	1	2	1	2
13	2	2	1	1	2	2	1	1	2	2	1	1	2	2	1
14	2	2	1	1	2	2	1	2	1	1	2	2	1	1	2
15	2	2	1	2	1	1	2	1	2	2	1	2	1	1	2
16	2	2	1	2	1	1	2	2	1	1	2	1	2	2	1

L_{16} $(8^2 \times 4^1 \times 2^2)$

No.	1	2	3	4	5
1	1	1	1	1	1
2	2	2	1	2	2
3	3	1	1	3	1
4	4	2	1	4	2
5	4	3	2	2	1
6	3	4	2	1	2
7	2	3	2	4	1
8	1	4	2	3	2
9	5	1	2	5	2
10	6	2	2	6	1
11	7	1	2	7	2
12	8	2	2	8	1
13	8	3	1	6	2
14	7	4	1	5	1
15	6	3	1	8	2
16	5	4	1	7	1

L_{27} (3^{13})

No.	1	2	3	4	5	6	7	8	9	10	11	12	13
1	1	1	1	1	1	1	1	1	1	1	1	1	1
2	1	1	1	1	2	2	2	2	2	2	2	2	2
3	1	1	1	1	3	3	3	3	3	3	3	3	3
4	1	2	2	2	1	1	1	2	2	2	3	3	3
5	1	2	2	2	2	2	2	3	3	3	1	1	1
6	1	2	2	2	3	3	3	1	1	1	2	2	2
7	1	3	3	3	1	1	1	3	3	3	2	2	2
8	1	3	3	3	2	2	2	1	1	1	3	3	3
9	1	3	3	3	3	3	3	2	2	2	1	1	1
10	2	1	2	3	1	2	3	1	2	3	1	2	3
11	2	1	2	3	2	3	1	2	3	1	2	3	1
12	2	1	2	3	3	1	2	3	1	2	3	1	2
13	2	2	3	1	1	2	3	2	3	1	3	1	2
14	2	2	3	1	2	3	1	3	1	2	1	2	3
15	2	2	3	1	3	1	2	1	2	3	2	3	1
16	2	3	1	2	1	2	3	3	1	2	2	3	1
17	2	3	1	2	2	3	1	1	2	3	3	1	2
18	2	3	1	2	3	1	2	2	3	1	1	2	3
19	3	1	3	2	1	3	2	1	3	2	1	3	2
20	3	1	3	2	2	1	3	2	1	3	2	1	3
21	3	1	3	2	3	2	1	3	2	1	3	2	1
22	3	2	1	3	1	3	2	2	1	3	3	2	1
23	3	2	1	3	2	1	3	3	2	1	1	3	2
24	3	2	1	3	3	2	1	1	3	2	2	1	3
25	3	3	2	1	1	3	2	3	2	1	2	1	3
26	3	3	2	1	2	1	3	1	3	2	3	2	1
27	3	3	2	1	3	2	1	2	1	3	1	3	2

L_{32} (2^{31})

No.	1	2	3	4	5	6	7	8	9	10	11	12	13	14	15	16	17	18	19	20	21	22	23	24	25	26	27	28	29	30	31
1	1	1	1	1	1	1	1	1	1	1	1	1	1	1	1	1	1	1	1	1	1	1	1	1	1	1	1	1	1	1	1
2	1	1	1	1	1	1	1	1	1	1	1	1	1	1	1	2	2	2	2	2	2	2	2	2	2	2	2	2	2	2	2
3	1	1	1	1	1	1	1	2	2	2	2	2	2	2	2	1	1	1	1	1	1	1	1	2	2	2	2	2	2	2	2
4	1	1	1	1	1	1	1	2	2	2	2	2	2	2	2	2	2	2	2	2	2	2	2	1	1	1	1	1	1	1	1
5	1	1	1	2	2	2	2	1	1	1	1	2	2	2	2	1	1	1	1	2	2	2	2	1	1	1	1	2	2	2	2
6	1	1	1	2	2	2	2	1	1	1	1	2	2	2	2	2	2	2	2	1	1	1	1	2	2	2	2	1	1	1	1
7	1	1	1	2	2	2	2	2	2	2	2	1	1	1	1	1	1	1	1	2	2	2	2	2	2	2	2	1	1	1	1
8	1	1	1	2	2	2	2	2	2	2	2	1	1	1	1	2	2	2	2	1	1	1	1	1	1	1	1	2	2	2	2
9	1	2	2	1	1	2	2	1	1	2	2	1	1	2	2	1	1	2	2	1	1	2	2	1	1	2	2	1	1	2	2
10	1	2	2	1	1	2	2	1	1	2	2	1	1	2	2	2	2	1	1	2	2	1	1	2	2	1	1	2	2	1	1
11	1	2	2	1	1	2	2	2	2	1	1	2	2	1	1	1	1	2	2	1	1	2	2	2	2	1	1	2	2	1	1
12	1	2	2	1	1	2	2	2	2	1	1	2	2	1	1	2	2	1	1	2	2	1	1	1	1	2	2	1	1	2	2
13	1	2	2	2	2	1	1	1	1	2	2	2	2	1	1	1	1	2	2	2	2	1	1	1	1	2	2	2	2	1	1
14	1	2	2	2	2	1	1	1	1	2	2	2	2	1	1	2	2	1	1	1	1	2	2	2	2	1	1	1	1	2	2
15	1	2	2	2	2	1	1	2	2	1	1	1	1	2	2	1	1	2	2	2	2	1	1	2	2	1	1	1	1	2	2
16	1	2	2	2	2	1	1	2	2	1	1	1	1	2	2	2	2	1	1	1	1	2	2	1	1	2	2	2	2	1	1

17	2	1	2	1	2	1	2	1	2	1	2	1	2	1	2	1	2	1	2	1	2	1	2	1	2	1	2	1	2	1	2
18	2	1	2	1	2	1	2	1	2	1	2	1	2	1	2	2	1	2	1	2	1	2	1	2	1	2	1	2	1	2	1
19	2	1	2	1	2	1	2	2	1	2	1	2	1	2	1	1	2	1	2	1	2	1	2	2	1	2	1	2	1	2	1
20	2	1	2	1	2	1	2	2	1	2	1	2	1	2	1	2	1	2	1	2	1	2	1	1	2	1	2	1	2	1	2
21	2	1	2	2	1	2	1	1	2	1	2	2	1	2	1	1	2	1	2	2	1	2	1	1	2	1	2	2	1	2	1
22	2	1	2	2	1	2	1	1	2	1	2	2	1	2	1	2	1	2	1	1	2	1	2	2	1	2	1	1	2	1	2
23	2	1	2	2	1	2	1	2	1	2	1	1	2	1	2	1	2	1	2	2	1	2	1	2	1	2	1	1	2	1	2
24	2	1	2	2	1	2	1	2	1	2	1	1	2	1	2	2	1	2	1	1	2	1	2	1	2	1	2	2	1	2	1
25	2	2	1	1	2	2	1	1	2	2	1	1	2	2	1	1	2	2	1	1	2	2	1	1	2	2	1	1	2	2	1
26	2	2	1	1	2	2	1	1	2	2	1	1	2	2	1	2	1	1	2	2	1	1	2	2	1	1	2	2	1	1	2
27	2	2	1	1	2	2	1	2	1	1	2	2	1	1	2	1	2	2	1	1	2	2	1	2	1	1	2	2	1	1	2
28	2	2	1	1	2	2	1	2	1	1	2	2	1	1	2	2	1	1	2	2	1	1	2	1	2	2	1	1	2	2	1
29	2	2	1	2	1	1	2	1	2	2	1	2	1	1	2	1	2	2	1	2	1	1	2	1	2	2	1	2	1	1	2
30	2	2	1	2	1	1	2	1	2	2	1	2	1	1	2	2	1	1	2	1	2	2	1	2	1	1	2	1	2	2	1
31	2	2	1	2	1	1	2	2	1	1	2	1	2	2	1	1	2	2	1	2	1	1	2	2	1	1	2	1	2	2	1
32	2	2	1	2	1	1	2	2	1	1	2	1	2	2	1	2	1	1	2	1	2	2	1	1	2	2	1	2	1	1	2

L_{64} (2^{63})

No.	1	2	3	4	5	6	7	8	9	10	11	12	13	14	15	16	17	18	19	20	21	22	23	24	25	26	27	28	29	30	31	32
1	1	1	1	1	1	1	1	1	1	1	1	1	1	1	1	1	1	1	1	1	1	1	1	1	1	1	1	1	1	1	1	1
2	1	1	1	1	1	1	1	1	1	1	1	1	1	1	1	1	1	1	1	1	1	1	1	1	1	1	1	1	1	1	1	2
3	1	1	1	1	1	1	1	1	1	1	1	1	1	1	1	2	2	2	2	2	2	2	2	2	2	2	2	2	2	2	2	1
4	1	1	1	1	1	1	1	1	1	1	1	1	1	1	1	2	2	2	2	2	2	2	2	2	2	2	2	2	2	2	2	2
5	1	1	1	1	1	1	1	2	2	2	2	2	2	2	2	1	1	1	1	1	1	1	1	2	2	2	2	2	2	2	2	1
6	1	1	1	1	1	1	1	2	2	2	2	2	2	2	2	1	1	1	1	1	1	1	1	2	2	2	2	2	2	2	2	2
7	1	1	1	1	1	1	1	2	2	2	2	2	2	2	2	2	2	2	2	2	2	2	2	1	1	1	1	1	1	1	1	1
8	1	1	1	1	1	1	1	2	2	2	2	2	2	2	2	2	2	2	2	2	2	2	2	1	1	1	1	1	1	1	1	2
9	1	1	1	2	2	2	2	1	1	1	1	2	2	2	2	1	1	1	1	2	2	2	2	1	1	1	1	2	2	2	2	1
10	1	1	1	2	2	2	2	1	1	1	1	2	2	2	2	1	1	1	1	2	2	2	2	1	1	1	1	2	2	2	2	2
11	1	1	1	2	2	2	2	1	1	1	1	2	2	2	2	2	2	2	2	1	1	1	1	2	2	2	2	1	1	1	1	1
12	1	1	1	2	2	2	2	1	1	1	1	2	2	2	2	2	2	2	2	1	1	1	1	2	2	2	2	1	1	1	1	2
13	1	1	1	2	2	2	2	2	2	2	2	1	1	1	1	1	1	1	1	2	2	2	2	2	2	2	2	1	1	1	1	1
14	1	1	1	2	2	2	2	2	2	2	2	1	1	1	1	1	1	1	1	2	2	2	2	2	2	2	2	1	1	1	1	2
15	1	1	1	2	2	2	2	2	2	2	2	1	1	1	1	2	2	2	2	1	1	1	1	1	1	1	1	2	2	2	2	1
16	1	1	1	2	2	2	2	2	2	2	2	1	1	1	1	2	2	2	2	1	1	1	1	1	1	1	1	2	2	2	2	2
17	1	2	2	1	1	2	2	1	1	2	2	1	1	2	2	1	1	2	2	1	1	2	2	1	1	2	2	1	1	2	2	1
18	1	2	2	1	1	2	2	1	1	2	2	1	1	2	2	1	1	2	2	1	1	2	2	1	1	2	2	1	1	2	2	2
19	1	2	2	1	1	2	2	1	1	2	2	1	1	2	2	2	2	1	1	2	2	1	1	2	2	1	1	2	2	1	1	1
20	1	2	2	1	1	2	2	1	1	2	2	1	1	2	2	2	2	1	1	2	2	1	1	2	2	1	1	2	2	1	1	2
21	1	2	2	1	1	2	2	2	2	1	1	2	2	1	1	1	1	2	2	1	1	2	2	2	2	1	1	2	2	1	1	1
22	1	2	2	1	1	2	2	2	2	1	1	2	2	1	1	1	1	2	2	1	1	2	2	2	2	1	1	2	2	1	1	2

23	1	2	2	1	1	2	2	2	2	1	1	2	2	1	1	2	2	1	1	2	2	1	1	1	1	2	2	1	1	2	2	1
24	1	2	2	1	1	2	2	2	2	1	1	2	2	1	1	2	2	1	1	2	2	1	1	1	1	2	2	1	1	2	2	2
25	1	2	2	2	2	1	1	1	1	2	2	2	2	1	1	1	1	2	2	2	2	1	1	1	1	2	2	2	2	1	1	1
26	1	2	2	2	2	1	1	1	1	2	2	2	2	1	1	1	1	2	2	2	2	1	1	1	1	2	2	2	2	1	1	2
27	1	2	2	2	2	1	1	1	1	2	2	2	2	1	1	2	2	1	1	1	1	2	2	2	2	1	1	1	1	2	2	1
28	1	2	2	2	2	1	1	1	1	2	2	2	2	1	1	2	2	1	1	1	1	2	2	2	2	1	1	1	1	2	2	2
29	1	2	2	2	2	1	1	2	2	1	1	1	1	2	2	1	1	2	2	2	2	1	1	2	2	1	1	1	1	2	2	1
30	1	2	2	2	2	1	1	2	2	1	1	1	1	2	2	1	1	2	2	2	2	1	1	2	2	1	1	1	1	2	2	2
31	1	2	2	2	2	1	1	2	2	1	1	1	1	2	2	2	2	1	1	1	1	2	2	1	1	2	2	2	2	1	1	1
32	1	2	2	2	2	1	1	2	2	1	1	1	1	2	2	2	2	1	1	1	1	2	2	1	1	2	2	2	2	1	1	2
33	2	1	2	1	2	1	2	1	2	1	2	1	2	1	2	1	2	1	2	1	2	1	2	1	2	1	2	1	2	1	2	1
34	2	1	2	1	2	1	2	1	2	1	2	1	2	1	2	1	2	1	2	1	2	1	2	1	2	1	2	1	2	1	2	2
35	2	1	2	1	2	1	2	1	2	1	2	1	2	1	2	2	1	2	1	2	1	2	1	2	1	2	1	2	1	2	1	1
36	2	1	2	1	2	1	2	1	2	1	2	1	2	1	2	2	1	2	1	2	1	2	1	2	1	2	1	2	1	2	1	2
37	2	1	2	1	2	1	2	2	1	2	1	2	1	2	1	1	2	1	2	1	2	1	2	2	1	2	1	2	1	2	1	1
38	2	1	2	1	2	1	2	2	1	2	1	2	1	2	1	1	2	1	2	1	2	1	2	2	1	2	1	2	1	2	1	2
39	2	1	2	1	2	1	2	2	1	2	1	2	1	2	1	2	1	2	1	2	1	2	1	1	2	1	2	1	2	1	2	1
40	2	1	2	1	2	1	2	2	1	2	1	2	1	2	1	2	1	2	1	2	1	2	1	1	2	1	2	1	2	1	2	2
41	2	1	2	2	1	2	1	1	2	1	2	2	1	2	1	1	2	1	2	2	1	2	1	1	2	1	2	2	1	2	1	1
42	2	1	2	2	1	2	1	1	2	1	2	2	1	2	1	1	2	1	2	2	1	2	1	1	2	1	2	2	1	2	1	2

(*Continued*)

L_{64} (2^{63}) (*Continued*)

No.	1	2	3	4	5	6	7	8	9	10	11	12	13	14	15	16	17	18	19	20	21	22	23	24	25	26	27	28	29	30	31	32
43	2	1	2	2	1	2	1	1	2	1	2	2	1	2	1	2	1	2	1	1	2	1	2	2	1	2	1	1	2	1	2	1
44	2	1	2	2	1	2	1	1	2	1	2	2	1	2	1	2	1	2	1	1	2	1	2	2	1	2	1	1	2	1	2	2
45	2	1	2	2	1	2	1	2	1	2	1	1	2	1	2	1	2	1	2	2	1	2	1	2	1	2	1	1	2	1	2	1
46	2	1	2	2	1	2	1	2	1	2	1	1	2	1	2	1	2	1	2	2	1	2	1	2	1	2	1	1	2	1	2	2
47	2	1	2	2	1	2	1	2	1	2	1	1	2	1	2	2	1	2	1	1	2	1	2	1	2	1	2	2	1	2	1	1
48	2	1	2	2	1	2	1	2	1	2	1	1	2	1	2	2	1	2	1	1	2	1	2	1	2	1	2	2	1	2	1	2
49	2	2	1	1	2	2	1	1	2	2	1	1	2	2	1	1	2	2	1	1	2	2	1	1	2	2	1	1	2	2	1	1
50	2	2	1	1	2	2	1	1	2	2	1	1	2	2	1	1	2	2	1	1	2	2	1	1	2	2	1	1	2	2	1	2
51	2	2	1	1	2	2	1	1	2	2	1	1	2	2	1	2	1	1	2	2	1	1	2	2	1	1	2	2	1	1	2	1
52	2	2	1	1	2	2	1	1	2	2	1	1	2	2	1	2	1	1	2	2	1	1	2	2	1	1	2	2	1	1	2	2
53	2	2	1	1	2	2	1	2	1	1	2	2	1	1	2	1	2	2	1	1	2	2	1	2	1	1	2	2	1	1	2	1
54	2	2	1	1	2	2	1	2	1	1	2	2	1	1	2	1	2	2	1	1	2	2	1	2	1	1	2	2	1	1	2	2
55	2	2	1	1	2	2	1	2	1	1	2	2	1	1	2	2	1	1	2	2	1	1	2	1	2	2	1	1	2	2	1	1
56	2	2	1	1	2	2	1	2	1	1	2	2	1	1	2	2	1	1	2	2	1	1	2	1	2	2	1	1	2	2	1	2
57	2	2	1	2	1	1	2	1	2	2	1	2	1	1	2	1	2	2	1	2	1	1	2	1	2	2	1	2	1	1	2	1
58	2	2	1	2	1	1	2	1	2	2	1	2	1	1	2	1	2	2	1	2	1	1	2	1	2	2	1	2	1	1	2	2
59	2	2	1	2	1	1	2	1	2	2	1	2	1	1	2	2	1	1	2	1	2	2	1	2	1	1	2	1	2	2	1	1
60	2	2	1	2	1	1	2	1	2	2	1	2	1	1	2	2	1	1	2	1	2	2	1	2	1	1	2	1	2	2	1	2
61	2	2	1	2	1	1	2	2	1	1	2	1	2	2	1	1	2	2	1	2	1	1	2	2	1	1	2	1	2	2	1	1
62	2	2	1	2	1	1	2	2	1	1	2	1	2	2	1	1	2	2	1	2	1	1	2	2	1	1	2	1	2	2	1	2
63	2	2	1	2	1	1	2	2	1	1	2	1	2	2	1	2	1	1	2	1	2	2	1	1	2	2	1	2	1	1	2	1
64	2	2	1	2	1	1	2	2	1	1	2	1	2	2	1	2	1	1	2	1	2	2	1	1	2	2	1	2	1	1	2	2

No.	33	34	35	36	37	38	39	40	41	42	43	44	45	46	47	48	49	50	51	52	53	54	55	56	57	58	59	60	61	62	63
1	1	1	1	1	1	1	1	1	1	1	1	1	1	1	1	1	1	1	1	1	1	1	1	1	1	1	1	1	1	1	1
2	2	2	2	2	2	2	2	2	2	2	2	2	2	2	2	2	2	2	2	2	2	2	2	2	2	2	2	2	2	2	2
3	1	1	1	1	1	1	1	1	1	1	1	1	1	1	1	2	2	2	2	2	2	2	2	2	2	2	2	2	2	2	2
4	2	2	2	2	2	2	2	2	2	2	2	2	2	2	2	1	1	1	1	1	1	1	1	1	1	1	1	1	1	1	1
5	1	1	1	1	1	1	1	2	2	2	2	2	2	2	2	1	1	1	1	1	1	1	1	2	2	2	2	2	2	2	2
6	2	2	2	2	2	2	2	1	1	1	1	1	1	1	1	2	2	2	2	2	2	2	2	1	1	1	1	1	1	1	1
7	1	1	1	1	1	1	1	2	2	2	2	2	2	2	2	2	2	2	2	2	2	2	2	1	1	1	1	1	1	1	1
8	2	2	2	2	2	2	2	1	1	1	1	1	1	1	1	1	1	1	1	1	1	1	1	2	2	2	2	2	2	2	2
9	1	1	1	2	2	2	2	1	1	1	1	2	2	2	2	1	1	1	1	2	2	2	2	1	1	1	1	2	2	2	2
10	2	2	2	1	1	1	1	2	2	2	2	1	1	1	1	2	2	2	2	1	1	1	1	2	2	2	2	1	1	1	1
11	1	1	1	2	2	2	2	1	1	1	1	2	2	2	2	2	2	2	2	1	1	1	1	2	2	2	2	1	1	1	1
12	2	2	2	1	1	1	1	2	2	2	2	1	1	1	1	1	1	1	1	2	2	2	2	1	1	1	1	2	2	2	2
13	1	1	1	2	2	2	2	2	2	2	2	1	1	1	1	1	1	1	1	2	2	2	2	2	2	2	2	1	1	1	1
14	2	2	2	1	1	1	1	1	1	1	1	2	2	2	2	2	2	2	2	1	1	1	1	1	1	1	1	2	2	2	2
15	1	1	1	2	2	2	2	2	2	2	2	1	1	1	1	2	2	2	2	1	1	1	1	1	1	1	1	2	2	2	2
16	2	2	2	1	1	1	1	1	1	1	1	2	2	2	2	1	1	1	1	2	2	2	2	2	2	2	2	1	1	1	1
17	1	2	2	1	1	2	2	1	1	2	2	1	1	2	2	1	1	2	2	1	1	2	2	1	1	2	2	1	1	2	2
18	2	1	1	2	2	1	1	2	2	1	1	2	2	1	1	2	2	1	1	2	2	1	1	2	2	1	1	2	2	1	1
19	1	2	2	1	1	2	2	1	1	2	2	1	1	2	2	2	2	1	1	2	2	1	1	2	2	1	1	2	2	1	1
20	2	1	1	2	2	1	1	2	2	1	1	2	2	1	1	1	1	2	2	1	1	2	2	1	1	2	2	1	1	2	2

(*Continued*)

L_{64} (2^{63}) (*Continued*)

No.	33	34	35	36	37	38	39	40	41	42	43	44	45	46	47	48	49	50	51	52	53	54	55	56	57	58	59	60	61	62	63
21	1	2	2	1	1	2	2	2	2	1	1	2	2	1	1	1	1	2	2	1	1	2	2	2	2	1	1	2	2	1	1
22	2	1	1	2	2	1	1	1	1	2	2	1	1	2	2	2	2	1	1	2	2	1	1	1	1	2	2	1	1	2	2
23	1	2	2	1	1	2	2	2	2	1	1	2	2	1	1	2	2	1	1	2	2	1	1	1	1	2	2	1	1	2	2
24	2	1	1	2	2	1	1	1	1	2	2	1	1	2	2	1	1	2	2	1	1	2	2	2	2	1	1	2	2	1	1
25	1	2	2	2	2	1	1	1	1	2	2	2	2	1	1	1	1	2	2	2	2	1	1	1	1	2	2	2	2	1	1
26	2	1	1	1	1	2	2	2	2	1	1	1	1	2	2	2	2	1	1	1	1	2	2	2	2	1	1	1	1	2	2
27	1	2	2	2	2	1	1	1	1	2	2	2	2	1	1	2	2	1	1	1	1	2	2	2	2	1	1	1	1	2	2
28	2	1	1	1	1	2	2	2	2	1	1	1	1	2	2	1	1	2	2	2	2	1	1	1	1	2	2	2	2	1	1
29	1	2	2	2	2	1	1	2	2	1	1	1	1	2	2	1	1	2	2	2	2	1	1	2	2	1	1	1	1	2	2
30	2	1	1	1	1	2	2	1	1	2	2	2	2	1	1	2	2	1	1	1	1	2	2	1	1	2	2	2	2	1	1
31	1	2	2	2	2	1	1	2	2	1	1	1	1	2	2	2	2	1	1	1	1	2	2	1	1	2	2	2	2	1	1
32	2	1	1	1	1	2	2	1	1	2	2	2	2	1	1	1	1	2	2	2	2	1	1	2	2	1	1	1	1	2	2
33	2	1	2	1	2	1	2	1	2	1	2	1	2	1	2	1	2	1	2	1	2	1	2	1	2	1	2	1	2	1	2
34	1	2	1	2	1	2	1	2	1	2	1	2	1	2	1	2	1	2	1	2	1	2	1	2	1	2	1	2	1	2	1
35	2	1	2	1	2	1	2	1	2	1	2	1	2	1	2	2	1	2	1	2	1	2	1	2	1	2	1	2	1	2	1
36	1	2	1	2	1	2	1	2	1	2	1	2	1	2	1	1	2	1	2	1	2	1	2	1	2	1	2	1	2	1	2
37	2	1	2	1	2	1	2	2	1	2	1	2	1	2	1	1	2	1	2	1	2	1	2	2	1	2	1	2	1	2	1
38	1	2	1	2	1	2	1	1	2	1	2	1	2	1	2	2	1	2	1	2	1	2	1	1	2	1	2	1	2	1	2
39	2	1	2	1	2	1	2	2	1	2	1	2	1	2	1	2	1	2	1	2	1	2	1	1	2	1	2	1	2	1	2
40	1	2	1	2	1	2	1	1	2	1	2	1	2	1	2	1	2	1	2	1	2	1	2	2	1	2	1	2	1	2	1
41	2	1	2	2	1	2	1	1	2	1	2	2	1	2	1	1	2	1	2	2	1	2	1	1	2	1	2	2	1	2	1
42	1	2	1	1	2	1	2	2	1	2	1	1	2	1	2	2	1	2	1	1	2	1	2	2	1	2	1	1	2	1	2

43	2	1	2	2	1	2	1	1	2	1	2	2	1	2	1	2	1	2	1	1	2	1	2	2	1	2	1	1	2	1	2
44	1	2	1	1	2	1	2	2	1	2	1	1	2	1	2	1	2	1	2	2	1	2	1	1	2	1	2	2	1	2	1
45	2	1	2	2	1	2	1	2	1	2	1	1	2	1	2	1	2	1	2	2	1	2	1	2	1	2	1	1	2	1	2
46	1	2	1	1	2	1	2	1	2	1	2	2	1	2	1	2	1	2	1	1	2	1	2	1	2	1	2	2	1	2	1
47	2	1	2	2	1	2	1	2	1	2	1	1	2	1	2	2	1	2	1	1	2	1	2	1	2	1	2	2	1	2	1
48	1	2	1	1	2	1	2	1	2	1	2	2	1	2	1	1	2	1	2	2	1	2	1	2	1	2	1	1	2	1	2
49	2	2	1	1	2	2	1	1	2	2	1	1	2	2	1	1	2	2	1	1	2	2	1	1	2	2	1	1	2	2	1
50	1	1	2	2	1	1	2	2	1	1	2	2	1	1	2	2	1	1	2	2	1	1	2	2	1	1	2	2	1	1	2
51	2	2	1	1	2	2	1	1	2	2	1	1	2	2	1	2	1	1	2	2	1	1	2	2	1	1	2	2	1	1	2
52	1	1	2	2	1	1	2	2	1	1	2	2	1	1	2	1	2	2	1	1	2	2	1	1	2	2	1	1	2	2	1
53	2	2	1	1	2	2	1	2	1	1	2	2	1	1	2	1	2	2	1	1	2	2	1	2	1	1	2	2	1	1	2
54	1	1	2	2	1	1	2	1	2	2	1	1	2	2	1	2	1	1	2	2	1	1	2	1	2	2	1	1	2	2	1
55	2	2	1	1	2	2	1	2	1	1	2	2	1	1	2	2	1	1	2	2	1	1	2	1	2	2	1	1	2	2	1
56	1	1	2	2	1	1	2	1	2	2	1	1	2	2	1	1	2	2	1	1	2	2	1	2	1	1	2	2	1	1	2
57	2	2	1	2	1	1	2	1	2	2	1	2	1	1	2	1	2	2	1	2	1	1	2	1	2	2	1	2	1	1	2
58	1	1	2	1	2	2	1	2	1	1	2	1	2	2	1	2	1	1	2	1	2	2	1	2	1	1	2	1	2	2	1
59	2	2	1	2	1	1	2	1	2	2	1	2	1	1	2	2	1	1	2	1	2	2	1	2	1	1	2	1	2	2	1
60	1	1	2	1	2	2	1	2	1	1	2	1	2	2	1	1	2	2	1	2	1	1	2	1	2	2	1	2	1	1	2
61	2	2	1	2	1	1	2	2	1	1	2	1	2	2	1	1	2	2	1	2	1	1	2	2	1	1	2	1	2	2	1
62	1	1	2	1	2	2	1	1	2	2	1	2	1	1	2	2	1	1	2	1	2	2	1	1	2	2	1	2	1	1	2
63	2	2	1	2	1	1	2	2	1	1	2	1	2	2	1	2	1	1	2	1	2	2	1	1	2	2	1	2	1	1	2
64	1	1	2	1	2	2	1	1	2	2	1	2	1	1	2	1	2	2	1	2	1	1	2	2	1	1	2	1	2	2	1

$L_{81}\ (3^{40})$

No.	1	2	3	4	5	6	7	8	9	10	11	12	13	14	15	16	17	18	19	20	21	22	23	24	25	26	27	28	29	30	31	32	33	34	35	36	37	38	39	40
1	1	1	1	1	1	1	1	1	1	1	1	1	1	1	1	1	1	1	1	1	1	1	1	1	1	1	1	1	1	1	1	1	1	1	1	1	1	1	1	1
2	1	1	1	1	1	1	1	1	1	1	1	1	1	2	2	2	2	2	2	2	2	2	2	2	2	2	2	2	2	2	2	2	2	2	2	2	2	2	2	2
3	1	1	1	1	1	1	1	1	1	1	1	1	1	3	3	3	3	3	3	3	3	3	3	3	3	3	3	3	3	3	3	3	3	3	3	3	3	3	3	3
4	1	1	1	1	2	2	2	2	2	2	2	2	2	1	1	1	1	1	1	1	1	1	2	2	2	2	2	2	2	2	2	3	3	3	3	3	3	3	3	3
5	1	1	1	1	2	2	2	2	2	2	2	2	2	2	2	2	2	2	2	2	2	2	3	3	3	3	3	3	3	3	3	1	1	1	1	1	1	1	1	1
6	1	1	1	1	2	2	2	2	2	2	2	2	2	3	3	3	3	3	3	3	3	3	1	1	1	1	1	1	1	1	1	2	2	2	2	2	2	2	2	2
7	1	1	1	1	3	3	3	3	3	3	3	3	3	1	1	1	1	1	1	1	1	1	3	3	3	3	3	3	3	3	3	2	2	2	2	2	2	2	2	2
8	1	1	1	1	3	3	3	3	3	3	3	3	3	2	2	2	2	2	2	2	2	2	1	1	1	1	1	1	1	1	1	3	3	3	3	3	3	3	3	3
9	1	1	1	1	3	3	3	3	3	3	3	3	3	3	3	3	3	3	3	3	3	3	2	2	2	2	2	2	2	2	2	1	1	1	1	1	1	1	1	1
10	1	2	2	2	1	1	1	2	2	2	3	3	3	1	1	1	2	2	2	3	3	3	1	1	1	2	2	2	3	3	3	1	1	1	2	2	2	3	3	3
11	1	2	2	2	1	1	1	2	2	2	3	3	3	2	2	2	3	3	3	1	1	1	2	2	2	3	3	3	1	1	1	2	2	2	3	3	3	1	1	1
12	1	2	2	2	1	1	1	2	2	2	3	3	3	3	3	3	1	1	1	2	2	2	3	3	3	1	1	1	2	2	2	3	3	3	1	1	1	2	2	2
13	1	2	2	2	2	2	2	3	3	3	1	1	1	1	1	1	2	2	2	3	3	3	2	2	2	3	3	3	1	1	1	3	3	3	1	1	1	2	2	2
14	1	2	2	2	2	2	2	3	3	3	1	1	1	2	2	2	3	3	3	1	1	1	3	3	3	1	1	1	2	2	2	1	1	1	2	2	2	3	3	3
15	1	2	2	2	2	2	2	3	3	3	1	1	1	3	3	3	1	1	1	2	2	2	1	1	1	2	2	2	3	3	3	2	2	2	3	3	3	1	1	1
16	1	2	2	2	3	3	3	1	1	1	2	2	2	1	1	1	2	2	2	3	3	3	3	3	3	1	1	1	2	2	2	2	2	2	3	3	3	1	1	1
17	1	2	2	2	3	3	3	1	1	1	2	2	2	2	2	2	3	3	3	1	1	1	1	1	1	2	2	2	3	3	3	3	3	3	1	1	1	2	2	2
18	1	2	2	2	3	3	3	1	1	1	2	2	2	3	3	3	1	1	1	2	2	2	2	2	2	3	3	3	1	1	1	1	1	1	2	2	2	3	3	3
19	1	3	3	3	1	1	1	3	3	3	2	2	2	1	1	1	3	3	3	2	2	2	1	1	1	3	3	3	2	2	2	1	1	1	3	3	3	2	2	2
20	1	3	3	3	1	1	1	3	3	3	2	2	2	2	2	2	1	1	1	3	3	3	2	2	2	1	1	1	3	3	3	2	2	2	1	1	1	3	3	3
21	1	3	3	3	1	1	1	3	3	3	2	2	2	3	3	3	2	2	2	1	1	1	3	3	3	2	2	2	1	1	1	3	3	3	2	2	2	1	1	1
22	1	3	3	3	2	2	2	1	1	1	3	3	3	1	1	1	3	3	3	2	2	2	2	2	2	1	1	1	3	3	3	3	3	3	2	2	2	1	1	1
23	1	3	3	3	2	2	2	1	1	1	3	3	3	2	2	2	1	1	1	3	3	3	3	3	3	2	2	2	1	1	1	1	1	1	3	3	3	2	2	2
24	1	3	3	3	2	2	2	1	1	1	3	3	3	3	3	3	2	2	2	1	1	1	1	1	1	3	3	3	2	2	2	2	2	2	1	1	1	3	3	3
25	1	3	3	3	3	3	3	2	2	2	1	1	1	1	1	1	3	3	3	2	2	2	3	3	3	2	2	2	1	1	1	2	2	2	1	1	1	3	3	3
26	1	3	3	3	3	3	3	2	2	2	1	1	1	2	2	2	1	1	1	3	3	3	1	1	1	3	3	3	2	2	2	3	3	3	2	2	2	1	1	1
27	1	3	3	3	3	3	3	2	2	2	1	1	1	3	3	3	2	2	2	1	1	1	2	2	2	1	1	1	3	3	3	1	1	1	3	3	3	2	2	2

28	2	1	2	3	1	2	3	1	2	3	1	2	3	1	2	3	1	2	3	1	2	3	1	2	3	1	2	3	1	2	3	1	2	3	1	2	3	1	2	3
29	2	1	2	3	1	2	3	1	2	3	1	2	3	2	3	1	2	3	1	2	3	1	2	3	1	2	3	1	2	3	1	2	3	1	2	3	1	2	3	1
30	2	1	2	3	1	2	3	1	2	3	1	2	3	3	1	2	3	1	2	3	1	2	3	1	2	3	1	2	3	1	2	3	1	2	3	1	2	3	1	2
31	2	1	2	3	2	3	1	2	3	1	2	3	1	1	2	3	1	2	3	1	2	3	2	3	1	2	3	1	2	3	1	3	1	2	3	1	2	3	1	2
32	2	1	2	3	2	3	1	2	3	1	2	3	1	2	3	1	2	3	1	2	3	1	3	1	2	3	1	2	3	1	2	1	2	3	1	2	3	1	2	3
33	2	1	2	3	2	3	1	2	3	1	2	3	1	3	1	2	3	1	2	3	1	2	1	2	3	1	2	3	1	2	3	2	3	1	2	3	1	2	3	1
34	2	1	2	3	3	1	2	3	1	2	3	1	2	1	2	3	1	2	3	1	2	3	3	1	2	3	1	2	3	1	2	2	3	1	2	3	1	2	3	1
35	2	1	2	3	3	1	2	3	1	2	3	1	2	2	3	1	2	3	1	2	3	1	1	2	3	1	2	3	1	2	3	3	1	2	3	1	2	3	1	2
36	2	1	2	3	3	1	2	3	1	2	3	1	2	3	1	2	3	1	2	3	1	2	2	3	1	2	3	1	2	3	1	1	2	3	1	2	3	1	2	3
37	2	2	3	1	1	2	3	2	3	1	3	1	2	1	2	3	2	3	1	3	1	2	1	2	3	2	3	1	3	1	2	1	2	3	2	3	1	3	1	2
38	2	2	3	1	1	2	3	2	3	1	3	1	2	2	3	1	3	1	2	1	2	3	2	3	1	3	1	2	1	2	3	2	3	1	3	1	2	1	2	3
39	2	2	3	1	1	2	3	2	3	1	3	1	2	3	1	2	1	2	3	2	3	1	3	1	2	1	2	3	2	3	1	3	1	2	1	2	3	2	3	1
40	2	2	3	1	2	3	1	3	1	2	1	2	3	1	2	3	2	3	1	3	1	2	2	3	1	3	1	2	1	2	3	3	1	2	1	2	3	2	3	1
41	2	2	3	1	2	3	1	3	1	2	1	2	3	2	3	1	3	1	2	1	2	3	3	1	2	1	2	3	2	3	1	1	2	3	2	3	1	3	1	2
42	2	2	3	1	2	3	1	3	1	2	1	2	3	3	1	2	1	2	3	2	3	1	1	2	3	2	3	1	3	1	2	2	3	1	3	1	2	1	2	3
43	2	2	3	1	3	1	2	1	2	3	2	3	1	1	2	3	2	3	1	3	1	2	3	1	2	1	2	3	2	3	1	2	3	1	3	1	2	1	2	3
44	2	2	3	1	3	1	2	1	2	3	2	3	1	2	3	1	3	1	2	1	2	3	1	2	3	2	3	1	3	1	2	3	1	2	1	2	3	2	3	1
45	2	2	3	1	3	1	2	1	2	3	2	3	1	3	1	2	1	2	3	2	3	1	2	3	1	3	1	2	1	2	3	1	2	3	2	3	1	3	1	2
46	2	3	1	2	1	2	3	3	1	2	2	3	1	1	2	3	3	1	2	2	3	1	1	2	3	3	1	2	2	3	1	1	2	3	3	1	2	2	3	1
47	2	3	1	2	1	2	3	3	1	2	2	3	1	2	3	1	1	2	3	3	1	2	2	3	1	1	2	3	3	1	2	2	3	1	1	2	3	3	1	2
48	2	3	1	2	1	2	3	3	1	2	2	3	1	3	1	2	2	3	1	1	2	3	3	1	2	2	3	1	1	2	3	3	1	2	2	3	1	1	2	3
49	2	3	1	2	2	3	1	1	2	3	3	1	2	1	2	3	3	1	2	2	3	1	2	3	1	1	2	3	3	1	2	3	1	2	2	3	1	1	2	3
50	2	3	1	2	2	3	1	1	2	3	3	1	2	2	3	1	1	2	3	3	1	2	3	1	2	2	3	1	1	2	3	1	2	3	3	1	2	2	3	1
51	2	3	1	2	2	3	1	1	2	3	3	1	2	3	1	2	2	3	1	1	2	3	1	2	3	3	1	2	2	3	1	2	3	1	1	2	3	3	1	2
52	2	3	1	2	3	1	2	2	3	1	1	2	3	1	2	3	3	1	2	2	3	1	3	1	2	2	3	1	1	2	3	2	3	1	1	2	3	3	1	2
53	2	3	1	2	3	1	2	2	3	1	1	2	3	2	3	1	1	2	3	3	1	2	1	2	3	3	1	2	2	3	1	3	1	2	2	3	1	1	2	3
54	2	3	1	2	3	1	2	2	3	1	1	2	3	3	1	2	2	3	1	1	2	3	2	3	1	1	2	3	3	1	2	1	2	3	3	1	2	2	3	1

(*Continued*)

L_{81} (3^{40}) (*Continued*)

No.	1	2	3	4	5	6	7	8	9	10	11	12	13	14	15	16	17	18	19	20	21	22	23	24	25	26	27	28	29	30	31	32	33	34	35	36	37	38	39	40
55	3	1	3	2	1	3	2	1	3	2	1	3	2	1	3	2	1	3	2	1	3	2	1	3	2	1	3	2	1	3	2	1	3	2	1	3	2	1	3	2
56	3	1	3	2	1	3	2	1	3	2	1	3	2	2	1	3	2	1	3	2	1	3	2	1	3	2	1	3	2	1	3	2	1	3	2	1	3	2	1	3
57	3	1	3	2	1	3	2	1	3	2	1	3	2	3	2	1	3	2	1	3	2	1	3	2	1	3	2	1	3	2	1	3	2	1	3	2	1	3	2	1
58	3	1	3	2	2	1	3	2	1	3	2	1	3	1	3	2	1	3	2	1	3	2	2	1	3	2	1	3	2	1	3	3	2	1	3	2	1	3	2	1
59	3	1	3	2	2	1	3	2	1	3	2	1	3	2	1	3	2	1	3	2	1	3	3	2	1	3	2	1	3	2	1	1	3	2	1	3	2	1	3	2
60	3	1	3	2	2	1	3	2	1	3	2	1	3	3	2	1	3	2	1	3	2	1	1	3	2	1	3	2	1	3	2	2	1	3	2	1	3	2	1	3
61	3	1	3	2	3	2	1	3	2	1	3	2	1	1	3	2	1	3	2	1	3	2	3	2	1	3	2	1	3	2	1	2	1	3	2	1	3	2	1	3
62	3	1	3	2	3	2	1	3	2	1	3	2	1	2	1	3	2	1	3	2	1	3	1	3	2	1	3	2	1	3	2	3	2	1	3	2	1	3	2	1
63	3	1	3	2	3	2	1	3	2	1	3	2	1	3	2	1	3	2	1	3	2	1	2	1	3	2	1	3	2	1	3	1	3	2	1	3	2	1	3	2
64	3	2	1	3	1	3	2	2	1	3	3	2	1	1	3	2	2	1	3	3	2	1	1	3	2	2	1	3	3	2	1	1	3	2	2	1	3	3	2	1
65	3	2	1	3	1	3	2	2	1	3	3	2	1	2	1	3	3	2	1	1	3	2	2	1	3	3	2	1	1	3	2	2	1	3	3	2	1	1	3	2
66	3	2	1	3	1	3	2	2	1	3	3	2	1	3	2	1	1	3	2	2	1	3	3	2	1	1	3	2	2	1	3	3	2	1	1	3	2	2	1	3
67	3	2	1	3	2	1	3	3	2	1	1	3	2	1	3	2	2	1	3	3	2	1	2	1	3	3	2	1	1	3	2	3	2	1	1	3	2	2	1	3
68	3	2	1	3	2	1	3	3	2	1	1	3	2	2	1	3	3	2	1	1	3	2	3	2	1	1	3	2	2	1	3	1	3	2	2	1	3	3	2	1
69	3	2	1	3	2	1	3	3	2	1	1	3	2	3	2	1	1	3	2	2	1	3	1	3	2	2	1	3	3	2	1	2	1	3	3	2	1	1	3	2
70	3	2	1	3	3	2	1	1	3	2	2	1	3	1	3	2	2	1	3	3	2	1	3	2	1	1	3	2	2	1	3	2	1	3	3	2	1	1	3	2
71	3	2	1	3	3	2	1	1	3	2	2	1	3	2	1	3	3	2	1	1	3	2	1	3	2	2	1	3	3	2	1	3	2	1	1	3	2	2	1	3
72	3	2	1	3	3	2	1	1	3	2	2	1	3	3	2	1	1	3	2	2	1	3	2	1	3	3	2	1	1	3	2	1	3	2	2	1	3	3	2	1
73	3	3	2	1	1	3	2	3	2	1	2	1	3	1	3	2	3	2	1	2	1	3	1	3	2	3	2	1	2	1	3	1	3	2	3	2	1	2	1	3
74	3	3	2	1	1	3	2	3	2	1	2	1	3	2	1	3	1	3	2	3	2	1	2	1	3	1	3	2	3	2	1	2	1	3	1	3	2	3	2	1
75	3	3	2	1	1	3	2	3	2	1	2	1	3	3	2	1	2	1	3	1	3	2	3	2	1	2	1	3	1	3	2	3	2	1	2	1	3	1	3	2
76	3	3	2	1	2	1	3	1	3	2	3	2	1	1	3	2	3	2	1	2	1	3	2	1	3	1	3	2	3	2	1	3	2	1	2	1	3	1	3	2
77	3	3	2	1	2	1	3	1	3	2	3	2	1	2	1	3	1	3	2	3	2	1	3	2	1	2	1	3	1	3	2	1	3	2	3	2	1	2	1	3
78	3	3	2	1	2	1	3	1	3	2	3	2	1	3	2	1	2	1	3	1	3	2	1	3	2	3	2	1	2	1	3	2	1	3	1	3	2	3	2	1
79	3	3	2	1	3	2	1	2	1	3	1	3	2	1	3	2	3	2	1	2	1	3	3	2	1	2	1	3	1	3	2	2	1	3	1	3	2	3	2	1
80	3	3	2	1	3	2	1	2	1	3	1	3	2	2	1	3	1	3	2	3	2	1	1	3	2	3	2	1	2	1	3	3	2	1	2	1	3	1	3	2
81	3	3	2	1	3	2	1	2	1	3	1	3	2	3	2	1	2	1	3	1	3	2	2	1	3	1	3	2	3	2	1	1	3	2	3	2	1	2	1	3

$L_{81}\ (9^8 \times 3^8)$

No.	1	2	3	4	5	6	7	8	9	10	11	12	13	14	15	16
1	1	1	1	1	1	1	1	1	1	1	1	1	1	1	1	1
2	1	2	2	2	2	2	2	2	1	1	2	2	2	2	2	2
3	1	3	3	3	3	3	3	3	1	1	3	3	3	3	3	3
4	2	1	2	3	4	4	4	4	2	2	1	1	2	2	3	3
5	2	2	3	1	5	5	5	5	2	2	2	2	3	3	1	1
6	2	3	1	2	6	6	6	6	2	2	3	3	1	1	2	2
7	3	1	3	2	7	7	7	7	3	3	1	1	3	3	2	2
8	3	2	1	3	8	8	8	8	3	3	2	2	1	1	3	3
9	3	3	2	1	9	9	9	9	3	3	3	3	2	2	1	1
10	2	4	4	4	1	2	3	8	3	3	1	3	2	3	1	2
11	2	5	5	5	2	3	1	9	3	3	2	1	3	1	2	3
12	2	6	6	6	3	1	2	7	3	3	3	2	1	2	3	1
13	3	4	5	6	4	5	6	2	1	1	1	3	3	1	3	1
14	3	5	6	4	5	6	4	3	1	1	2	1	1	2	1	2
15	3	6	4	5	6	4	5	1	1	1	3	2	2	3	2	3
16	1	4	6	5	7	8	9	5	2	2	1	3	1	2	2	3
17	1	5	4	6	8	9	7	6	2	2	2	1	2	3	3	1
18	1	6	5	4	9	7	8	4	2	2	3	2	3	1	1	2
19	3	7	7	7	1	3	2	6	2	2	1	2	3	2	1	3
20	3	8	8	8	2	1	3	4	2	2	2	3	1	3	2	1
21	3	9	9	9	3	2	1	5	2	2	3	1	2	1	3	2
22	1	7	8	9	4	6	5	9	3	3	1	2	1	3	3	2
23	1	8	9	7	5	4	6	7	3	3	2	3	2	1	1	3
24	1	9	7	8	6	5	4	8	3	3	3	1	3	2	2	1
25	2	7	9	8	7	9	8	3	1	1	1	2	2	1	2	1
26	2	8	7	9	8	7	9	1	1	1	2	3	3	2	3	2
27	2	9	8	7	9	8	7	2	1	1	3	1	1	3	1	3
28	4	1	4	7	2	5	9	3	2	3	3	2	1	1	3	2
29	4	2	5	8	3	6	7	1	2	3	1	3	2	2	1	3
30	4	3	6	9	1	4	8	2	2	3	2	1	3	3	2	1
31	5	1	5	9	5	8	3	6	3	1	3	2	2	2	2	1
32	5	2	6	7	6	9	1	4	3	1	1	3	3	3	3	2
33	5	3	4	8	4	7	2	5	3	1	2	1	1	1	1	3
34	6	1	6	8	8	2	6	9	1	2	3	2	3	3	1	3
35	6	2	4	9	9	3	4	7	1	2	1	3	1	1	2	1
36	6	3	5	7	7	1	5	8	1	2	2	1	2	2	3	2
37	5	4	7	1	2	6	8	7	1	2	3	1	2	3	3	3
38	5	5	8	2	3	4	9	8	1	2	1	2	3	1	1	1
39	5	6	9	3	1	5	7	9	1	2	2	3	1	2	2	2
40	6	4	8	3	5	9	2	1	2	3	3	1	3	1	2	2
41	6	5	9	1	6	7	3	2	2	3	1	2	1	2	3	3
42	6	6	7	2	4	8	1	3	2	3	2	3	2	3	1	1

(*Continued*)

L_{81} ($9^8 \times 3^8$) (*Continued*)

No.	1	2	3	4	5	6	7	8	9	10	11	12	13	14	15	16
43	4	4	9	2	8	3	5	4	3	1	3	1	1	2	1	1
44	4	5	7	3	9	1	6	5	3	1	1	2	2	3	2	2
45	4	6	8	1	7	2	4	6	3	1	2	3	3	1	3	3
46	6	7	1	4	2	4	7	5	3	1	3	3	3	2	3	1
47	6	8	2	5	3	5	8	6	3	1	1	1	1	3	1	2
48	6	9	3	6	1	6	9	4	3	1	2	2	2	1	2	3
49	4	7	2	6	5	7	1	8	1	2	3	3	1	3	2	3
50	4	8	3	4	6	8	2	9	1	2	1	1	2	1	3	1
51	4	9	1	5	4	9	3	7	1	2	2	2	3	2	1	2
52	5	7	3	5	8	1	4	2	2	3	3	3	2	1	1	2
53	5	8	1	6	9	2	5	3	2	3	1	1	3	2	2	3
54	5	9	2	4	7	3	6	1	2	3	2	2	1	3	3	1
55	7	1	7	4	3	9	5	2	3	2	2	3	1	1	2	3
56	7	2	8	5	1	7	6	3	3	2	3	1	2	2	3	1
57	7	3	9	6	2	8	4	1	3	2	1	2	3	3	1	2
58	8	1	8	6	6	3	8	5	1	3	2	3	2	2	1	2
59	8	2	9	4	4	1	9	6	1	3	3	1	3	3	2	3
60	8	3	7	5	5	2	7	4	1	3	1	2	1	1	3	1
61	9	1	9	5	9	6	2	8	2	1	2	3	3	3	3	1
62	9	2	7	6	7	4	3	9	2	1	3	1	1	1	1	2
63	9	3	8	4	8	5	1	7	2	1	1	2	2	2	2	3
64	8	4	1	7	3	7	4	9	2	1	2	2	2	3	2	1
65	8	5	2	8	1	8	5	7	2	1	3	3	3	1	3	2
66	8	6	3	9	2	9	6	8	2	1	1	1	1	2	1	3
67	9	4	2	9	6	1	7	3	3	2	2	2	3	1	1	3
68	9	5	3	7	4	2	8	1	3	2	3	3	1	2	2	1
69	9	6	1	8	5	3	9	2	3	2	1	1	2	3	3	2
70	7	4	3	8	9	4	1	6	1	3	2	2	1	2	3	2
71	7	5	1	9	7	5	2	4	1	3	3	3	2	3	1	3
72	7	6	2	7	8	6	3	5	1	3	1	1	3	1	2	1
73	9	7	4	1	3	8	6	4	1	3	2	1	3	2	2	2
74	9	8	5	2	1	9	4	5	1	3	3	2	1	3	3	3
75	9	9	6	3	2	7	5	6	1	3	1	3	2	1	1	1
76	7	7	5	3	6	2	9	7	2	1	2	1	1	3	1	1
77	7	8	6	1	4	3	7	8	2	1	3	2	2	1	2	2
78	7	9	4	2	5	1	8	9	2	1	1	3	3	2	3	3
79	8	7	6	2	9	5	3	1	3	2	2	1	2	1	3	3
80	8	8	4	3	7	6	1	2	3	2	3	2	3	2	1	1
81	8	9	5	1	8	4	2	3	3	2	1	3	1	3	2	2

$L_{81}(9^{10})$

No.	1	2	3	4	5	6	7	8	9	10
1	1	1	1	1	1	1	1	1	1	1
2	1	2	2	2	2	2	2	2	2	2
3	1	3	3	3	3	3	3	3	3	3
4	2	1	2	3	4	4	4	4	4	4
5	2	2	3	1	5	5	5	5	5	5
6	2	3	1	2	6	6	6	6	6	6
7	3	1	3	2	7	7	7	7	7	7
8	3	2	1	3	8	8	8	8	8	8
9	3	3	2	1	9	9	9	9	9	9
10	2	4	4	4	1	2	3	8	7	9
11	2	5	5	5	2	3	1	9	8	7
12	2	6	6	6	3	1	2	7	9	8
13	3	4	5	6	4	5	6	2	1	3
14	3	5	6	4	5	6	4	3	2	1
15	3	6	4	5	6	4	5	1	3	2
16	1	4	6	5	7	8	9	5	4	6
17	1	5	4	6	8	9	7	6	5	4
18	1	6	5	4	9	7	8	4	6	5
19	3	7	7	7	1	3	2	6	4	5
20	3	8	8	8	2	1	3	4	5	6
21	3	9	9	9	3	2	1	5	6	4
22	1	7	8	9	4	6	5	9	7	8
23	1	8	9	7	5	4	6	7	8	9
24	1	9	7	8	6	5	4	8	9	7
25	2	7	9	8	7	9	8	3	1	2
26	2	8	7	9	8	7	9	1	2	3
27	2	9	8	7	9	8	7	2	3	1
28	4	1	4	7	2	5	9	3	6	8
29	4	2	5	8	3	6	7	1	4	9
30	4	3	6	9	1	4	8	2	5	7
31	5	1	5	9	5	8	3	6	9	2
32	5	2	6	7	6	9	1	4	7	3
33	5	3	4	8	4	7	2	5	8	1
34	6	1	6	8	8	2	6	9	3	5
35	6	2	4	9	9	3	4	7	1	6
36	6	3	5	7	7	1	5	8	2	4
37	5	4	7	1	2	6	8	7	3	4
38	5	5	8	2	3	4	9	8	1	5
39	5	6	9	3	1	5	7	9	2	6
40	6	4	8	3	5	9	2	1	6	7
41	6	5	9	1	6	7	3	2	4	8
42	6	6	7	2	4	8	1	3	5	9

(*Continued*)

L_{81} (9^{10}) (*Continued*)

No.	1	2	3	4	5	6	7	8	9	10
43	4	4	9	2	8	3	5	4	9	1
44	4	5	7	3	9	1	6	5	7	2
45	4	6	8	1	7	2	4	6	8	3
46	6	7	1	4	2	4	7	5	9	3
47	6	8	2	5	3	5	8	6	7	1
48	6	9	3	6	1	6	9	4	8	2
49	4	7	2	6	5	7	1	8	3	6
50	4	8	3	4	6	8	2	9	1	4
51	4	9	1	5	4	9	3	7	2	5
52	5	7	3	5	8	1	4	2	6	9
53	5	8	1	6	9	2	5	3	4	7
54	5	9	2	4	7	3	6	1	5	8
55	7	1	7	4	3	9	5	2	8	6
56	7	2	8	5	1	7	6	3	9	4
57	7	3	9	6	2	8	4	1	7	5
58	8	1	8	6	6	3	8	5	2	9
59	8	2	9	4	4	1	9	6	3	7
60	8	3	7	5	5	2	7	4	1	8
61	9	1	9	5	9	6	2	8	5	3
62	9	2	7	6	7	4	3	9	6	1
63	9	3	8	4	8	5	1	7	4	2
64	8	4	1	7	3	7	4	9	5	2
65	8	5	2	8	1	8	5	7	6	3
66	8	6	3	9	2	9	6	8	4	1
67	9	4	2	9	6	1	7	3	8	5
68	9	5	3	7	4	2	8	1	9	6
69	9	6	1	8	5	3	9	2	7	4
70	7	4	3	8	9	4	1	6	2	8
71	7	5	1	9	7	5	2	4	3	9
72	7	6	2	7	8	6	3	5	1	7
73	9	7	4	1	3	8	6	4	2	7
74	9	8	5	2	1	9	4	5	3	8
75	9	9	6	3	2	7	5	6	1	9
76	7	7	5	3	6	2	9	7	5	1
77	7	8	6	1	4	3	7	8	6	2
78	7	9	4	2	5	1	8	9	4	3
79	8	7	6	2	9	5	3	1	8	4
80	8	8	4	3	7	6	1	2	9	5
81	8	9	5	1	8	4	2	3	7	6

REFERENCES

Akao, Y. (1972), New product development and quality assurance: System of quality function deployment (in Japanese), *Standardization and Quality Control*, Vol. 25, No. 4, pp. 9–14.

Akao, Y. (ed.) (1988), *The Practice of Applying Quality Deployment to New Product Development* (in Japanese), Japanese Standards Association.

Akao, Y. (ed.) (1990a), *Quality Function Deployment: Integrating Customer Requirements into Product Design*, Productivity Press.

Akao Y. (1990b), *Introduction to Quality Deployment* (in Japanese), JUSE Press.

Akao Y. (1996), Quality function deployment on total quality management and future subject (QFD and TQM Series No. 1) (in Japanese), *Quality Management*, Vol. 47, No. 8, pp. 55–64.

Akao, Y. (1997), QFD: Past, present, and future, in *Proceedings of the 3rd International Symposium on Quality Function Deployment*, pp. 19–29.

Akao, Y. (2001a), TQM quality system by QFD based on ISO 9000, in *Proceedings of the 6th International Conference on ISO 9000 & TQM*, pp. 137–146.

Akao, Y. (2001b), Quality management system by quality function deployment, in *Proceedings of the 7th International Symposium on Quality Function Deployment*, pp. 1–6.

Akao, Y. (2002a), QFD and knowledge management, *Proceedings of the 7th International Conference on ISO 9000 & TQM*, pp. 83–84.

Akao, Y. (2002b), QFD and knowledge management on health care service, in *Proceedings of the 8th International Symposium on Quality Function Deployment*.

Akao, Y. (2004), Product quality and work quality, in *Proceedings of the 9th International Conference on ISO 9000 & TQM*, pp. 246–253.

Akao, Y. (2010), *Quality Function Deployment for Product Development: SECI Model of Knowledge Conversion and QFD (in Japanese)*, Japanese Standards Association.

Akao, Y. (2012), The method for motivation by quality function deployment, *Nang Yan Business Journal*, No. 1-01, pp. 1–9.

Akao, Y. and Inayoshi, K. (2003), QFD and administrative knowledge management, in *Proceedings of the 9th International Symposium on Quality Function Deployment*.

Quality Strategy for Research and Development, First Edition. Ming-Li Shiu, Jui-Chin Jiang, and Mao-Hsiung Tu.

Akao, Y. and Kamimura, T. (2005), QFD and knowledge management: A QFD case study on the development of a large outboard motor, in *Proceedings of the 7th Management International Congress*.

Akao, Y. and Kozawa, Y. (2005), QFD and knowledge management: A QFD case study on a survey of surgery patient satisfaction, in *Proceedings of the 10th International Conference on ISO 9000 & TQM*.

Akao, Y. and Mazur, G. H. (2003), The leading edge in QFD: Past, present and future, *International Journal of Quality & Reliability Management*, Vol. 20, No. 1, pp. 20–35.

Akao, Y., Naoi, T., and Ohfuji, T. (1989), Survey of the status of quality deployment—Report from the Quality Deployment Research Committee (in Japanese), *Quality Control*, Vol. 19, No. 1, pp. 35–44.

Akao Y., Ohfuji, T., and Kaneko, N. (1990), Methods for collecting customer information and translating demanded-quality (Application of QFD Series No. 1) (in Japanese), *Quality Management*, Vol. 41, No. 7, pp. 77–86.

Akao, Y., Ono, T., Harada, A., Tanaka, H., and Iwasawa, K. (1983), Quality deployment including Cost, Reliability and Technology (Part I)—Design of Quality, Cost and Reliability (in Japanese), *Quality*, Vol. 13, No. 3, pp. 61–70.

Akao, Y. and Takaomoto, M. (2005), QFD and knowledge management: A QFD case study on the development of substation equipment, in *Proceedings of the 11th International Symposium on Quality Function Deployment*.

Akao, Y. and Tsugawa, M. (2004), QFD and knowledge management: A QFD case study on the development of a commercial scanner, in *Proceedings of the 10th International Symposium on Quality Function Deployment*.

Allen, C. W. (ed.) (1990), *Simultaneous Engineering: Integrating Manufacturing and Design*, Society of Manufacturing Engineers.

Antony, J. and Banuelas, R. (2002), Design for six sigma, *Manufacturing Engineer*, February, pp. 24–26.

Armacost, R. L., Componation, P. J., Mullens, M. A., and Swart, W. W. (1994), An AHP framework for prioritizing customer requirements in QFD: An industrialized housing application, *IIE Transactions*, Vol. 26, No. 4, pp. 72–79.

Beckford, J. (1998), *Quality: A Critical Introduction*, Routledge.

Bendell, T., Boulter, L., and Goodstadt, P. (1993), *Benchmarking for Competitive Advantage*, Pitman Publishing.

Booker, J. D. (2003), Industrial practice in designing for quality, in *International Journal of Quality & Reliability Management*, Vol. 20, No. 3, pp. 288–303.

Bouchereau, V. and Rowlands, H. (2000), Methods and techniques to help quality function deployment, *Benchmarking*, Vol. 7, No. 1, pp. 8–19.

Bralla, J. G. (1996), *Design for Excellence*, McGraw-Hill.

Burke, E., Kloeber, J. M., and Deckro, R. F. (2002), Using and abusing QFD scores, *Quality Engineering*, Vol. 15, No. 1, pp. 9–21.

Büyüközkan, G., Dereli, T., and Baykasoğlu, A. (2004a), A survey on the methods and tools of concurrent new product development and agile manufacturing, *Journal of Intelligent Manufacturing*, Vol. 15, pp. 731–751.

Büyüközkan, G., Ertay, T., Kahraman, C., and Ruan, D. (2004b), Determining the importance weights for the design requirements in the house of quality using the fuzzy analytic network approach, *International Journal of Intelligent Systems*, Vol. 19, pp. 443–461.

Carter, D. E. and Baker, B. S. (1992), *Concurrent Engineering: The Product Development Environment for the 1990s*, Addison-Wesley.

Chan, L. K. and Wu, M. L. (2002), Quality function deployment: A comprehensive review of its concepts and methods, *Quality Engineering*, Vol. 15, No. 1, pp. 23–35.

Chen, Y., Fung, R. Y. K., and Tang, J. (2005), Fuzzy expected value modeling approach for determining target values of engineering characteristics in QFD, *International Journal of Production Research*, Vol. 43, No. 17, pp. 3583–3604.

Cheng, L. C. (2003), QFD in product development: Methodological characteristics and a guide for intervention, *International Journal of Quality & Reliability Management*, Vol. 20, No. 1, pp. 107–122.

Clausing, D. (1988), Taguchi methods to improve the development process, in *Proceedings of IEEE Conference on Communications*, Vol. 2, pp. 826–832.

Clausing, D. (1993), *Total Quality Development: A Step-by-Step Guide to World-Class Concurrent Engineering*, ASME Press.

Clausing, D. and Pugh, S. (1991), Enhanced quality function deployment, in *Proceedings of the Design Productivity International Conference*, pp. 15–25.

Clausing, D. and Simpson, B. H. (1990), Quality by design, *Quality Progress*, Vol. 23, No. 1, pp. 41–44.

Cohen, L. (1988), Quality function deployment: An application perspective from Digital Equipment Corporation, *National Productivity Review*, Vol. 7, No. 3, pp. 197–208.

Cohen, L. (1995), *Quality Function Deployment: How to Make QFD Work for You*, Addison-Wesley.

Conti, T. (1989), *Process Management and Quality Function Deployment*, *Quality Progress*, Vol. 22, No. 12, pp. 45–48.

Coombs, C. F. (ed.) (2001), *Printed Circuits Handbook*, fifth edition, McGraw-Hill.

Coughlan, P. and Coghlan, D. (2002), Action research for operations management, *International Journal of Operations & Production Management*, Vol. 22, No. 2, pp. 220–240.

Cox, C. A. (1992), Keys to success in quality function deployment, *APICS—The Performance Advantage*, Vol. 2, No. 4, pp. 25–29.

Creveling, C. M., Slutsky, J. L., and Antis, D. (2002), *Design for Six Sigma in Technology and Product Development*, Prentice Hall PTR.

Cristiano, J. J., Liker, J. K., and White, C. C. III (2000), Customer-Driven Product Development through Quality Function Deployment in the U.S. and Japan, *Journal of Product Innovation Management*, Vol. 17, No. 4, pp. 286–308.

Cristiano, J. J., Liker, J. K., and White, C. C. III (2001), Key factors in the successful application of quality function deployment (QFD), *IEEE Transactions on Engineering Management*, Vol. 48, No. 1, pp. 81–95.

Crowe, T. J. and Cheng, C. C. (1996), Using quality function deployment in manufacturing strategic planning, *International Journal of Operations & Production Management*, Vol. 16, No. 4, pp. 35–48.

Dawson, D. and Askin, R. G. (1999), Optimal new product design using quality function deployment with empirical value functions, *Quality and Reliability Engineering International*, Vol. 15, pp. 17–32.

Deming, W. E. (1982), *Out of the Crisis*, MIT Press.

Elliott, J. (1991), *Action Research for Educational Change*, Open University Press.

Eureka, W. E. and Ryan, N. E. (eds.) (1995), *Quality Up, Costs Down: A Manager's Guide to Taguchi Methods and QFD*, ASI Press.

Flatt, M. O. (1997), *Printed Circuit Board Basics*, Miller Freeman Inc.

Fortuna, R. M. (1988), Beyond quality: Taking SPC upstream, *Quality Progress*, Vol. 21, No. 6, pp. 23–28.

Franceschini, F. and Rossetto, S. (1997), Design for quality: Selecting product's technical features, *Quality Engineering*, Vol. 9, No. 4, pp. 681–688.

Franceschini, F. (2002), *Advanced Quality Function Deployment*, CRC Press LLC.

Garvin, D. A. (1987), Competing on the eight dimensions of quality, *Harvard Business Review, November–December*, No. 6, pp. 101–109.

Ghobadian, A. and Terry, A. J. (1995), How alitalia improves service quality through quality function deployment, *Managing Service Quality*, Vol. 5, No. 5, pp. 25–30.

Hahn, G. J., Doganaksoy, N., and Hoerl, R. (2000), The evolution of six sigma, *Quality Engineering*, Vol. 12, No. 3, pp. 317–326.

Hauser, J. R. (1993), How Puritan-Bennett used the house of quality, *Sloan Management Review*, Vol. 30, No. 3, pp. 61–70.

Hauser, J. R. and Clausing, D. (1988), The house of quality, *Harvard Business Review*, Vol. 66, No. 3, pp. 63–73.

Hunt, R. A. and Xavier, F. B. (2003), The leading edge in strategic QFD, *International Journal of Quality & Reliability Management*, Vol. 20, No. 1, pp. 56–73.

Jiang, J. C., Shiu, M. L., and Tu, M. H. (2007a), DFX and DFSS: How QFD integrates them, *Quality Progress*, Vol. 40, No. 10, pp. 45–51.

Jiang, J. C., Shiu, M. L., and Tu, M. H. (2007b), QFD's evolution in Japan and the West, *Quality Progress*, Vol. 40, No. 7, pp. 30–37.

Jiang, J. C., Shiu, M. L., and Tu, M. H. (2007c), Quality function deployment (QFD) technology designed for contract manufacturing, *TQM Magazine*, Vol. 19, No. 4, pp. 291–307.

Juran, J. M. (1992), *Juran on Quality by Design: The New Steps for Planning Quality into Goods and Services*, Free Press.

Kanda, N. (1995), *Seven Tools for New Product Planning*, JUSE Press.

Kanji, G. K. (2002), Performance measurement system, *Total Quality Management*, Vol. 13, No. 5, pp. 715–728.

Kano, N. (1987), Quality management in management engineering (in Japanese), *Quality*, Vol. 17, No.1, pp. 23–39.

Kano, N. (1994), Downsizing through reengineering and upsizing through attractive quality creation, *Proceeding of ASQC Annual Quality Conference*, pp. 1–8.

Kano, N. (ed.) (1996), *Guide to TQM in Service Industries*, Asian Productivity Organization.

Kano, N. (2002), Attractive quality creation under globalization, in *The 8th Asia Pacific Quality Organization Conference Proceedings*, Vol. 1, pp. K-18–K-21.

Kano, N., Seraku, N., Takahashi, F., and Tsuji, S. (1984), Attractive quality and must-be quality (in Japanese), *Quality*, Vol. 14, No. 2, pp. 147–156.

Kathawala, Y. and Motwani, J. (1994), Implementing quality function deployment: A systems approach, *TQM Magazine*, Vol. 6, No. 6, pp. 31–37.

Khoo, L. P. and Ho, N. C. (1996), Framework of a fuzzy quality function deployment system, *International Journal of Production Research*, Vol. 34, No. 2, pp. 299–311.

Killen, C. P., Walker, M., and Hunt, R. A. (2005), Strategic planning using QFD, *International Journal of Quality & Reliability Management*, Vol. 22, No. 1, pp. 17–29.

Kim, W. C. and Mauborgne, R. (2005), *Blue Ocean Strategy: How to Create Uncontested Market Space and Make Competition Irrelevant*, Harvard Business Review Press.

Kim, K. J., Moskowitz, H., Dhingra, A., and Evans, G. (2000), Fuzzy multicriteria models for quality function deployment, *European Journal of Operational Research*, Vol. 121, No. 3, pp. 504–518.

King, B. (1987), *Better Designs in Half the Time: Implementing QFD Quality Function Deployment in America*, GOAL/QPC.

Kogure, M. and Akao, Y. (1983), Quality function deployment and CWQC in Japan, *Quality Progress*, Vol. 16, No. 10, pp. 25–29.

Kondo, Y. (1995), *Companywide Quality Control*, 3A Corporation.

Kondo, Y. (2002), Quality is the center of integrated management, *Managerial Auditing Journal*, Vol. 17, No. 6, pp. 298–303.

Kotler, P. (2000), *Marketing Management*, millennium edition, Prentice-Hall.

Kusiak, A. (ed.) (1993), *Concurrent Engineering: Automation, Tools and Techniques*, John Wiley & Sons.

Lai, X., Xie, M., and Tan, K. C. (2005), Dynamic programming for QFD optimization, *Quality and Reliability Engineering International*, Vol. 21, pp. 769–780.

LePrevost, J. and Mazur, G. (2005), Quality infrastructure improvement: Using QFD to manage project priorities and project management resources, *International Journal of Quality & Reliability Management*, Vol. 22, No. 1, pp. 10–16.

Lockamy, A. III and Khurana, A. (1995), Quality function deployment: Total quality management for new product design, *International Journal of Quality & Reliability Management*, Vol. 12, No. 6, pp. 73–84.

Lu, M. H., Madu, C. N., Kuei, C. H., and Winokur, D. (1994), Integrating QFD, AHP and benchmarking in strategic marketing, *Journal of Business & Industrial Marketing*, Vol. 9, No. 1, pp. 41–50.

Maddux, G. A., Amos, R. W., and Wyskids, A. R. (1991), Organizations can apply quality function deployment as strategic planning tool, *Industrial Engineering*, Vol. 23, No. 9, pp. 33–37.

Mader, D. P. (2002), Design for six sigma, *Quality Progress*, Vol. 35, No. 7, pp. 82–86.

Magrab, E. B. (1997), *Integrated Product and Process Design and Development: The Product Realization Process*, CRC Press LLC.

Marks, L. and Caterina, J. (2000), *Printed Circuit Assembly Design*, McGraw-Hill.

Masud, A. S. and Dean, E. B. (1993), Using fuzzy sets in quality function deployment, in *Proceedings of the 2nd Industrial Engineering Research Conference*, pp. 270–274.

Miguel, P. A. C. (2005), Evidence of QFD best practices for product development: A multiple case study, *International Journal of Quality & Reliability Management*, Vol. 22, No. 1, pp. 72–82.

Mizuno, S. and Akao, Y. (eds.) (1978), *Quality Function Deployment: A Company-wide Quality Approach* (in Japanese), JUSE Press.

Mochimoto, T. (1996), A study on product value, quality, price and the ability to market competitiveness (in Japanese), *Quality*, Vol. 26, No. 2, pp. 81–85.

Mochimoto, T. (1997), Reconsideration of VE equation and deployment of quality concept (in Japanese), *Quality*, Vol. 27, No. 4, pp. 182–193.

Monplaisir, L. and Singh, N. (eds.) (2002), *Collaborative Engineering for Product Design and Development*, American Scientific Publishers.

Moskowitz, H. and Kim, K. J. (1997), QFD optimizer: A novice friendly quality function deployment decision support system for optimizing product designs, *Computers & Industrial Engineering*, Vol. 32, No. 3, pp. 641–655.

Mrad, F. (1997), An industrial workstation characterization and selection using quality function deployment, *Quality and Reliability Engineering International*, Vol. 13, pp. 261–268.

Nagai, K. and Ohfuji, T. (2008), *The Third Generation of QFD: Quality Function Deployment for Development Process Management* (in Japanese), JUSE Press.

Nonaka, I. and Takeuchi, H. (1995), *The Knowledge-Creating Company: How Japanese Companies Create the Dynamics of Innovation*, Oxford University Press.

Ohfuji, T. (1993), Quality function deployment: The basics of QFD (in Japanese), *Societas Qualitatis*, Vol. 7, No. 2, pp. 2–3.

Ohfuji, T. (2010), *QFD: Approach to Quality Assurance at the Planning Stage* (in Japanese), Japanese Standards Association.

Ohfuji, T., Cristiano, J. J., and White, C. C. III (1996), Comparison of QFD status in Japan and the USA (in Japanese), in *Proceedings of the JSQC 25th Anniversary and 52nd Research Presentations*, pp. 1–4.

Ohfuji, T., Ono, M., and Akao, Y. (1990), *Quality Deployment (1): Construction and Exercise of Quality Chart* (in Japanese), JUSE Press.

Ohfuji, T., Ono, M., and Akao, Y. (1994), *Quality Deployment (2): Total Deployment including Technology, Reliability and Cost* (in Japanese), JUSE Press.

Osada, H. (1998), Strategic management by policy in total quality management, *Strategic Change*, Vol. 7, No. 5, pp. 277–287.

Özgener, S. (2003), Quality function deployment: A teamwork approach, *TQM & Business Excellence*, Vol. 14, No. 9, pp. 969–979.

Park, T. and Kim, K. J. (1998), Determination of an optimal set of design requirements using house of quality, *Journal of Operations Management*, Vol. 16, No. 5, pp. 569–581.

Parsaei, H. R. and Sullivan, W. G. (eds.) (1993), *Concurrent Engineering: Contemporary Issues and Modern Design Tools*, Chapman & Hall.

Partovi, F. Y. (1999), A quality function deployment approach to strategic capital budgeting, *Engineering Economist*, Vol. 44, No. 3, pp. 239–260.

Porter, M. E. (1985), *Competitive Advantage: Creating and Sustaining Superior Performance*, Free Press.

Prahalad, C. K. and Hamel, G. (1990), The core competence of the corporation, *Harvard Business Review*, May–June, pp. 79–91.

Prasad, B. (1996), *Concurrent Engineering Fundamentals: Integrated Product and Process Organization*, volume 1, Prentice Hall PTR.

Prasad, B. (1998), Review of QFD and related deployment techniques, *Journal of Manufacturing Systems*, Vol. 17, No. 3, pp. 221–234.

Ramasamy, N. R. and Selladurai, V. (2004), Fuzzy logic approach to prioritise engineering characteristics in quality function deployment (FL-QFD), *International Journal of Quality & Reliability Management*, Vol. 21, No. 9, pp. 1012–1023.

ReVelle, J. B., Moran, J. W., and Cox, C. A. (1998), *The QFD Handbook*, John Wiley & Sons.

Ross, P. J. (1988), The role of taguchi methods and design of experiments in QFD, *Quality Progress*, Vol. 21, No. 6, pp. 41–47.

Schmidt, R. (1997), The implementation of simultaneous engineering in the stage of product concept development: A process oriented improvement of quality function deployment, *European Journal of Operational Research*, Vol. 100, pp. 293–314.

Shin, J. S., Kim, K. J., and Chandra, M. J. (2002), Consistency check of a house of quality chart, *International Journal of Quality & Reliability Management*, Vol. 19, No. 4, pp. 471–484.

Shina, S. G. (1991), *Concurrent Engineering and Design for Manufacture of Electronics Products*, Van Nostrand Reinhold.

Shina, S. G. (ed.) (1994), *Successful Implementation of Concurrent Engineering Products and Processes*, Van Nostrand Reinhold.

Shindo, H. (ed.) (1998), *Advanced QFD Technology for Value Creation* (in Japanese), JUSE Press.

Shiu, M. L., Jiang, J. C., and Tu, M. H. (2007), Reconstruct QFD for integrated product and process development management, *TQM Magazine*, Vol. 19, No. 5, pp. 403–418.

Snee, R. D. (2003), Eight essential tools, *Quality Progress*, Vol. 36, No. 12, pp. 86–88.

Suh, N. P. (2005), *Complexity: Theory and Applications*, Oxford University Press.

Sullivan, L. P. (1986), Quality function deployment, *Quality Progress*, Vol. 19, No. 6, pp. 39–50.

Sullivan, L. P. (1988), Policy management through quality function deployment, *Quality Progress*, Vol. 21, No. 6, pp. 18–20.

Swink, M. (2002), Product development—Faster, on-time, *Research Technology Management*, Vol. 45, No. 4, pp. 50–58.

Syan, C. S. and Menon, U. (eds.) (1994), *Concurrent Engineering: Concepts, Implementation and Practice*, Chapman & Hall.

Taguchi, G., Chowdhury, S., and Taguchi, S. (2000), *Robust Engineering: Learn How to Boost Quality While Reducing Costs & Time to Market*, McGraw-Hill.

Taguchi, G., Chowdhury, S., and Wu Y. (2005), *Taguchi's Quality Engineering Handbook*, John Wiley & Sons.

Temponi, C., Yen, J., and Tiao, W. A. (1999), House of quality: A fuzzy logic-based requirements analysis, *European Journal of Operational Research*, Vol. 117, No. 2, pp. 340–354.

Terninko, J. (1997), *Step-by-Step QFD: Customer-Driven Product Design*, second edition, CRC Press LLC.

Turino, J. (1992), *Managing Concurrent Engineering: Buying Time to Market*, Van Nostrand Reinhold.

Usher, J. M., Roy, U., and Parsaei, H. R. (eds.) (1998), *Integrated Product and Process Development: Methods, Tools and Technologies*, John Wiley & Sons.

Wang, B. (ed.) (1997), *Integrated Product, Process and Enterprise Design*, Chapman & Hall.

Wasserman, G. S. (1993), On how to prioritize design requirements during the QFD planning process, *IIE Transactions*, Vol. 25, No. 3, pp. 59–65.

Wasserman, G. S., Sudjianto, A., Mohanty, G., and Sanrow, C. W. (1993), Using fuzzy set theory to derive an overall customer satisfaction index, in *Proceedings of the 5th Symposium on Quality Function Deployment*, pp. 36–54.

Wheelwright, S. C. and Clark, K. B. (1992), *Revolutionizing Product Development: Quantum Leaps in Speed*, Efficiency and Quality, Free Press.

Yan, W., Khoo, L. P., and Chen, C. H. (2005), A QFD-enabled product conceptualization approach via design knowledge hierarchy and RCE neural network, *Knowledge-Based Systems*, Vol. 18, pp. 279–293.

Yoshizawa, T., Nagai, K., and Ohfuji, T. (2004), *Quality Function Deployment for Sustainable Growth: Effective Utilization and Case Studies of JIS Q 9025* (in Japanese), Japanese Standards Association.

Zairi, M. and Youssef, M. A. (1995), Quality function deployment: A main pillar for successful total quality management and product development, *International Journal of Quality & Reliability Management*, Vol. 12, No. 6, pp. 9–23.

Zhang, X., Bode, J., and Ren, S. (1996), Neural networks in quality function deployment, *Computers & Industrial Engineering*, Vol. 31, No. 3/4, pp. 669–673.

Zhou, M. (1998), Fuzzy logic and optimization models for implementing QFD, *Computers & Industrial Engineering*, Vol. 35, No. 1/2, pp. 237–240.

INDEX

Quality Strategy for Research and Development, First Edition. Ming-Li Shiu, Jui-Chin Jiang, and Mao-Hsiung Tu.

WILEY SERIES IN SYSTEMS ENGINEERING AND MANAGEMENT

Andrew P. Sage, Editor

ANDREW P. SAGE and JAMES D. PALMER
Software Systems Engineering

WILLIAM B. ROUSE
Design for Success: A Human-Centered Approach to Designing Successful Products and Systems

LEONARD ADELMAN
Evaluating Decision Support and Expert System Technology

ANDREW P. SAGE
Decision Support Systems Engineering

YEFIM FASSER and DONALD BRETTNER
Process Improvement in the Electronics Industry, Second Edition

WILLIAM B. ROUSE
Strategies for Innovation

ANDREW P. SAGE
Systems Engineering

HORST TEMPELMEIER and HEINRICH KUHN
Flexible Manufacturing Systems: Decision Support for Design and Operation

WILLIAM B. ROUSE
Catalysts for Change: Concepts and Principles for Enabling Innovation

LIPING FANG, KEITH W. HIPEL, and D. MARC KILGOUR
Interactive Decision Making: The Graph Model for Conflict Resolution

DAVID A. SCHUM
Evidential Foundations of Probabilistic Reasoning

JENS RASMUSSEN, ANNELISE MARK PEJTERSEN, and LEONARD P. GOODSTEIN
Cognitive Systems Engineering

ANDREW P. SAGE
Systems Management for Information Technology and Software Engineering

ALPHONSE CHAPANIS
Human Factors in Systems Engineering

YACOV Y. HAIMES
Risk Modeling, Assessment, and Management, Third Edition

DENNIS M. BUEDE
The Engineering Design of Systems: Models and Methods, Second Edition

ANDREW P. SAGE and JAMES E. ARMSTRONG, Jr.
Introduction to Systems Engineering

WILLIAM B. ROUSE
Essential Challenges of Strategic Management

YEFIM FASSER and DONALD BRETTNER
Management for Quality in High-Technology Enterprises

THOMAS B. SHERIDAN
Humans and Automation: System Design and Research Issues

ALEXANDER KOSSIAKOFF and WILLIAM N. SWEET
Systems Engineering Principles and Practice

HAROLD R. BOOHER
Handbook of Human Systems Integration

JEFFREY T. POLLOCK and RALPH HODGSON
Adaptive Information: Improving Business Through Semantic Interoperability, Grid Computing, and Enterprise Integration

ALAN L. PORTER and SCOTT W. CUNNINGHAM
Tech Mining: Exploiting New Technologies for Competitive Advantage

REX BROWN
Rational Choice and Judgment: Decision Analysis for the Decider

WILLIAM B. ROUSE and KENNETH R. BOFF (editors)
Organizational Simulation

HOWARD EISNER
Managing Complex Systems: Thinking Outside the Box

STEVE BELL
Lean Enterprise Systems: Using IT for Continuous Improvement

J. JERRY KAUFMAN and ROY WOODHEAD
Stimulating Innovation in Products and Services: With Function Analysis and Mapping

WILLIAM B. ROUSE
Enterprise Tranformation: Understanding and Enabling Fundamental Change

JOHN E. GIBSON, WILLIAM T. SCHERER, and WILLAM F. GIBSON
How to Do Systems Analysis

WILLIAM F. CHRISTOPHER
Holistic Management: Managing What Matters for Company Success

WILLIAM B. ROUSE
People and Organizations: Explorations of Human-Centered Design

MO JAMSHIDI
System of Systems Engineering: Innovations for the Twenty-First Century

ANDREW P. SAGE and WILLIAM B. ROUSE
Handbook of Systems Engineering and Management, Second Edition

JOHN R. CLYMER
Simulation-Based Engineering of Complex Systems, Second Edition

KRAG BROTBY
Information Security Governance: A Practical Development and Implementation Approach

JULIAN TALBOT and MILES JAKEMAN
Security Risk Management Body of Knowledge

SCOTT JACKSON
Architecting Resilient Systems: Accident Avoidance and Survival and Recovery from Disruptions

JAMES A. GEORGE and JAMES A. RODGER
Smart Data: Enterprise Performance Optimization Strategy

YORAM KOREN
The Global Manufacturing Revolution: Product-Process-Business Integration and Reconfigurable Systems

AVNER ENGEL
Verification, Validation, and Testing of Engineered Systems

WILLIAM B. ROUSE (editor)
The Economics of Human Systems Integration: Valuation of Investments in People's Training and Education, Safety and Health, and Work Productivity

ALEXANDER KOSSIAKOFF, WILLIAM N. SWEET, SAM SEYMOUR, and STEVEN M. BIEMER
Systems Engineering Principles and Practice, Second Edition

GREGORY S. PARNELL, PATRICK J. DRISCOLL, and DALE L. HENDERSON (editors)
Decision Making in Systems Engineering and Management, Second Edition

ANDREW P. SAGE and WILLIAM B. ROUSE
Economic Systems Analysis and Assessment: Intensive Systems, Organizations, and Enterprises

BOHDAN W. OPPENHEIM
Lean for Systems Engineering with Lean Enablers for Systems Engineering

LEV M. KLYATIS
Accelerated Reliability and Durability Testing Technology

BJOERN BARTELS , ULRICH ERMEL, MICHAEL PECHT, and PETER SANDBORN
Strategies to the Prediction, Mitigation, and Management of Product Obsolescence

LEVANT YILMAS and TUNCER ÖREN
Agent-Directed Simulation and Systems Engineering

ELSAYED A. ELSAYED
Reliability Engineering, Second Edition

BEHNAM MALKOOTI
Operations and Production Systems with Multiple Objectives

MING-LI SHIU, JUI-CHIN JIANG, and MAO-HSIUNG TU
Quality Strategy for Research and Development

Printed in the United States
By Bookmasters